Societies in Motion

NEW HORIZONS IN REGIONAL SCIENCE

Series Editor: Philip McCann, *Professor of Economic Geography, University of Groningen, The Netherlands and Professor of Economics, University of Waikato, New Zealand*

Regional science analyses important issues surrounding the growth and development of urban and regional systems and is emerging as a major social science discipline. This series provides an invaluable forum for the publication of high quality scholarly work on urban and regional studies, industrial location economics, transport systems, economic geography and networks.

New Horizons in Regional Science aims to publish the best work by economists, geographers, urban and regional planners and other researchers from throughout the world. It is intended to serve a wide readership including academics, students and policymakers.

Titles in the series include:

Migration and Human Capital
Edited by Jacques Poot, Brigitte Waldorf and Leo van Wissen

Universities, Knowledge Transfer and Regional Development
Geography, Entrepreneurship and Policy
Edited by Attila Varga

International Knowledge and Innovation Networks
Knowledge Creation and Innovation in Medium Technology Clusters
Riccardo Cappellin and Rüdiger Wink

Leadership and Institutions in Regional Endogenous Development
Robert Stimson and Roger R. Stough with Maria Salazar

Entrepreneurship and Regional Development
Local Processes and Global Patterns
Edited by Charlie Karlsson, Börje Johansson and Roger R. Stough

Endogenous Regional Development
Perspectives, Measurement and Empirical Investigation
Edited by Robert Stimson, Roger R. Stough and Peter Nijkamp

Media Clusters
Spatial Agglomeration and Content Capabilities
Edited by Charlie Karlsson and Robert G. Picard

Spatial Scenarios in a Global Perspective
Europe and the Latin Arc Countries
Edited by Roberto Camagni and Roberta Capello

Creative Knowledge Cities
Myths, Visions and Realities
Edited by Marina van Geenhuizen and Peter Nijkamp

Societies in Motion
Innovation, Migration and Regional Transformation
Edited by Amnon Frenkel, Peter Nijkamp and Philip McCann

Innovation, Global Change and Territorial Resilience
Edited by Philip Cooke, Mario Davide Parrilli and José Luis Curbelo

Societies in Motion

Innovation, Migration and Regional Transformation

Essays in Honor of Dani Shefer

Edited by

Amnon Frenkel

Associate Professor of Urban and Regional Planning, Technion - Israel Institute of Technology, Israel

Peter Nijkamp

Professor of Regional, Urban and Environmental Economics, Free University of Amsterdam, The Netherlands

Philip McCann

Professor of Economic Geography, University of Groningen, The Netherlands and Professor of Economics, University of Waikato, New Zealand

NEW HORIZONS IN REGIONAL SCIENCE

Edward Elgar

Cheltenham, UK • Northampton, MA, USA

Published by
Edward Elgar Publishing Limited
The Lypiatts
15 Lansdown Road
Cheltenham
Glos GL50 2JA
UK

Edward Elgar Publishing, Inc.
William Pratt House
9 Dewey Court
Northampton
Massachusetts 01060
USA

A catalogue record for this book
is available from the British Library

Library of Congress Control Number: 2011932945

ISBN 978 0 85793 993 7

Printed and bound by MPG Books Group Ltd, UK

Contents

Figures

Tables

Contributors

Patricio Aroca Professor and Head of the Institute of Regional Applied Economics and Faculty of Economics and Management, Universidad Catolica del Norte, Antofagasia, Chile.

Haim Aviram Associate Professor of Economics, Kinneret College on the Lake of Galilee, Israel.

Gil Avnimelech Lecturer of Business Administration, Ono Academic College, Kiryat Ono, Israel.

David Banister Professor and Director of Transport Studies, School of Geography and the Environment, University of Oxford, Oxford, United Kingdom.

Raphael Bar-El Professor of Regional Economic Development at Department of Public Policy and Administration, Ben-Gurion University of the Negev, Beer-Sheva, Israel.

Hillel Bar-Gera Senior Lecturer, Department of Industrial Engineering and Management, Ben-Gurion University of the Negev, Beer-Sheva, Israel.

David Boyce Adjunct Professor, Department of Civil and Environmental Engineering, Northwestern University, Evanston, Illinois, USA.

Daniela Constantin Professor, Academy of Economic Studies, Bucharest, Romania.

Jonathan Corcoran Senior Lecture, School of Geography, Planning and Environmental Management, Faculty of Science, University of Queensland, Brisbane, Australia.

Henri L.F. de Groot Professor of Regional Economic Dynamics, Free University of Amsterdam, Department of Spatial Economics, The Netherlands.

Alessandra Faggian Researcher in Economic Geography, School of Geography, University of Southampton, Highfield, Southampton, United Kingdom.

Daniel Felsenstein Associate Professor at the Department of Geography and Institute of Urban and Regional Studies, Hebrew University of Jerusalem, Jerusalem, Israel.

Amnon Frenkel Associate Professor of Urban and Regional Planning at the Faculty of Architecture and Town Planning and Center for Urban and Regional Studies, Technion-Israel Institute of Technology, Haifa, Israel.

Moshe Givoni Senior Researcher, School of Geography and the Environment, Centre of Research, Outreach, and Graduate Studies in Transport, University of Oxford, Oxford, United Kingdom.

Zizi Goschin Professor, Bucharest Academy of Economic Studies, Bucharest, Romania.

Daniel A. Griffith Ashbel Smith Professor, School of Economic, Political and Policy Sciences, University of Texas-Dallas, Dallas, Texas, USA.

Shlomie Hazam Researcher, Department of Geography and Institute of Urban and Regional Studies, Hebrew University of Jerusalem, Jerusalem, Israel.

Geoffrey J.D. Hewings Professor of Geography, University of Illinois at Urbana-Champaign, Urbana, Illinois, USA.

Joris Knoben Assisant Professor, Faculty of Social and Behavioral Sciences, Department of Organisation Studies, Tilburg University, Tilburg, The Netherlands.

Miki Malul Senior Lecturer, Department of Public Policy and Administration, Ben-Gurion University of the Negev, Beer Sheva, Israel.

Philip McCann Endowed Chair of Economic Geography, Faculty of Spatial Sciences, University of Groningen, Groningen, The Netherlands and Professor of Economics, Department of Economics, University of Waikato, Hamilton, New Zealand.

Paavo Monkkonen Department of City and Regional Planning, University of California Berkeley, Berkeley, California, USA.

Peter Nijkamp Professor of Regional, Urban and Environmental Economics, Department of Spatial Economics, Free University of Amsterdam, The Netherlands.

Roberto Patuelli Assistant Professor, Department of Economics, University of Bologna, Bologna, Italy.

Piet H. Pellenbarg Professor of Economic Geography, Department of Spatial Economics, Free University of Amsterdam, The Netherlands.

John M. Quigley I. Donald Terner Distinguished Professor, Berkeley Programme on Housing and Urban Policy, University of California Berkeley, California, USA.

Masagus M. Ridhwan Bureau of Economic Research, Central Bank of Indonesia and Free University of Amsterdam, Department of Spatial Economics, The Netherlands.

Piet Rietveld Professor of Transport Economics, Free University of Amsterdam, The Netherlands.

Dafna Schwartz Associate Professor, Department of Business Administration, Ben-Gurion University of the Negev, Beer-Sheva, Israel.

Daniel Shefer Professor Emeritus, Faculty of Architecture and Town Planning, Technion-Israel Institute of Technology, Haifa, Israel.

Yoram Shiftan Associate Professor, Faculty of Civil and Environmental Engineering, Technion-Israel Institute of Technology, Haifa, Israel.

Michael Sonis Professor Emeritus of Geography, Bar Ilan University, Israel, Distinguished Research Professor, University of Illinois Urbana-Champaign.

Michael Tiefelsdorf Associate Professor, School of Economic, Political and Policy Sciences, University of Texas-Dallas, Dallas, Texas, USA.

Preface

Compared to 2000 years ago, our world has displayed a surprisingly dynamic panorama. Spatial mobility has increased dramatically from an average of some 15,000 km over someone's lifetime to an almost limitless spatial action radius. This megatrend was mainly instigated by modern technological innovations and logistics revolutions, by the rise in advanced information and communication technologies, and by far-reaching global demographic transitions (including migration and ageing). Consequently, our modern society is in a permanent state of spatial flux, with dynamic changes in land use patterns, urban and regional settlement developments (including urban sprawl and mega-city development) global (spatial-)economic and financial transitions, and rising geographic connectivity patterns.

These developments already have a long history and date sometimes back to the pre-industrial revolution age. But in the past decades we have observed an unprecedented acceleration in these dynamic space–time processes. Hence, the title of this volume *Societies in Motion*. It aims to map out and analyse the complexity involved with regional development, industrial innovation and spatial mobility patterns in the first decades of our 21st century.

This volume was composed by several colleagues from all over the world. Clearly, their goal was to offer new scientific insights into the driving forces and spatial-economic consequences of our modern dynamic societies. But an additional intention was to pay a scientific tribute to one of the leading scholars in regional science, Daniel Shefer. He has really been a global ambassador of advanced regional science analysis, with a broad interest in many regional, urban, environmental and transportation issues. His pioneering work in many fields of regional science deserves the highest appreciation from scholars all over the world. The present book honours a man of a great scientific and social calibre. Dani: chapeau!

Amnon Frenkel
Peter Nijkamp
Philip McCann

Introduction

Amnon Frenkel, Peter Nijkamp and Philip McCann

In our global world, modern society is characterized by rapid transitions in space that are contributed by the technological development of the previous decades. This book, *Societies in Motion*, contains 15 chapters dealing with several aspects that are affected by these changes that are expressed as regional development, technological innovation and its distribution in space, and spatial mobility.

The book is organized into four parts. The first part describes factors that affect regional development. The opening chapter, contributed by Paavo Monkkonen and John M. Quigley, presents an empirical examination of the spatial distribution of specific categories of land-use regulations based on original data collected for the San Francisco Bay Area. The authors document the strong positive autocorrelation, which characterizes regulations enacted independently by local governments in nearby cities and look to find and analyse the significance of a spatial relationship among land-use rules and strategic interaction in the enactment of land-use regulations.

Chapter 2, written by Daniel Felsenstein and Shlomie Hazam, tests whether increasing metropolitanization impacts on metropolitan core area density functions in the Tel Aviv metropolitan area over the last decade. It uses census and model-simulated data as well as GIS-based and statistical tools. The authors observed a minimal shift in the centre of gravity of the metropolitan core over time, identified the emergence of sub-centres and found that their results upheld the expected negative impact of distance on density for both population and economic activity.

Chapter 3, written by Masagus M. Ridhwan, Peter Nijkamp, Piet Rietveld and Henri L.F. de Groot, provides a critical review of theoretical arguments and empirical evidence on the assumption made by standard economic theory that money is neutral and is driven by interregional arbitrage and perfect capital mobility. It devotes special attention to asymmetric information problems caused by geographical factors and concludes that monetary policy and financial markets can have a potentially important role to play in promoting regional development, especially in less-developed countries.

The second part looks at the spatial distribution of industrial and technological innovation. The starting chapter, Chapter 4, written by Piet H. Pellenbarg and Joris Knoben, presents the spatial dimension of firm adjustment processes and analyses the basic form of firm relocation. The chapter provides an overview of the firm relocation literature and identifies the remaining gaps regarding this topic. It brings together almost six decades of theoretical and empirical work and identifies the most promising paths for future research on this topic.

Chapter 5, written by Dafna Schwartz, Gil Avnimelech and Raphael Bar-El, analyses the location considerations of various categories of knowledge-intensive and high-tech sectors in the Israeli experience. This analysis is based on a brief examination of the geographical patterns of employment in six different sectorial categories in Israel during the years 1995–2006. The chapter argues that as the activity of a company becomes more advanced (or technology-based) and more knowledge-intensive, the geographical location pattern tends to become more concentrated in metropolitan locations and extends this by arguing that the more innovative the company is, the greater the concentration tendency.

Chapter 6, by Amnon Frenkel and Daniel Shefer, deals with factors and programmes that act to foster or to hinder the rate of technology transfer between academia and industry and presents the basic concepts of a technology-transfer production-function model in which human capital, investment capital and a technology transfer office (TTO) staff interact to produce innovations or patent registration. It analyses a field study involving twelve universities in the USA and Israel whose technology-transfer policies were thoroughly analysed and looks at government programmes designed to encourage research cooperation between academia and industry.

The third part covers human capital, migration and labour force. The seventh chapter, written by Patricio Aroca, Geoffrey J.D. Hewings and Michael Sonis begins this part by addressing migration decision-making in developing economies from the perspective of labour force status. The chapter clarifies the role of assets and credit constraints in the decision to migrate from an efficiency point of view. A dynamic model is formulated to analyse this intrinsically dynamic problem of migration. The model considers an imperfect capital market, that is introduced as the unemployed workers lack of access to borrowing from financial institutions.

In Chapter 8, Alessandra Faggian, Jonathan Corcoran and Philip McCann examine the spatial shape and skewness of labour migration patterns using an empirical analysis based on a unique dataset detailing employment mobility behaviour of university graduates in Great Britain, introduce the concept of circular variance into migration modelling and demonstrate how this allows for the simplification of complex two-dimensional phenomena into one-dimensional scalar indices suitable for econometric modelling.

The ninth chapter, written by Raphael Bar-El and Miki Malul, attempts to analyse labour demand and supply, as well as commuting and migration patterns of two regions in Israel, the centre and the suburbs. For this purpose they use an empirical analysis to see if the "spatial mismatch hypothesis" can be supported or rejected. This "spatial mismatch hypothesis" basically states that the process of job dispersal from the centre to the suburbs induces a "people follow jobs" effect, leading to migration of the labour force to the suburbs.

Daniela Constantin, Zizi Goschin and Philip McCann, in Chapter 10, discuss the issues raised by flows of foreign direct investment (FDI) into the Central and Eastern European Countries (CEECs), review the major themes of globalization as they impact on all parts of the world, discuss the recent experience of the CEECs and shifts of FDI eastward within the EU, as well as analyse the periodic fundamental reforms on the EU Cohesion Policy. Although the analysis in this chapter refers to the period before the 2008 global financial crisis in the EU and the CEECs in particular, the countries which are perceived to be more risky investments have suffered significantly in the crisis and the urgency of the need for appropriate institutional reform has increased within the EU.

Chapter 11, written by Roberto Patuelli, Daniel A. Griffith, Michael Tiefelsdorf and Peter Nijkamp, aims to map out German regional unemployment patterns by means of spatial filtering techniques. The authors provide an assessment of how important spatial effects are in explaining unemployment levels in Germany and, particularly, to show that these patterns are consistent over time. The authors present statistical analyses carried out by means of a semi-parametric "spatial filtering" technique based on the decomposition of spatial weights matrices and employ data at a refined level of disaggregation enabling for a highly detailed analysis of the underlying spatial patterns. The empirical study examines data from 439 German districts on both topological and distance-based criteria – such as shared boundaries or centroid distance – and economic flows.

Part four deals with spatial mobility and commences with David Boyce and Hillel Bar-Gera's retrospective in Chapter 12 on the role of computing in urban travel forecasting. The chapter traces the evolution of computer use, and especially computer programmes and software, from the early urban transportation studies, government in-house computer programmeming activities and innovations by consultants to the present software systems supplied by commercial developers. It examines the roles of these actors in advancing as well as deterring the application of methods evolving from research and practice.

Authoring the thirteenth chapter, Yoram Shiftan and Daniel Shefer highlight the important role of understanding travel behaviour and the complex relations between the various travel externalities to the development of sustainable transport policies. They introduce the complex relations between congestion, speed variance, vehicle mix and road safety, present their view of the required actions for the transition toward sustainable transportation-planning practice and the role of better understanding of travel behaviour in this process. The chapter provides examples and an overview of activity-based models and their contribution to travel behaviour in support of sustainable transportation planning.

Chapter 14, written by Moshe Givoni and David Banister, reviews trends over the last 35 years of passenger rail travel in the European Union, and questions the sustainability of rail travel, arguing for better integration between rail and other services so that the total journey is considered, and not just the main part of it. The chapter contrasts the benefits of rail travel with the income levels of the users of rail services and provides some guidelines for a sustainable development of the rail network.

The final and fifteenth chapter, composed by Haim Aviram and Daniel Shefer, asks the question, "Are transportation benefits fully absorbed by land values?" and then goes on to analyse the possible impact of urban transport improvement projects particularly with regard to two forces: the classic indirect impact of transport projects on land values and land area and the less frequently employed direct impact of a reduction in transport costs on the transport sector itself.

The collection of chapters in this book expresses the dynamic forces that work in the field and affect the design of society, the economy and the environment for a better and sustainable future. The various methods and approaches presented are designed to expand our understanding of the connections between the various issues raised within it. All of these issues have a strong relationship to the possibilities of regional development. The chapters provide answers to various questions that were raised for analysis, but at the same time raise more questions of their own that require continued research as are specified in the various chapters.

PART ONE

Regional Development

1. The spatial consequences of autarky in land-use regulation: strategic interaction or simply parallelism?

Paavo Monkkonen and John M. Quigley

INTRODUCTION

The regulation of land use in most urban areas of the United States does not occur at the metropolitan level. Local jurisdictions in most states have almost complete control over land-use regulation within their boundaries. Since metropolitan areas are typically composed of tens or hundreds of jurisdictions of varying sizes, it is not surprising that these regions exhibit great variation in the stringency of land-use regulation within their borders. Moreover, it is generally supposed that the lack of coordination in land-use regulation, combined with strategic implementation of ordinances that limit growth or density, has untoward effects on population growth patterns, leads to the exclusion of low-income households from some communities, and results in higher housing prices overall.

Metropolitan regions in California provide an extreme example of the uncoordinated spatial distribution of land-use regulation, as their constituent jurisdictions are almost completely autarchic. Previous research on land-use regulations in California has demonstrated that the stringency of regulation at the city level has significant and important effects on housing prices and demographics throughout the state. In this analysis, we focus on the spatial patterns of land-use regulation within the San Francisco Bay Area. We test whether the observed spatial pattern of land-use regulations across cities is the result of the spatial autocorrelation of their demographic characteristics, or is a consequence of the interaction among the regulatory policies adopted by city governments.

Initially, we estimate models similar to those previously applied to investigate the interaction between local government policies. We find that the number of development caps enacted by cities is positively correlated over space, and the level of density limitation imposed by a city, measured by the minimum lot-size for development, is negatively correlated over space. We consider two other regulatory variables: the number of approvals or reviews needed to obtain a permit to build and the number needed to secure a zoning change, but find that there is no significant autocorrelation over space when the demographic characteristics of the city and its neighbours are controlled for appropriately. Additionally, we examine the spatial patterns of building permits issued during the period from 1990 to 2006 for the same cities and find that there is significant, positive spatial correlation in permits issued for single-family houses and for all residential units when controlling for each city's own demographic characteristics.

However, in a series of models that test the impact of neighbours' policy decisions on the regulations enacted by a city, while controlling for that city's demographic characteristics as well as the demographics of neighbouring cities, spatial autocorrelation is not significant. This implies that the influence of policy decisions or the permitting activity of neighbouring jurisdictions on the number of land-use regulations enacted or building permits issued by a city cannot be isolated and identified, due to the strong spatial autocorrelation in the determinants of land-use regulation. Thus, despite the spatial autocorrelation in the regulatory decisions of cities, we find no credible evidence of strategic interaction among governments in the region.

In the following section, we examine the link between political fragmentation, land-use regulation and spatial outcomes. We synthesize theories and findings by economists and planners linking these phenomena. The third section considers these issues in the specific context of California's San Francisco Bay Area. We exploit newly assembled data detailing the land-use regulations enacted by more than 75 jurisdictions in the Bay Area, the building permits issued by these jurisdictions, and their demographic characteristics. The fourth section presents a series of empirical models describing the spatial pattern of land-use regulations adopted independently by nearby jurisdictions. We distinguish between strategic interactions in the adoption of regulation by nearby cities and the influence of the similar demographic conditions observed in neighbouring jurisdictions. The fifth section is a brief conclusion that discusses directions for future research.

THE IMPACTS OF POLITICALLY FRAGMENTED LAND-USE REGULATION ON URBAN GROWTH, SEGREGATION AND HOUSING PRICES

A growing body of empirical research focuses on the impact of land-use regulation on three outcomes: urban growth patterns, the spatial distribution of demographic groups and housing prices. The politically fragmented nature of land-use regulation in most metropolitan areas is central to much of this analysis, as it is the variation in the regulatory stringency in different places (metropolitan areas or the jurisdictions within them) that is linked to variations in housing prices, growth patterns and levels of segregation. Much of this research has been conducted at the metropolitan level. As noted, however, land-use regulation is quite often controlled by city governments, and there are typically a large number of local governments in any given metropolitan area. Moreover, this intra-metropolitan political fragmentation in land-use regulation gives rise to some of the more important and complicated impacts of the regulatory system.

The study of 'sprawl' and its causes has addressed the question of political fragmentation most explicitly. For example, Fulton and his colleagues found that, in metropolitan areas with more fragmented local governments, more land was converted to urban use to accommodate a given level of population growth (Fulton et al. 2001). Carruthers (2003) has developed a conceptual model indicating how the politically fragmented landscape leads to less dense patterns of growth. Land-use regulation in cities in the interior of a metropolitan area pushes new growth to the peripheral areas. After the passage of time, these outlying areas incorporate as cities with the power to regulate land use and thus to push growth further out again. In a test of this model, Carruthers finds significant, positive effects of municipal fragmentation on the percentage of metropolitan population change that occurs at the urban fringe.

The connection between land-use regulation and high housing prices is most apparent in metropolitan level comparisons, and is evident using simple graphical analysis (Malpezzi 1996; Quigley 2007). In two recent studies which focus on cities in Florida and the San Francisco Bay Area, it has been shown that variation in land-use regulation at the city level within metropolitan areas is linked to higher housing prices. Both studies use instrumental variables techniques to account for the endogeneity of regulation in a hedonic model of housing prices. Both find a significant and positive relationship between indicators of regulatory restrictiveness and housing prices (Ihlanfeldt 2007; Quigley, Raphael and Rosenthal 2007).

Three quantitative studies of California cities provide evidence of 'exclusion' attributable to strict land-use regulation, which leads to increased levels of segregation of low-income and minority households (Donovan and Neiman 1992; Levine 1999; Quigley, Raphael, and Rosenthal 2004). Using a survey of 147 California cities, Donovan and Neiman (1992) show that an increase in the fraction of Black residents in a city is negatively associated with the number of regulations enacted. Using a similar but more comprehensive survey with a larger sample of California cities, Levine (1999) reports similar results, and he also extends the analysis to address the distribution of income and the share of a city's population that is Hispanic. In a more recent analysis of the issue, Quigley et al. (2004) show that land-use regulation which favours the development of single-family housing is associated with decreases in the proportion of the minority population in cities in the Los Angeles Metropolitan Area.

Only one study has analysed the spatial association of land-use regulation across cities. Brueckner (1998) uses data on cities across California to test for an interaction between the growth control measures adopted by one city and those adopted by neighbouring cities. Using a spatial autoregressive (SAR) model, Brueckner finds that – under reasonable conditions – a variable measuring local growth control limitations has a significant effect upon similar measures adopted by other cities.

There are two major problems in interpreting Brueckner's results. First, the distance thresholds used for the spatial analysis are quite high – 50 and 100 miles. The idea that the decision by one city to impose growth controls is influenced by the decisions of cities 100 miles away is implausible – the land-use regulation decisions made by the city of San Francisco are not influenced by those made by the city of Sacramento. Second, Brueckner's model does not recognize the spatial autocorrelation of the demographic characteristics of cities, and the possibility that the observed spatial correlation of the regulations adopted by neighbouring cities may arise simply from the similarity of their socio-demographic characteristics and underlying citizen preferences. We address these two problems in the empirical analysis reported below, using new data on several distinct aspects of land-use regulations imposed by cities of the San Francisco Bay Area.

THE METROPOLITAN CONTEXT AND THE DATA

The San Francisco Bay Area includes over one hundred separate and independent land-use jurisdictions. The nine counties that define the Bay Area have authority over land-use decisions in their unincorporated areas, while the 101 incorporated cities control the land use within their boundaries. The San Francisco Bay Area has received considerable attention in the study of land-use regulation because it is one of the most heavily regulated metropolitan landscapes in the world. It also contains some of the most expensive housing in the United States.[1] The Bay Area is also notable in that it has more open space and preserved greenbelt than any other metropolitan area in the United States, due both to its unique geography and its history as home of the conservation movement (Walker 2008).

The regulatory data analysed below were collected as part of a survey conducted in 2006/2007.[2] The survey was administered to public officials in the planning offices of city and county governments, and the responses were corroborated by developers, builders and environmental consultants in the local area. The survey asked a variety of questions about the political influences on land-use regulation, the process of project approvals and zoning changes, the enactment of specific ordinances to control growth or to restrict development, and the average rate of delay and rejection of proposed-development projects. The response rate of public officials on the survey was quite high (79 percent).

We use responses from 76 of the 77 cities for which complete data are available.[3] Figure 1.1 shows the cities and counties of the San Francisco Bay Area, and indicates those cities for which data are not available. The figure also shows distance boundaries of 20, 30 and 40 kilometres around the city of Berkeley in order to illustrate the different spatial neighbourhood thresholds that will be used later.

The survey instrument represents one of the most comprehensive attempts to measure land-use regulation to date. Nevertheless, transforming qualitative survey responses on regulation into quantitative measures of restrictiveness is challenging (see, for example, Gyourko, Saiz and Summers 2008). The state of the art in this enterprise remains the simple summation of ordinances in some category of regulation, such as growth control. This yields a meaningful measure, e.g. the number of growth control regulations, which can be used to test the impact of growth controls. However, some categories of regulation have proven more difficult to quantify meaningfully, and a consistent method for combining categories into an overall index of regulatory restrictiveness remains elusive.[4]

In our empirical analysis, we focus on four subsets of questions in this survey: the approvals needed for obtaining a permit for a new housing project; the approvals needed for a change in the zoning code; the presence of caps on residential development; and the restriction of density through minimum lot-size ordinances. Tables 1.1, 1.2 and 1.3 describe the responses to these questions. Table 1.1 presents the responses to questions about reviews needed for project approval and for zoning changes. The vast majority of cities require approval from the Planning Commission, the Building and Fire Departments, and a California Environmental Quality Act (CEQA) review for all projects. In addition, slightly more than half require some sort of architectural or design review. The most prevalent type of approval needed for a zoning change is from the Planning Commission and City Council, required in almost every city surveyed. In addition, a surprisingly high number of cities require a growth management analysis and an approval from the Health Department for a zoning change. Other reviews or approvals required by cities include a public benefits review, Police Department approval, a geotechnical assessment and approval from the California Coastal Commission.

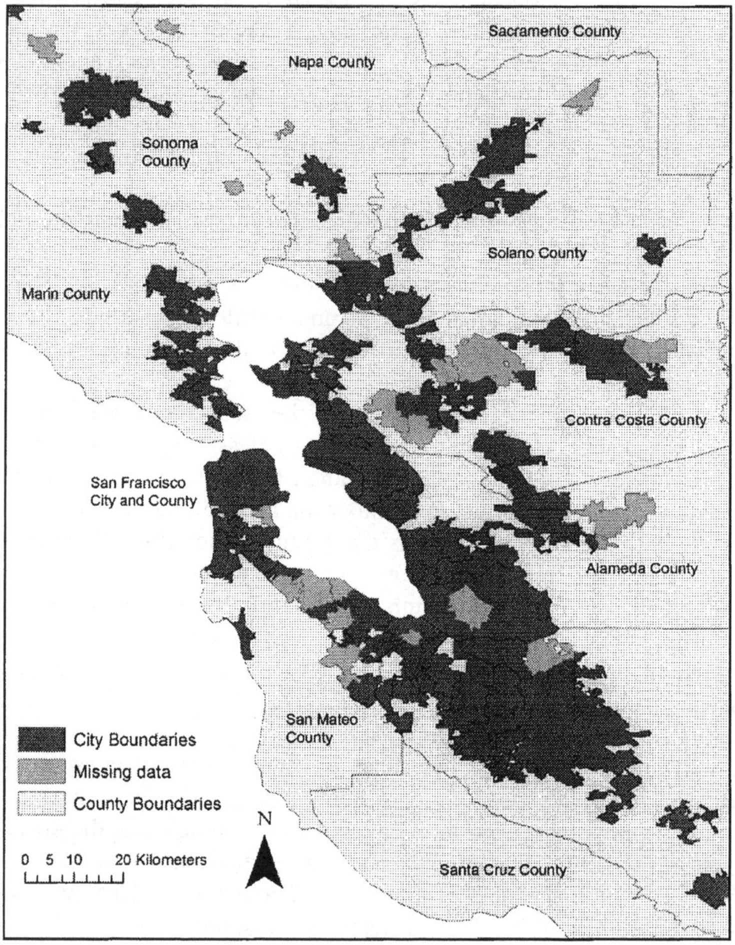

Figure 1.1 Cities and counties surveyed in the San Francisco Bay Area

Tables 1.3, 1.4 and 1.5 describe the quantitative variables derived from these survey responses. As noted, Table 1.3 outlines the minimum lot-size variable,

Table 1.1 Number and percent of cities requiring project approval or review by various public bodies

Approval or Review by:	A. No zoning change		B. Zoning change	
	Number	Percent	Number	Percent
Planning Commission	57	75	72	95
City Council	18	24	74	97
Landmarks/Historical Commission	11	14	1	1
Architectural/Design Review	45	59	7	9
Building Department	64	84	42	55
Fire Department	64	84	59	78
Health Department	19	25	60	79
Parking/Transportation	21	28	20	26
CEQA review	62	82	26	34
Growth management analysis	12	16	67	88
Other	16	21	16	21

Table 1.2 Number and percent of cities by type of development cap enacted

Category of Caps	Number of Jurisdictions	Percent of Jurisdictions
No development cap	58	76
Single-family building permits granted	12	16
Multi-family building permits granted	11	14
New single-family units	10	13
New multi-family units	10	13
Population growth	4	5

obtained by categorizing increasing sizes from 0 to 4. Tables 1.4 and 1.5 describe the cumulative number of approvals or reviews required and the number of development caps enacted by the local government. Not surprisingly, cities generally require more reviews and approvals for zoning changes than for projects that do not require a zoning change. Most cities in the San Francisco Bay Area have not enacted any development caps, but those that have enacted these caps generally adopt two or more.

Table 1.3 Number and percent of cities by lot-size restrictions enacted

Lot-Size Restrictions	Variable Coding	Number of Jurisdictions	Percent of Jurisdictions
None	0	9	12
Less than ½ acre	1	36	47
½ acre up to 1 acre	2	13	17
1 acre to 2 acres	3	5	7
More than 2 acres	4	13	17

Table 1.4 Number and percent of cities by cumulative number of approvals required

Number of Approvals Required	A. No zoning change		B. Zoning Change	
	Number	Percent	Number	Percent
0	2	3	1	1
1	4	5	2	3
2	3	4	3	4
3	5	7	7	9
4	10	13	5	7
5	18	24	11	14
6	18	24	13	17
7	9	12	11	14
8	3	4	16	21
9	4	5	5	7
10	0	0	1	1

The indexes summarized in Tables 1.3, 1.4 and 1.5 suffer from a problem inherent to the construction of quantitative indicators – they do not describe the stringency of a regulation exactly, rather the number of components it has. More ordinances need not mean stricter regulation; however, this is as good a proxy as is available. This limitation should be taken into consideration when interpreting results.

Table 1.5 Number and percent of cities by cumulative development caps enacted

Number of Caps	Number of Jurisdictions	Percent of Jurisdictions
0	58	76
1	2	3
2	10	13
3	1	1
4	3	4
5	2	3

Table 1.6 Summary of demographic data for cities surveyed in the San Francisco Bay Area (Year 2000)

Variable	Mean	Std. Dev.	Min.	Max.
Median household income (thousands of dollars)	75.83	31.42	37.18	200.00
Percent of adult residents with a college degree	21.78	10.25	5.23	41.98
Percent of dwellings that are owner-occupied	64.99	13.43	34.98	98.34
Percent of population under 18 years of age	25.43	4.82	7.00	34.00
Percent of population that is White	68.32	19.27	26.00	97.00
Single-family housing permits issued 1990–2006/(Housing stock in 1990)	0.14	0.20	0.00	1.44
All permits issued 1990–2006/(Housing stock in 1990)	0.19	0.23	0.00	1.46

Source: United States Census Bureau, 2002; California Building Institute, 2006

In addition to the data on regulations imposed by the cities of the San Francisco Bay Area, data on the number of building permits issued by these cities from 1990 to 2006 is available from the California building institute (2006). Table 1.6 summarizes descriptive information on selected demographic characteristics of the sample of 76 cities.

THE SPATIAL DISTRIBUTION OF LAND-USE REGULATIONS IN THE SAN FRANCISCO BAY AREA

Figures 1.2 to 1.5 present the spatial distribution of the measures of regulatory restrictiveness for cities in the San Francisco Bay Area. Figure 1.2 reports the number of reviews required for the approval of a new development project in each of the 76 cities. Figure 1.3 reports analogous information for the approval of a development project requiring a zoning change. Figure 1.4 documents the spatial distribution of the caps on development enacted by those same cities, and Figure 1.5 indicates the spatial distribution of density restrictions imposed by San Francisco Bay Area cities. Figures 1.6 and 1.7 present information on the housing permits issued by the same cities.

In Figures 1.2 to 1.7 there appears to be a spatial association in the geographical distribution of regulations and permits across these cities. Clusters of cities with a high number of reviews required for project approval and zoning changes are evident north of San Jose, on the peninsula near San Francisco, in Sonoma County and in central Contra Costa County. Clusters of cities with development caps are even more apparent in Sonoma and Marin County and in southern Santa Clara County. Cities with strict density restrictions are clearly clustered in the East Bay, and in Marin and Sonoma counties. As expected, high numbers of building permits for single-family housing, and all residential construction, relative to the pre-existing stock of residences are found throughout the peripheral cities of the metropolitan area. To investigate the extent to which land-use regulation is affected by the spatial relationships among these variables and the demographic characteristics of cities, we conduct four sets of tests. First, we analyse the spatial autocorrelation of the regulatory and permit variables, as well as that of the demographic variables. Then, we analyse simple autoregressive models using spatial lags and compare these to ordinary least squares statistical models of the determinants of regulatory stringency. Finally, based on the results of the autoregressive models, we incorporate spatial lags of some of the demographic variables in models known as spatial Durbin models (LeSage 1999).

First Order Spatial Autoregressive Models

The basic empirical model of spatial autocorrelation takes the following form:

$$y = \rho^* W^* y + \varepsilon, \tag{1.1}$$

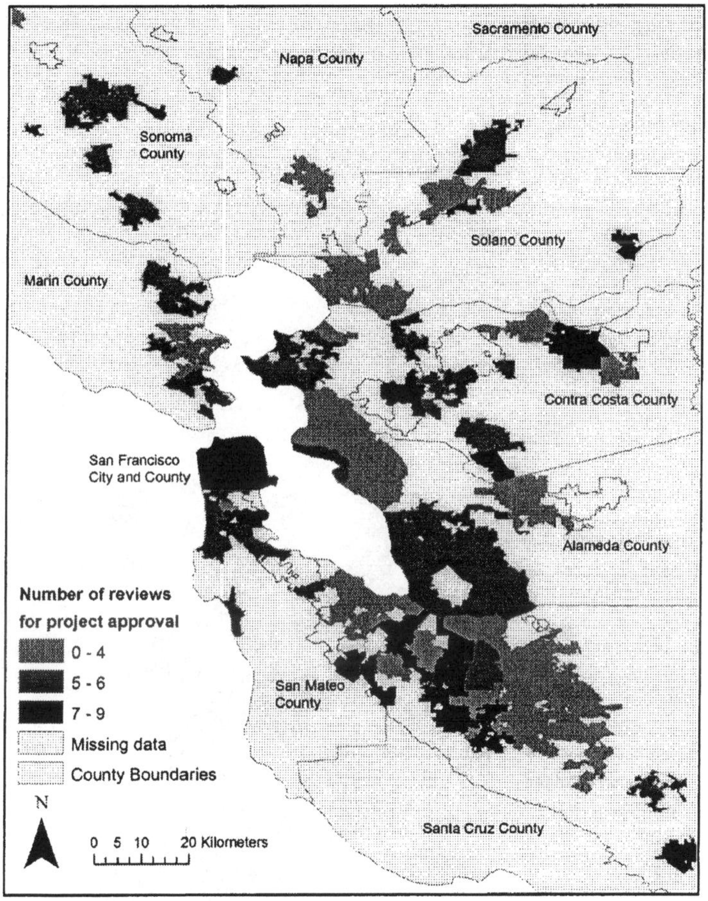

Figure 1.2 The number of reviews or approvals required for housing development projects in San Francisco Bay Area cities

Where W is a matrix of spatial weights, standardized so that its rows sum to one, and y is a quantitative variable, e.g., one of the regulation measures described above (expressed in deviations from the mean to eliminate the constant term in the model). The ρ term, a simple correlation coefficient, is estimated using maximum likelihood m ethods. We test this spatial autoregressive model using four different spatial weights matrixes: a contiguity matrix, in which the elements of the weight matrix, W_{ij}, have a value of one one if $city_i$ and $city_j$ share a border and zero otherwise, and three different inverse distance matrices. The elements of each inverse distance matrix are $W_{ij} = 1/d_{ij}$ for $d_{ij} \leq d^*$, and $d_{ij} = 0$ for $d_{ij} > d^*$. d_{ij} is the distance between $city_i$ and $city_j$, and d^* is the distance cut-off point. In the simple autocorrelation tests, we consider cut-off points at 20, 30 and 40 kilometres, as illustrated in Figure 1.1. Beyond the cut-off point, cities i and j do not influence each other.

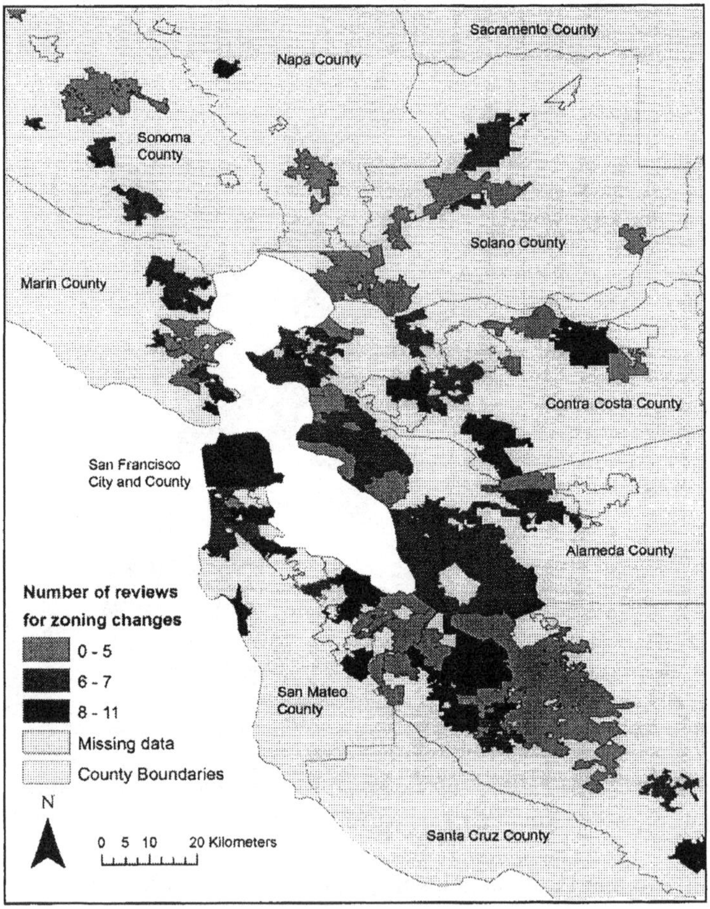

Figure 1.3 The number of approvals required for zoning change in San Francisco Bay Area Cities

Table 1.7 presents spatial autocorrelation measures for the 76 cities in the sample, using these four spatial weight matrices. All of the demographic variables are highly positively spatially correlated – cities are similar to their neighbours – and increasingly so as the definition of neighbour becomes broader. The regulatory variables are also consistently serially correlated over space, although much less so than the demographic variables, and not very highly when only considering immediate neighbours. The number of development caps and density restrictions are less spatially correlated than the number of reviews needed for project approval or zoning changes. Permits for single-family homes and all residential permits combined are also less spatially correlated than the demographic variables, especially at low distance thresholds.

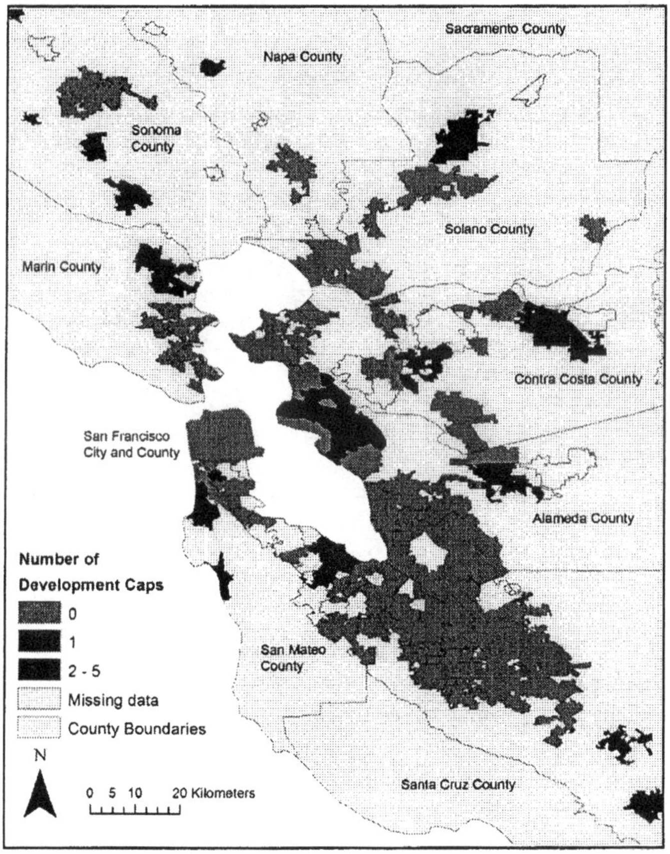

Figure 1.4 The number of development caps enacted in San Francisco Bay Area cities

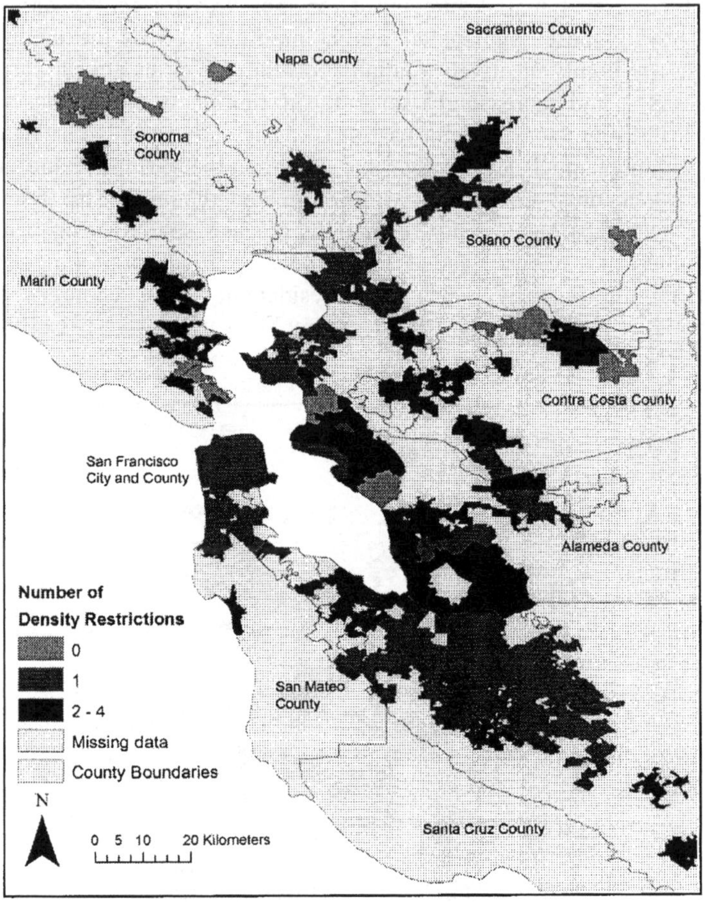

Figure 1.5 The number of density restrictions enacted in San Francisco Bay Area cities

It is striking that all the correlation coefficients reported in Table 1.7 are positive, and all coefficients are statistically significant at the 0.05 level. Indeed, of the 44 correlation coefficients reported in Table 1.7, 42 are significant at the 0.01 level. Demographic characteristics and local regulations are significantly and positively correlated over space, and the propensity of cities to issue building permits is also highly correlated.

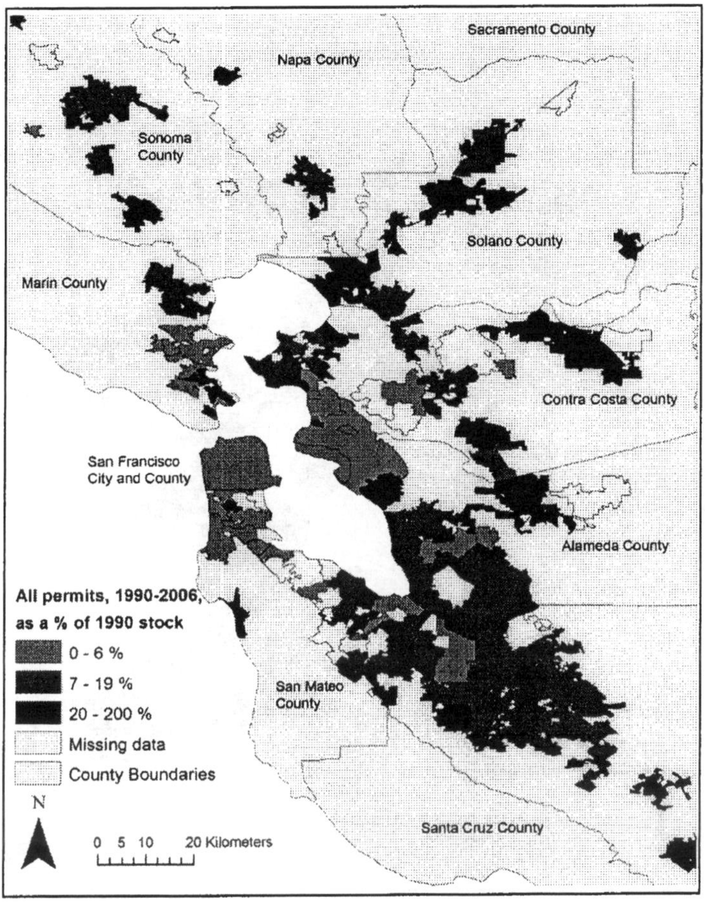

Figure 1.6 The number of single-family housing permits issued by San Francisco Bay Area cities, 1990–2006 (as a fraction of housing stock in 1990)

Based on the above measures of spatial autocorrelation, we use the 30 km inverse distance weights matrix for the remainder of the analysis. While the 'right' spatial neighbourhood is always difficult to define, considering a 30 km boundary is appropriate for the influence on regulation. Cities are influenced by more than just those cities they border, or those very close by. Restricting the distance boundary to 30 km also simplifies the analysis.

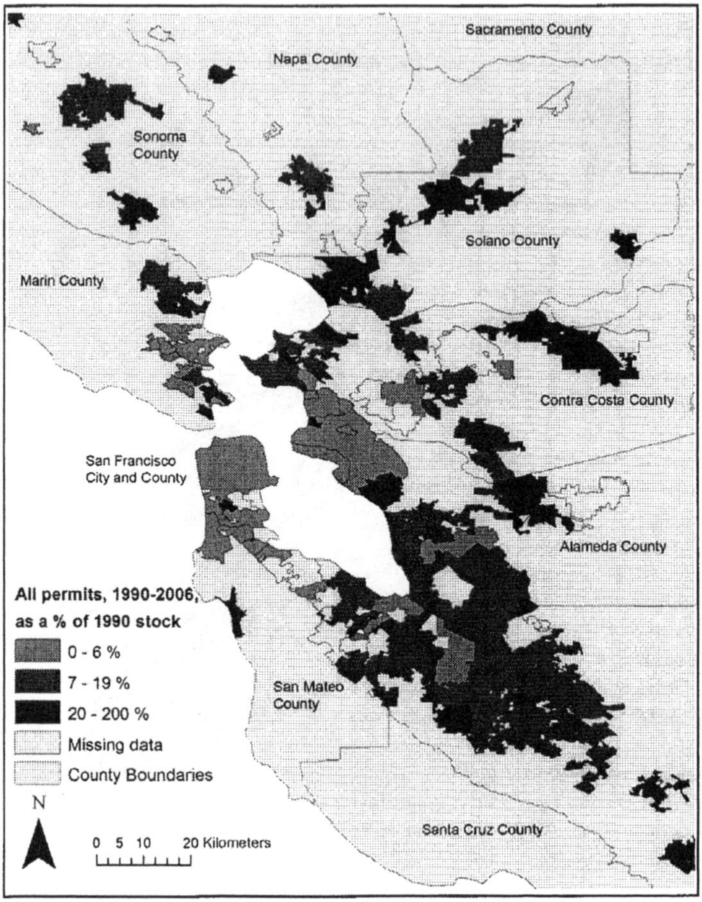

Figure 1.7 The number of all housing permits issued by San Francisco Bay Area cities, 1990–2006 (as a fraction of housing stock in 1990)

OLS and Spatial Autoregressive (SAR) Models

From Table 1.7, it is clear that a positive spatial autocorrelation exists between regulations and permits. But does this depend simply on the spatial autocorrelation of demographic variables? In order to test this, we compare simple ordinary least squares models of the determinants of regulation with results obtained using the SAR model, which includes a spatial lag of the dependent variable. SAR models take the form:

$$y = \rho^* W^* y + X^* \beta + \varepsilon\,, \qquad (1.2)$$

where again, the W matrix is a spatial weight matrix and the ρ term is the spatial correlation coefficient. Note that in this formulation, the spatial autocorrelation is conditional on the matrix of explanatory variables in X. As with other empirical work on regulation, we assume that city demographic characteristics are not endogenous in these models; the demographic information refers to conditions seven years earlier than the regulatory data, and is thus relatively unaffected by the regulation.

Table 1.8 reports the results from the OLS models relating demographic characteristics of the cities to the regulations they have adopted and the building permits they have issued. There is essentially no impact of demographic characteristics on the approvals process for new housing projects or for zoning changes. However, several demographic characteristics (the percent of residents with a college education, the percent of the residents under 18 years of age, and the percent of residents that are White), have a significant impact on the number of development caps enacted by a jurisdiction. The percent of residents that are White is positively associated with the level of density restrictions. Overall, the influence of demographic characteristics on the regulations is not large. Several demographic variables – the percent of adults with a college education, the percent of dwellings that are owner-occupied, and the percent of residents that are White – have a weakly significant impact on the number of residential building permits issued by a jurisdiction for single-family housing between 1990 and 2006, measured as a fraction of the total residential housing stock in 1990.

The SAR models reported in Table 1.9 include a spatial lag of the dependent variable, making it possible to test the effect of the level of regulation in neighbouring cities on the regulation adopted in any city, while controlling for the demographic characteristics of that place. The estimate of ρ, the spatial lag, is statistically significant for the models of development caps, density restrictions, single-family housing permits and all permits. It indicates, for example, that the number of development caps enacted by neighbouring cities positively influences the number of development caps in any city, controlling for the demographic characteristics of that city. The spatial correlation is negative for density restrictions, which is consistent with the conventional wisdom that cities restrict density within their own jurisdiction in reaction to the absence of density control in neighbouring cities.

Table 1.7 Spatial autoregression coefficients for San Francisco Bay Area cities

		Inverse distance with cut off at:		
Demographic Characteristics	Adjacent	20 km	30 km	40 km
Median household income (thousands of dollars)	0.78** (12.07)	0.89** (15.71)	0.95** (20.99)	0.96** (28.12)
Percent of adult residents with a college degree	0.81** (14.10)	0.89** (15.85)	0.92** (14.16)	0.94** (16.92)
Percent of dwellings that are owner-ccupied	0.76** (11.05)	0.91** (18.89)	0.99** (135.72)	0.99** (70.38)
Percent of population under 18 years of age	0.76** (11.27)	0.92** (21.13)	0.99** (83.15)	0.99** (106.45)
Percent of population that is White	0.72** (9.29)	0.89** (15.57)	0.95** (23.05)	0.97** (38.73)
B. Measures of Regulation				
Approvals required for projects with no zoning change	0.66** (7.22)	0.82** (9.55)	0.93** (15.39)	0.96** (25.35)
Approvals required for projects with a zoning change	0.62** (6.35)	0.82** (9.67)	0.93** (16.64)	0.95** (21.59)
Development caps	0.28 (1.93)	0.63** (4.17)	0.73** (4.49)	0.77** (4.33)
Density restrictions	0.44** (3.51)	0.73** (6.05)	0.77** (5.42)	0.82** (5.66)
C. Permits Issued				
Single-family housing permits issued 1990-2006 / (Housing stock in 1990)	0.30* (2.16)	0.59** (3.69)	0.89** (10.89)	0.92** (12.09)
All permits issued 1990-2006 / (Housing stock in 1990)	0.38** (2.83)	0.64** (4.29)	0.84** (7.58)	0.87** (8.02)

Notes: Asymptotic t-ratios are in parenthesis.
* indicates significance at the 0.05 level and ** indicates significance at the 0.01 level.

Table 1.8 OLS models of the determinants of regulatory stringency

	Project	Zoning	Caps	Density	SFH	All
Intercept	-0.72 (-0.04)	-12.82 (-0.68)	-5.63 (-0.62)	-12.56 (-1.24)	0.87 (0.40)	0.85 (0.29)
Log (Median household income)	0.64 (0.37)	1.97 (1.03)	0.46 (0.52)	1.20 (1.16)	-0.10 (-0.45)	-0.07 (-0.23)
Percent of adult residents with a college degree	2.23 (0.45)	-2.88 (-0.46)	-5.31** (-2.37)	-3.50 (-1.16)	-0.86** (-2.61)	-1.05 (-1.42)
Percent of dwellings that are owner-occupied	-2.85 (-0.88)	-3.78 (-1.03)	-2.67 (-1.53)	-1.20 (-0.69)	0.53 (1.07)	0.46 (0.86)
Percent of population under 18 years of age	-3.15 (-0.42)	-3.71 (-0.47)	7.40** (2.01)	3.55 (0.93)	0.15 (0.19)	-0.31 (-0.30)
Percent of population that is White	1.21 (0.93)	1.13 (0.82)	3.16** (3.76)	2.28** (2.99)	0.22* (1.71)	0.15 (0.87)
R-squared	0.08	0.04	0.26	0.12	0.29	0.19
F	1.21	0.65	3.67**	2.86**	7.32**	3.93**

Notes: T-ratios are in parenthesis, generated using White robust standard errors.
* indicates significance at the 0.10 level
** indicates significance at the 0.05 level.

Each of the SAR models reported in Table 1.9 is highly significant, as indicated by their associated χ^2 values. It is tempting to interpret the autocorrelation parameters, significantly different from zero in four of the models, as evidence of strategic interaction among jurisdictions. Interpreted literally, the results in Table 1.9 state that cities are likely to take neighbours' land-use regulation into account when adopting their own rules. Indeed, in Brueckner's (1998) analysis, a test for strategic interaction among local governments is merely a test of the significance of the spatial lag term. But these SAR models suffer from a potentially spurious indication of spatial influence. It is certainly possible that the significant spatial association observed arises merely from the spatial correlation in demographic variables that determine regulatory policies and permitting outcomes. Thus, although a city's proclivity to restrict density or impose development caps is associated with that of its neighbours, this may arise from the similarity in the demographic characteristics of neighbouring cities rather than from some strategic interaction or reaction. We test this below.

Table 1.9 SAR models of the determinants of regulatory stringency

	Project	Zoning	Caps	Density	SFH	All
Intercept	-1.35	-12.62	-6.74	-10.47	0.64	0.63
	(-0.08)	(-0.65)	(-0.72)	(-1.02)	(0.46)	(0.35)
Log (Median household income)	0.83	2.12	0.58	1.05	-0.07	-0.04
	(0.48)	(1.08)	(0.61)	(1.01)	(-0.48)	(-0.23)
Percent of adult residents with a college degree	2.28	-2.90	-4.62*	-3.16	-0.75*	-0.99*
	(0.44)	(-0.49)	(-1.64)	(-1.02)	(-1.82)	(-1.83)
Percent of dwellings that are owner-occupied	-3.21	-4.06	-2.26	-1.08	0.54**	0.46
	(-0.98)	(-1.09)	(-1.28)	(-0.55)	(2.06)	(1.34)
Percent of population under 18 years of age	-2.93	-4.07	5.90	4.21	-0.31	-0.67
	(-0.43)	(-0.53)	(1.58)	(1.03)	(-0.57)	(-0.94)
Percent of population that is White	1.15	1.10	2.46**	2.18**	0.12	0.06
	(0.79)	(0.67)	(3.02)	(2.54)	(1.04)	(0.42)
P	-0.26	-0.27	0.36**	-0.41*	0.57**	0.39**
	(-1.07)	(-1.10)	(2.00)	(-1.79)	(3.95)	(2.09)
χ^2	263.38	282.46	170.5	185.9	118.76	79.4

Notes: Asymptotic t-ratios are in parenthesis.
* indicates significance at the 0.10 level
** indicates significance at the 0.05 level.

Spatial Durbin Models

In order to separate the influence of the regulations adopted by neighbours from the fact that cities are demographically similar over space, we estimate a series of spatial Durbin models.[5] By including spatial lags of demographic characteristics of neighbouring cities on the right hand side of the regression model, spatial Durbin models test whether the demographic composition of neighbours influence a city's decision to adopt regulations. More importantly, however, the significance of the spatial lag of regulation, ρ, now represents a test of whether cities recognize and react to their neighbours' regulatory decisions, controlling for the demographic similarity of cities across space.

The spatial Durbin models take the form:

$$y = \rho^*W^*y + X^*\beta_1 + W^*X^*\beta_2 + \varepsilon\,, \qquad (1.3)$$

Where y, ρ, W and X are the same as before, but now there is an additional matrix X of independent variables that is lagged over space with the weight matrix W. The vector of coefficients β_2 measures the influence of demographic characteristics of neighbouring cities on the regulatory outcomes observed in a city.

Table 1.10 reports the results of spatial Durbin models that include all the independent variables considered previously and the spatial lag of the most important right-hand-side variables from the SAR model (the percent of residents with a college education, the percent of dwellings that are owner-occupied, and the percent of residents that are White). All the models are significant as indicated by the log-likelihood ratios and their associated χ^2. Some of the independent variables in the models are statistically significant at conventional levels, including some of the spatial lags of demographic characteristics.

However, the spatial lags of development caps, density restrictions and the permits for single-family housing, significant in the SAR models, are insignificantly different from zero when controlling for the demographic characteristics of neighbouring cities. Thus, we cannot reject the hypothesis that there is no association between cities' regulatory decisions and those of their neighbours. Similarly, in the models for the number of approvals or reviews needed for development projects and zoning changes, and all residential permits issued from 1990 to 2006, we are unable to reject the null hypothesis for ρ. There is apparently no interaction among local governments' regulatory decisions.

To test the robustness of the relationship, we estimate a series of parsimonious spatial Durbin models, which only include the three variables that are most significant in the SAR models and their spatial lag. Table 1.11 reports these results. The significance of ρ for models of regulatory variables and permits issued does not change under this specification, indicating that it is not possible to distinguish the spatial autocorrelation in regulations enacted by cities from the spatial autocorrelation in their demographic characteristics.

Table 1.10 Spatial Durbin models of the determinants of regulatory stringency

	Project	Zoning	Caps	Density	SFH	All
Intercept	-11.24	-20.00	-14.05	-6.70	0.00	-0.77
	(-0.61)	(-0.95)	(-1.41)	(-0.60)	(0.00)	(-0.44)
Log (Median household income)	1.50 (0.81)	2.82 (1.34)	1.14 (1.14)	0.69 (0.62)	-0.07 (-0.49)	-0.01 (-0.03)
Percent of adult residents with a college degree	2.34 (0.45)	-2.85 (-0.49)	-4.89* (-1.75)	-3.43 (-1.10)	-0.78** (-2.02)	-1.00** (-2.05)
Percent of dwellings that are owner-occupied	-3.51 (-1.08)	-4.45 (-1.21)	-2.78 (-1.58)	-0.88 (-0.45)	0.53** (2.17)	0.44 (1.42)
Percent of population under 18 years of age	0.73 (0.41)	0.83 (0.41)	2.46** (2.53)	2.63** (2.46)	0.09 (0.69)	0.01 (0.06)
Percent of population that is White	-6.08 (-0.85)	-5.56 (-0.69)	4.55 (1.18)	5.11 (1.19)	-0.42 (-0.78)	-1.03 (-1.53)
*W**(Percent of adult residents with a college degree)	-4.32 (-0.70)	-2.06 (-0.30)	-3.27 (-0.92)	2.92 (0.80)	-1.02* (-1.92)	-1.56** (-2.48)
*W**(Percent of dwellings that are owner-occupied)	3.18 (0.55)	-3.02 (-0.46)	0.51 (0.17)	-0.51 (-0.15)	1.81** (3.72)	2.58** (4.49)
*W**(Percent of population that is White)	3.84 (1.29)	4.47 (1.32)	3.54* (1.93)	-1.18 (-0.65)	-0.28 (-1.25)	-0.12 (-0.42)
P	-0.26	-0.30	0.17	-0.37	0.10	-0.13
	(-1.04)	(-1.23)	(0.81)	(-1.52)	(0.45)	(-0.56)
χ^2	261.23	280.57	166.74	185.07	134.08	98.88

Notes: Asymptotic t-ratios are in parenthesis.
* indicates significance at the 0.10 level
** indicates significance at the 0.05 level.

Table 1.11 Limited spatial Durbin models of the determinants of regulatory stringency

	Project	Zoning	Caps	Density	SFH	All
Intercept	3.24	7.36	-2.78	0.20	-0.70**	-0.88**
	(0.78)	(1.52)	(-1.27)	(0.08)	(-2.25)	(-2.31)
Percent of adult residents with a college degree	6.40*	3.55	-3.66*	-3.19	-0.83**	-0.81**
	(1.85)	(0.90)	(-1.95)	(-1.52)	(-3.19)	(-2.46)
Percent of dwellings that are owner-occupied	-3.09*	-2.30	-0.31	1.16	0.35**	0.20
	(-1.64)	(-1.07)	(-0.30)	(1.01)	(2.47)	(1.09)
Percent of population that is White	1.14	1.28	2.20**	2.38**	0.11	0.07
	(0.66)	(0.65)	(2.30)	(2.26)	(0.88)	(0.44)
*W**(Percent of adult residents with a college degree)	-2.04	1.62	-2.25	3.12	-1.10**	-1.54**
	(-0.35)	(0.25)	(--0.66)	(0.91)	(-2.15)	(-2.53)
*W**(Percent of Dwellings that are owner-occupied)	2.41	-3.33	1.79	0.69	1.76**	2.46**
	(0.43)	(-0.52)	(0.58)	(0.20)	(3.62)	(4.26)
*W**(Percent of population that is White)	2.83	3.03	3.21*	-1.03	-0.28	-0.19
	(0.99)	(0.93)	(1.80)	(-0.58)	(-1.32)	(-0.69)
P	-0.26	-0.30	0.23	-0.38	0.05	-0.21
	(-1.04)	(-1.23)	(1.07)	(-1.56)	(0.23)	(-0.89)
χ^2	262.48	282.65	169.60	187.05	133.17	96.61

Notes: t-ratios are in parenthesis,
* indicates significance at the 0.10 level
** indicates significance at the 0.05 level.

CONCLUSION

This chapter analyses the geographical pattern of land-use regulation and permitting policies which are adopted independently by local governments and made by local governments about the kinds of reviews required to obtain permits for new residential construction. We also analyse decisions about the reviews required by local governments for a change in zoning. Beyond these, we analyse the number and type of development caps enacted by local governments as well as the lot-size restrictions enacted. Finally, we analyse construction of single-family and all housing.

Our analysis documents the high degree of similarity among these regulations over space. Maps suggest that cities that adopt more restrictive regulations tend to be close to other cities that have adopted similar policies. More precise measures confirm that regulations are spatially autocorrelated and that these correlations are highly statistically significant. Measures of the demographic characteristics of cities, presumably the determinants of local regulatory policy, are highly correlated over space as well. In our quantitative analysis, we seek to determine whether the observed autocorrelation of cities' decisions regarding land-use regulation across space represents strategic interaction or whether it arises simply from the strong demographic similarity of neighbouring cities.

Previous research based on SAR models has found that the number of growth control ordinances in a city is positively influenced by the number of growth control ordinances in its neighbouring cities. We present analogous tests, clarifying the definition of neighbouring cities and expanding the analysis to several regulatory categories: reviews needed for project and zoning change approval, development caps and density restrictions. We find that for some categories (project and zoning change approvals or reviews), there is no significant impact of a spatial lag term, for others (development caps) there is a positive impact, and still others (density restrictions), there is a negative impact. It is tempting to interpret the results of these models as evidence of recognition or interaction between cities' regulatory policy.

However, the interpretation of these findings based on SAR models is confounded by the strong spatial autocorrelation of the demographic characteristics of cities. We test explicitly for the possibility that the observed impacts of regulatory decisions by a city's neighbours are spurious, arising from demographic similarity of cities across space. We thus estimate statistical models (spatial Durbin models) that control for the demographic characteristics of neighbouring cities. The impact of neighbouring cities' regulatory decisions is never statistically significant when controlling for their demographic characteristics. These models do not disprove interaction or recognition of regulatory decisions by neighbouring cities, but they provide no evidence of any form of interaction. These results also provide a simple alternative explanation for the so-called strategic interaction of local regulation describe elsewhere.

The analysis suggests several directions for empirical research on the spatial aspects of land-use regulation. First, it would be useful to compare results from similar models in different metropolitan areas in order to understand the importance of region-specific factors on the distribution of regulatory decisions across space. Second, a method for incorporating the influence of county level land-use regulation of unincorporated areas is necessary to complete the understanding of local land-use regulation over space in metropolitan areas. Finally, it seems that cross-sectional analysis will not be sufficient to show evidence of interaction, and it is necessary to investigate explicitly the dynamics of regulation and regulatory interaction.

NOTES

1. In the terminology of Gyourko, Mayer and Sinai (2006), San Francisco is the prototypical 'superstar city', characterized by high housing prices, land scarcity and the in-migration of high-income families.
2. For detailed information on the methodology, see Calfee et al. (2007).
3. One city, Brentwood, granted five times the number of residential building permits between 1990 and 2006 than it had dwellings in 1990. This was four times more than the next highest city. We exclude it from the analysis reported below.
4. In an earlier paper on the house price impacts of land-use regulation that used the same survey data, such an index was constructed, the Berkeley Land Use Regulation Index (Quigley et al. 2008). However, it was subsequently demonstrated that sub-indexes of different categories of regulation were preferable as measures of regulatory strictness.
5. See Brasington and Hite (2005) for a clear exposition of the spatial Durbin model and an application to housing markets.

REFERENCES

Brasington, David M. and Diane Hite (2005), Demand for environmental quality: A spatial hedonic analysis, *Regional Science and Urban Economics*, 35: 57–82.

Brueckner, Jan K. (1998), Testing for strategic interaction among local governments: The case of growth controls, *Journal of Urban Economics*, 44(3): 438–467.

Calfee, Corie, Paavo Monkkonen, John M. Quigley, Stephen Raphael, Larry A. Rosenthal and Joseph Wright (2007), *Measuring Land-Use Regulation: Report to the MacArthur Foundation*, Berkeley Programme on Housing and Urban Policy Professional Report No. P07-002 Part 1.

California Building Institute (2006), *Building Permit Data: 1990–2006.* Burbank, CA.

Carruthers, John I. (2003), Growth at the fringe: The influence of political fragmentation in United States metropolitan areas.,*Papers in Regional Science*, 82: 475–499.

Donovan, Todd and Max Neiman (1992), Community social status, suburban growth and local government restrictions on residential development, *Urban Affairs Quarterly*, 28: 232–336.

Fulton, William, Rolf Pendall, Mai Nguyen and Alicia Harrison (2001), *Who Sprawls the Most? How Growth Patterns Differ Across the US*, The Brookings Institution, Center on Urban and Metropolitan Policy.

Gyourko, Joseph E., Christopher Mayer and Todd Sinai (2006), Superstar Cities, NBER Working Paper No. W12355.

Gyourko, Joseph E., Albert Saiz and Anita A. Summers (2008), A new measure of the local regulatory environment for housing markets: The Wharton Residential Land Use Regulatory Index, *Urban Studies* 45(3): 693–729.

Ihlanfeldt, Keith R. (2007), The effect of land use regulation on housing and land prices, *Journal of Urban Economics*, 61: 420–435.

LeSage, James P. (1999), *The Theory and Practice of Spatial Econometrics*, Department of Economics, University of Toledo, February 1999.

Levine, Ned (1999), The effects of local growth controls on regional housing production and population redistribution in California, *Urban Studies*, 36(12): 2047–2068.

Malpezzi, Stephen (1996), Housing prices, externalities, and regulation in US metropolitan areas, *Journal of Housing Research*, 7(2): 209–241.

Quigley, John M. (2007), Regulation and Property Values in the United States: The High Cost of Monopoly, *Land Policies and their Outcomes*, G. Ingram and Y.H. Hong, eds, Lincoln Institute: 46–66.

Quigley, John M., Steven Raphael and Larry A. Rosenthal (2004), Local land-use controls and demographic outcomes in a booming economy, *Urban Studies*, 41(2): 389–421.

Quigley, John M., Steven Raphael and Larry A. Rosenthal (2007), Measuring Land-Use Regulation: An Examination of the San Francisco Bay Area, 1992–2007. Berkeley Programme on Housing and Urban Policy Professional Report No. P07-002 Part 1.

Quigley, John M., Steven Raphael and Larry A. Rosenthal (2008), Measuring Land-Use Regulation, *Housing Markets and the Economy: Risk, Regulation, Policies*, E.L. Glaeser and J.M. Quigley, eds, Lincoln Institute.

United States Census Bureau (2002), *Census 2000 Summary File 3 – California*.

Walker, Richard (2008), *The Country in the City: The Greening of the San Francisco Bay Area*, Seattle, WA: University of Washington Press.

2. Intra-urban mobility and changing density functions in Tel Aviv, 1995–2006

Daniel Felsenstein and Shlomie Hazam

INTRODUCTION

While the impact of the CBD in the modern metropolis is rather different from that posited in the classic Alonso–Mills–Muth (AMM) models of urban structure and polycentricity has become a fact of life in many large urban areas, the measurement of this dispersal is still predicated on the existence of an urban focal point. The focal point is important because the estimation of change in house prices or urban densities with distance needs to be measured in relation to an existing condition which is invariably the CBD. Even in the presence of sub-centres, many studies have replicated the simple monocentric distance decay function of about 8 percent per mile for house prices (Glaeser and Kahn 2001), 12–16 percent for employment deconcentration (Macauley 1985; Glaeser and Kahn 2001) and a similar figure for population (McDonald and Bowman 1976). This surprising consistency illustrates the robustness of the density function as a tool for describing urban development.

Obviously the complex intra-mobility patterns of the modern metropolis serve to undermine some of the simplicities of the monocentric assumptions. While some authors extend the monocentric model to include secondary centres of employment or housing (Henderson and Mitra 1994), the distance decay relationship will necessarily be flatter. However, the basic underlying utility function that implies a spatial equilibrium in the trade-off between central location and housing/transportation costs will still hold irrespective of whether urban form is mono or polycentric. Metropolitan development is simply an extension in scale and scope of polycentric development. This study tests whether increasing metropolitanization impacts on the density functions of the metropolitan core area. Ostensibly, increasing the scale of the metro area should decrease core area densities unless other urban processes such as infilling and land incorporation are occurring simultaneously. In terms of dynamics, previous work indicates that this impact declines over time. We test these hypotheses with respect to the Tel Aviv metropolitan area over the last decade and in relation to densities of both population and economic activity.

The objective of the chapter is to take a fresh look at density functions for Tel Aviv using simulated data at a much finer level of analysis than has hitherto been used. In addition, we incorporate spatial estimation that takes direct account of the role of contagion effects in the estimation of density functions and the identification of incipient subcentres in the metropolitan area. While earlier studies have departed from the standard monocentric analysis by trying to identify clusters of emerging locations through examining significant positive residuals in the density functions, correlated or not across space (McDonald and Prather 1994; McMillen 2001), to our knowledge this has not been done in an explicitly spatial framework in the past. In the second section, we proceed to synthesize some of the findings relating to earlier work on the Tel Aviv metropolitan area. The third section then describes the method employed, both descriptive and analytical. We attempt to make the case for an estimation procedure that takes direct account of spatial spillover effects across neighbouring measurement units and accounts for the bias that spatial dependence can cause. We argue that the spatial autoregressive (SAR) model offers an appropriate form of estimation. Due to the synthetic nature of the data used, the process behind its generation is described in the fourth section. Empirical results are presented in the fifth section, and some insights relating to the ongoing metropolitan processes fashioning the development of Tel Aviv are discussed in the conclusions.

LITERATURE REVIEW

The use of the negative exponential density function to represent the urban structure for the monocentric city was pioneered by Clark (1951). This seminal contribution spawned considerable empirical work including Muth (1969) who tested this form on nearly 50 cities and McDonald and Bowman (1976) who went on to test the suitability of other functional forms such as the quadratic and higher order polynomials such as cubic spline (Mahmassani, Hadibaaj and Tong 1988). These forms were posited as more suitable for describing the realities of the monocentric city punctuated by discontinuities, emerging subcentres, peripheral housing estates and statutory land use controls such as growth boundaries or green belts. The argument was that all these complexities required more flexible functional forms for describing urban densities in population, employment and economic activity. However, whatever the form of estimation, the empirical results on the effect of polycentricity on population densities are far from equivocal. Some studies find very little effect (Griffith 1981). Others find evidence in classic polycentric cities only such as Los Angeles (Gordon, Richardson and Wong 1986) while others still relate population density patterns to the influence of exogenously defined subcentres (Small and Song 1994). Other work looks at the effect of dual centres within a single city and their competing impacts on population densities (Alperovich and Deutsch 1996).

In the context of the Tel Aviv metropolitan area, the earliest work grounded in the monocentric model is Shachar (1975). He fitted the standard negative exponential relationship for population density although his focus was really on assorted indices in the concentration of commercial activity. Subsequent work on density functions for both population and economic activity in Tel Aviv (Alperovich 1980; Stern 1994; Krakover and Adler 2009) reiterated this seemingly universal pattern. Much of this work has examined density functions for two or more periods. These dynamics show, as expected, an inter-temporal decline in the strength of the negative coefficient indicating that the effect of distance-decay decreases over time. This result is also replicated in other studies (Macauley 1985). In Table 2.1 we have calculated this inter-temporal change in the distance coefficients for Tel Aviv studies since Shachar (1975). To allow for comparison we have related to a uniform definition of the Tel Aviv metropolitan area that includes the 'metropolitan core' area defined here as encompassing the cities of Tel Aviv, Ramat Gan, Givatayim, Bnei Brak, Holon Bat Yam and Bnei Brak. This was the metropolitan area as defined in Shachar (1975) and serves at the centre of the contemporary metropolitan area. This is also the area of reference for the current study (Figure 2.1).

Despite the fact the distance decay effects relate to different time periods and that the studies all use different spatial units of analysis, a clear pattern emerges of a significant negative effect of distance that reduces over time. The size of this reduction varies by time period and form of estimation but their relative magnitudes remain comparable. The first comprehensive analysis of density functions (Shachar 1975) fitted two exponential functions and used 167 administrative units (statistical areas, SAs). In contrast, Alperovich's study (1980) that looked at the period 1961–76 and used only 25 SAs examined the factors impact distance decay (income, metropolitan size) and the temporal change in their effect. Stern's (1994) study attempted to estimate the differences in the intensity of the negative distance parameter

Table 2.1 Review of previous studies

	Study				
	(1) Shachar (1975)	(2) Alperovich (1980)	(3) Stern (1994)	(4) Krakover and Adler (2009)	(5) Current Chapter
% Change					
Period I	-14	-33	-38	-39	-25
Period II	-27	-28	–	-30	-10
Function	Geometric	Geometric	Geometric	OLS	OLS
Units	167 Stat. Areas	24 Stat. Areas	219 Stat. Areas	232–325 Stat. Areas	1832 Grid Cells

Notes
Study (1) I 1961–1967; II 1967–1972
Study (2) I 1961–1972; II 1972–1978
Study (3) I 1972–1983; II –
Study (4) I 1972–1983; II 1983–1995
Study (5) I 1995–2001; II 2001–2006

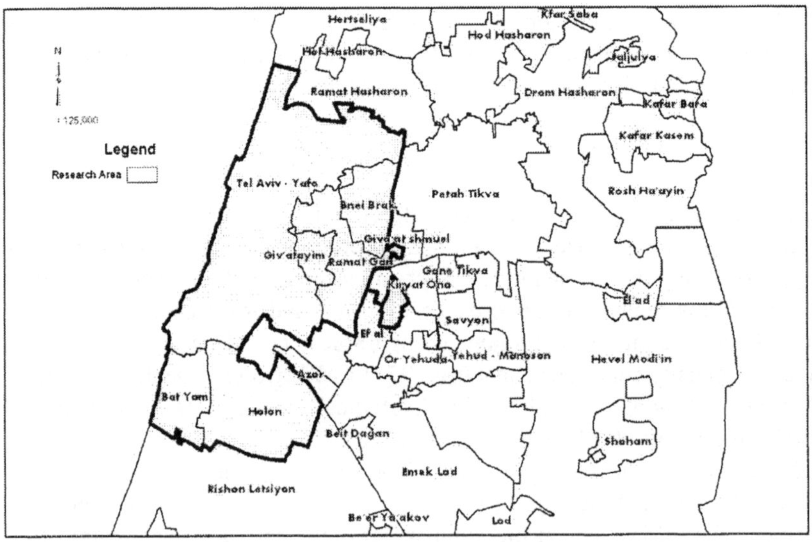

Figure 2.1 Tel Aviv metropolitan core area and wider region

when measured from different points within the metropolitan core (CBD, historic centre etc.) and by using various functions (standard Clark-Newling, root, lognormal etc.). Recently, Krakover and Adler (2009) have used linear and exponential functions for testing the hypothesis that Tel Aviv's urban structure has evolved from monocentric to polycentricity, over the period 1972–2000. To support this, they use standard functional forms and a combination of census data for SAs and synthetic data for travel area zone (TAZs) generated by a transportation model.

METHOD

The method used for this study consists of two main parts. This first is descriptive and deals with mapping the dynamics of metropolitan population density and tracing the changing centre of gravity of the metropolitan core over time. Despite the focus in this chapter on the metropolitan core area, we are cognizant of the fact that metropolitan Tel Aviv in the 21st century is a polycentric system characterized by subcentres (Krakover and Adler 2009). Thus, using the 'neighbourhood statistics' GIS function we endeavour to identify emergent metropolitan subcentres for the periods 1995–2006 both in terms of population and commercial activity. This procedure computes an output raster where the value at each location is a function of the input cells in some specified neighbourhood of the location. For each cell in the input raster, the neighbourhood function computes a statistic based on the value of the processing cells within a specified neighbourhood and then sends this value to the corresponding cell location on the output raster. In the present case, the neighbourhood comprises the seven cells adjacent to the target cell. The resultant maps depict the density nodes that emerge from the cumulative cell values.

An additional descriptive device relates to defining the centre for gravity of the study area and locating the weighted centre of the metropolitan core area. This exercise is important given the fact that this study estimates density functions over time. As such the measurement of the distance effect may need to change across time periods as the centre of gravity of the area of observation may shift. We define the geometric centre of the study area at each point in time as the weighted mean centre of a cluster of points as follows:

$$\bar{x}_{wc} = \sum_j f_i x_i / \sum_j f_i \,;\, \bar{y}_{wc} = \sum_j f_i y_i / \sum_j f_i \qquad (2.1)$$

where wc = the weighted centre and f = the frequency (or weighting) factor.

The analytic component of this analysis deals with estimating the density functions for the metropolitan core area for the years 1995, 2001 and 2006. This is done in two stages. In the first stage a simple estimate of the effect of distance on ln density is derived.[1] This is done both for population density and commercial activity density. However, given the nature of the data being used (continuous surfaces divided into grid cells), we suspect that spatial dependence inherent in the data is likely to result in biased estimation. On this basis, in the second stage we estimate the OLS model with spatial dependence. Two options are available for this. One is that spatial effects are expressed as a serial correlation in the model's residuals (spatial autocorrelation, SAC). The other is that spatial dependence exists between the observations themselves (spatial lag or autoregression, SAR). Estimating a spatial regression calls for the use of a weight matrix (*W*) that gives more weight to closer observations. In this study we use inverse distance weighting i.e. $\omega_{ij} = 1/d_{ij}$ to construct the weight matrix (Dubin 1998),[2] which seems suitable due to the grid-like structure of the data. We measure the inverse distance between the centroids of the grid cells as the data is point-oriented and not a continuous surface. First-order neighbours are defined as all points within a 250m radius.

An obvious first indication of spatial dependence in the data is provided by the Moran's *I* (*MI*) statistic for spatial autocorrelation.[3] Given the presence of a significant *MI*, it can be shown that the choice of spatial regression estimation in the univariate case results in nearly identical estimators. The reason for this is that the SAR model is formally a restricted version of the SAC model. The unrestricted SAC model can be expressed as follows:

$$\begin{aligned} Y_i &= \alpha_i + \beta X_i + \rho W Y_i + u_i \\ u_i &= \phi W u_i + v_i \end{aligned} \tag{2.2}$$

where:

$$\begin{aligned} u_i &= B v_i \\ B &= (I - \phi W)^{-1} \end{aligned} \tag{2.3}$$

Substituting for u_i:

$$\begin{aligned} Y_i &= \alpha_i + \beta X_i + (I - \phi W)^{-1} v_i \\ (I - \phi W)Y &= \alpha (I - \phi W) X + v \end{aligned} \tag{2.4}$$

Equation (2.4) can reduce to:

$$Y_i = \alpha_i + \phi W Y_i + \beta X_i + \phi W X_i + v_i \tag{2.5}$$

Equation (2.5) is the mixed SAR model also called the 'spatial Dubin' model in the literature (Anselin et al. 1996). In this model spatial effects are expressed both in the spatial autoregressive parameter and in the spatial lag on the independent variable ϕWX_i. We can therefore see overlap between the SAC and SAR models where the only difference is the spatial lag on the independent variable in the SAC model. Therefore the SAR model is a restricted version of the SAC model and our SAR estimates are presented below.[4]

DATA AND VARIABLES

Data Preparation

In contrast to previous work on intra-urban density patters in Israel, the data preparation for the present study was not straightforward. Two primary data sources have been used. We use observed (real) data on population and commercial activity densities from the National Census 1995 and Travel Survey 1996, conducted by the Israel Central Bureau of Statistics (CBS). These data are used at the statistical areas (SA) level for Tel Aviv and its five satellite cities, i.e. the metropolitan core as defined by Shachar (1976). In the absence of census data post 1995, density data for 2001 and 2006 is 'synthetic', i.e. derived from the UrbanSim land use microsimulation model that has been calibrated for the Tel Aviv metropolitan area (Felsenstein, Ashbel and Be-nun 2007). In similar fashion, Krakover and Adler (2009) used estimated data from a transportation model to estimate density functions for Tel Aviv for the year 2000.

The derived data is generated by letting the UrbanSim model run from its base year of 1995 till the target years of 2001 and 2006. The model simulates the activities of the principal agents affecting land use such as firms, households, developers and governments, and their resultant impacts on land use via density, land prices and mobility patterns. Data is organized on a 250m grid and each grid cell is allocated a SA to which it corresponds (Figure 2.2). Data from each census tract is transferred to the grid using GIS and where an SA covers more than one grid, data is divided equally between the grid cells. Tabular data is incorporated in a two-stage process. In the first stage, data is allocated to a GIS vector layer of SAs. In the second stage this is converted to raster using zonal statistics. Wherever there is more than one grid cell per SA, the data is allocated proportionally to the cells. This is based on the assumption that SAs are homogenous units.

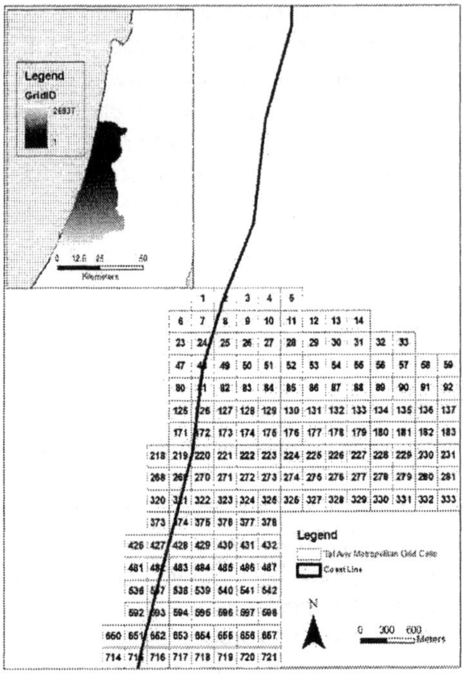

Figure 2.2 Grid cell data preparation in UrbanSim

Variables

This study relates to two density variables. The first relates to population density that is measured as the number of people residing in a given grid cell (*Res*) when all cells are of equal size.[5] The second variable relates to density of economic activity. This is measured by the share of the grid cell occupied by commercial activity (*Com*). The first variable therefore uses volume of people while the second relates to land area coverage.

Distance is measured as the aerial distance from the centre of each cell to the CBD of the city of Tel Aviv. This is defined above (in the third section) in respect to the geometric centre of the metropolitan core area. Using the same definition for the city, we combine the geometric centres for population and commercial activity as follows, $(\sqrt{\mathrm{Res}^{0.5}} + \sqrt{Com^{0.5}})$. Over the three periods, this centre is empirically found to be close to the Ichilov Hospital in central Tel Aviv with slight deviations across the years. To arrive at a point estimator, we delimited a polygon around the three weighted centres and used the centroid of the resultant polygon to represent the CBD for the city. The other variables in this study (weight matrix, lags etc.) are generated in the estimation process using the GeoDa software package (Anselin 2003).

EMPIRICAL RESULTS

Description of Trends over Time

Ostensibly, the shift in the centre of gravity over time in the study area should reflect the deconcentration of population and economic activity. Remarkably, the movement of the geometric centre has been minimal. Across all three time periods the weighted centre of the metropolitan core has remained on the east side of the Ayalon freeway in the area of Yad Eliyahu (Figure 2.3). This area is nearer to the new CBD of Tel Aviv clustered around the office towers of Ramat Gan than it is to the old CBD centred on the banking district around Allenby and Yehuda Halevi streets. It lies just east of the projected (planned) CBD that is expected to develop in the future along the Ayalon freeway. The estimated centre of gravity therefore reflects the equilibrium created by the centripetal forces pulling the population to the popular residential locations north and east of the metropolitan area and the centrifugal forces that entrench existing economic activity in the current CBD areas.

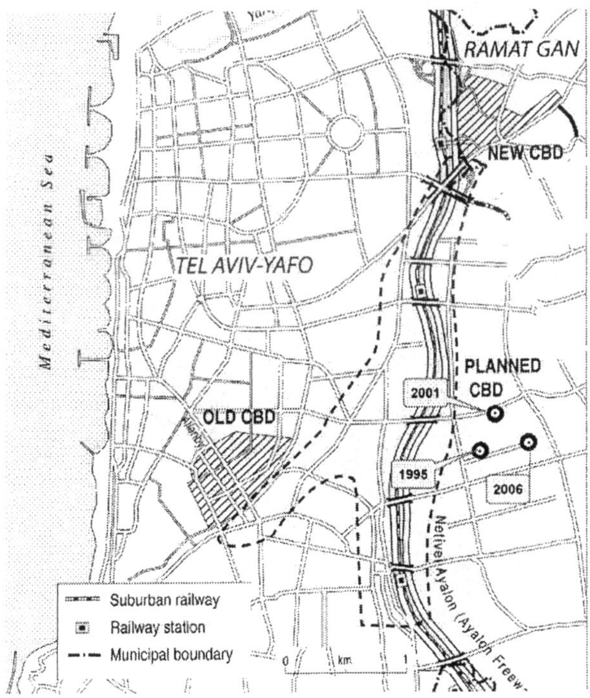

Figure 2.3 Estimated weighted centre of gravity for Tel Aviv metropolitan core in relation to old, new and planned CBD areas

From a metropolitan-wide perspective, two important questions arise. First, can we identify the emergence of incipient centres of either population or economic activity that are likely to compete with current centres? Second, do these centres develop on the basis of spatial spillover from existing centres? If so, contagion may be an issue in understanding density patterns. We investigate the first question using neighbourhood statistics. Figure 2.4 shows the analysis is relation to population density for the metropolitan core area (outlined) and wider region.

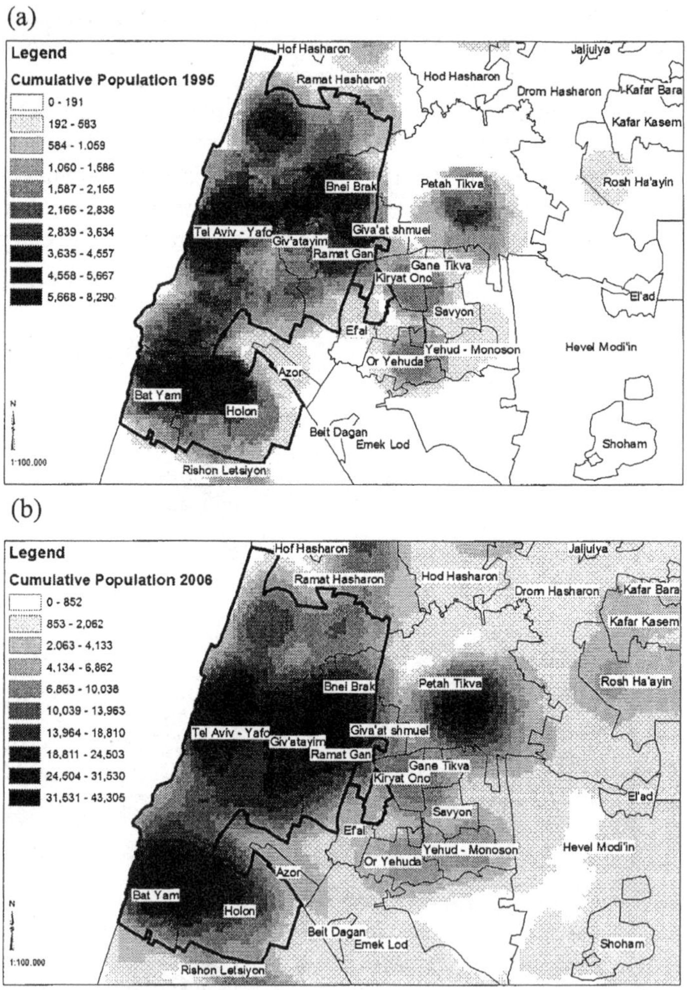

Figure 2.4 Identifying population subcentres using neighbourhood statistics (a) 1995 and (b) 2006

Each cell value incorporates the individual cell value and the density level of the neighbouring cells. Thus the data surface for the entire metro area gives an interpolative effect of emerging concentrations. As can be seen by comparing 1995 and 2006, within the confines of the metropolitan core there is a population entrenchment in existing centres (Tel Aviv, Bat Yam, Holon and Bnei Brak) while seemingly emergent concentrations in 1995 such as Herzliya do not to materialize by 2006. In the wider metropolitan area, Petach Tikva develops as a subcentre while the incipient growth of Or Yehuda and Kiryat Ono in 1995 does not seem to take root by 2006. The overall picture for population growth is one of self-sustaining growth of the centres within the metropolitan core and only selective subcentre development outside this area.

Commercial development over this same period displays a more diffuse picture (Figure 2.5). At the beginning of the period, initial subcentres exist in Bnei Brak in the north and Holon/Bat Yam in the south. By 2006, these have grown and have been supplemented by other centres in the city of Tel Aviv and on the perimeter of the metropolitan core in Rishon Letzion and Givat Shmuel. In the wider metropolitan area, we can identify a pattern of further non-continuous diffusion, which could be either leading or following residential movement. This situation could be indicative of disequilibrium in the metropolitan system with unsynchronized development of population and commercial subcentres.

Density Functions for the 2000s: Spatial Estimation

Table 2.2 presents the results of the estimated density functions for 1995, 2001 and 2006. The estimations relate to population and commercial activity and relate to the metropolitan core area. Each function is estimated twice: once using OLS and once using the SAR model (see Method section above). The unit of observation is the 250m sq. grid cell and 1832 such units are generated for the study area.

The first consistent finding is the (expected) significant inverse relationship between density and distance. This relationship is more accentuated for population density but nevertheless still characterizes the density pattern of commercial activity. The magnitudes of the OLS coefficients are slightly larger than those found in similar studies (Shachar 1975; Krakover and Adler 2009) but without noticeable bias. Coefficients for population density are roughly double those for employment density indicating a faster distance decay pattern. In all cases, however, distance itself contributes a small fraction of the variance in density as evidenced by the low R^2 values and this contribution is smaller for commercial activity than it is for population.

The spatial correlation coefficient (*MI*) is positive and significant for all periods and indicates spatial dependence in the data. As a result we run a series of SAR regressions that all produce positive and significant autoregressive coefficients for all activity in all time periods. This seems to indicate the potency of contagion effects in the creation of subcentres and in the emergence of clusters of residential and commercial activity (McMillen 2001).

(a)

(b)

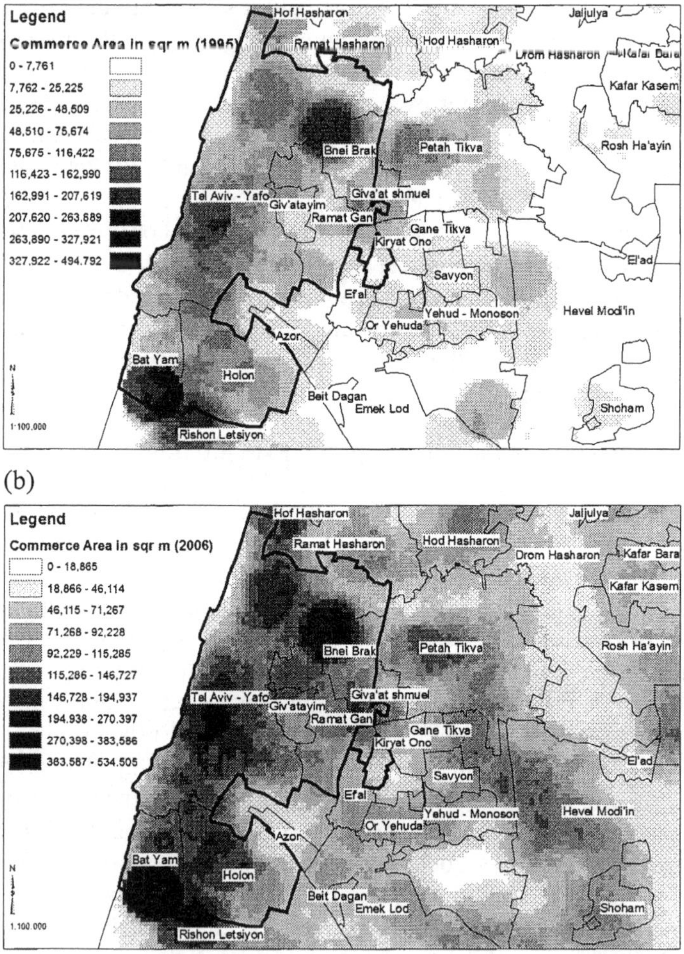

Figure 2.5 Identifying commercial activity subcentres using neighbourhood statistics (a) 1995 and (b) 2006

Table 2.2 Results of OLS and spatial regression

		Density Functions							
		Population				Economic Activity			
Period	Form of Estimation	LnDist	ρWY	R^2	*MI*	LnDist	ρWY	R^2	*MI*
1995	OLS	-0.096**		0.06	0.682**	-0.05**		0.01	0.211**
	SAR	-0.018**	0.082**	0.68		-0.031**	0.037**	0.12	
2001	OLS	-0.072**		0.04	0.641**	-0.054**		0.01	0.184**
	SAR	-0.016**	0.079**	0.63		-0.037**	0.031**	0.09	
2006	OLS	-0.064**		0.03	0.625**	-0.038**		0.01	0.309**
	SAR	-0.015**	0.078**	0.61		-0.021*	0.047**	0.20	

Notes:
$n = 1832$; SAR = Spatial Autoregressive Model; *MI* = Spatial Correlation Coefficient (Moran's I) .
ρWY = Spatially weighted autoregressive parameter = $p < 0.05^{**}$; $p < 0.001$.

Despite these general patterns we can discern some differences across the models. First, as noted earlier, population density functions become flatter over time. The impact of distance becomes attenuated over the period 1995-2006, a finding that is replicated other studies (Literature Review section above). This effect occurs in both the OLS and SAR estimations. The neighbourhood or contagion effect also becomes more moderate over time. In the case of commercial activity this finding is less equivocal. Over the period 1995–2001 density patterns in commercial activity seem to strengthen and the effect of distance increases, all pointing to growing concentration in commercial activity. However, these trends are reversed over the period 2001–06. The effect of neighbours is also inconsistent over this period.

In sum, the analytic findings point to a clear pattern of continuing concentration in metropolitan Tel Aviv although the impact of this trend seems to decline over time. The density patterns discovered would seem to include no small measure of the spatial spillover and reciprocal effects between neighbouring spatial units. These findings seem to show that spatial entrenchment and spatial spillover can occur simultaneously.

CONCLUSIONS

What new insights can we derive from the above results with respect to the continuing metropolinization of Tel Aviv? While many of the processes of deconcentration, both continuous and leap-frogging, are similar to those occurring elsewhere (Benguigi, Czamanski and Marinov 2001), in Tel Aviv this has not resulted in a weakened central city. The estimated density functions above hint at the continued dominance of the metropolitan centre, even if not to the same extent as a decade ago. In this respect, the metropolitan processes fashioning Tel Aviv's development are closer to the European form of decentralisation than that of the US (Dijst and Vazquez 2007). In contrast to the latter, they do not result in a hollow metropolitan centre or population density craters so often identified elsewhere (Newling 1969). This testifies to Tel Aviv's power in reinventing itself. Despite a population downturn in the 1980s and early 1990s, an unfavourable demographic balance and attempts to capture CBD activities by neighbouring jurisdictions, the estimated weighted centre of gravity for the city has remained remarkably stable over the decade 1995–2006.

Our use of spatial estimation also points to the existence of incipient subcentres at the metropolitan level. While the window of examination here is only a decade and not enough time to really observe the germination and growth of a subcentre, the empirical results are indicative of what can be done using tests of proximity to identify new clusters. It is true that these new concentrations do not display the sharp breaks in continuity often associated with subcentres (McMillen and Lester 2003) and that their form of graphic representation through neighbourhood statistics serves to wash out some of the mini-peaks that can exist in the data. However, it cannot be denied that if density functions have spatially correlated residuals or are spatially autoregressive, then something to do with contagion must be affecting the observed patterns. Our data show this to be the case for Tel Aviv. Subcentre development is one of the identifying characteristics of US metropolitan areas (McMillen and Lester 2003). Tel Aviv thus shows a dualism in its metropolitan development that positions it in some respects close to European model and in others, closer to the North American pattern.

NOTES

1. Krakover and Adler (2009) show that when estimating density functions for a monocentric city, no real difference arises from linear or geometric estimation.
2. Alternative weighting schemes of course exist. Common border weighting where element $\omega_{ij}=1$, units $_i$ and $_j$ share a common border and $\omega_{ij} = 0$ is otherwise (Cliff and Ord 1981) was not considered appropriate due to the grid structure of the data. Exponential distance decay weighting, i.e. $\omega_{ij} = \exp(-d_{ij})$ (Fik et al. 2003) did not yield very different outcomes.
3. Morans I (MI) is defined as:

$$MI = N / S_0 \sum_i \sum_j W_{ij} * (x_i - \mu) * (x_j - \mu) / \sum_i (x_i - \mu)^2$$

where $S_0 = \sum_i \sum_j w_{ij}$ (constant weight); N = Number of observations, w_{ij} = spatial weights based on aerial distances between centroids of the cells, x_i and x_j are observations i,j with average value μ .

4. Equation (2.5) was also estimated empirically in this study but since the ϕWX_i parameter did not yield significant results, the results presented relate to the SAR model only.
5. Since population density relates to the density of people at place of residence, population density and residential density are synonymous.

REFERENCES

Alperovich G. (1980) Determinants of Population Density Gradient in the Tel Aviv Metropolitan Area, *Urban Studies*, 17, 185–192.

Alperovich G. and Deutsch Y. (1996) Urban Structure with Two Co-existing and Almost Completely Separate Populations, *Regional Science and Urban Economics*, 26, 171–187.

Anselin L., Bera A., Florax R. and Yoon M. (1996) Simple Diagnostic Tests for Spatial Dependence, *Regional Science and Urban Economics*, 26, 77–104.

Anselin L. (2003) *GeoDa™ 0.9 User's Guide*, Spatial Analysis Laboratory.

Benguigui L., Czamanski D. and Marinov M. (2001) City Growth as a Leap-frogging Process: An Application to the Tel-Aviv Metropolis, *Urban Studies*, 38 (10), 1819–1839

Clark C. (1951) Urban Population Densities, *Journal of the Royal Statistical Society, Series A*, 114, 490–496.

Cliff A.D. and Ord J.K. (1981) *Spatial Processes: Models and Applications*, Pion, London.

Dijst M. and Vazquez C. (2007) Employment Deconcentration in European Metropolitan Areas: A Comprehensive Comparison and Policy Implications, pp. 265–292 in Razin E, Dijst and Vazquez (eds) *Employment Deconcentration in European Metropolitan Areas: Market Forces and Planning Regulations*, Springer, Berlin.

Dubin R.A. (1998) Spatial Autocorrelation: A Primer, *Journal of Housing Economics*, 7, 304–327.

Felsenstein D., Ashbel E. and Ben-Nun A. (2007) Microsimulation of Metropolitan Employment Deconcentration, pp. 199–218 in Koomen E., Stillwell J., Bakema A. and Schloten H. (eds) *Modelling Land Use Change: Progress and Applications*, Springer, Dordrecht, The Netherlands.

Fik T.J., Ling D.C. and Mulligan G.F. (2003) Modelling Spatial Variation in Housing Prices: A Variable Interaction Approach, *Real Estate Economics*, 31 (4), 623–646.

Glaeser E.L. and Kahn M. (2001) Decentralized Employment and the Transformation of the American City, *Brookings-Wharton Papers on Urban Affairs*, 2, 2–63.

Gordon P., Richardson H. and Wong H. (1986) The Distribution of Population and Employment in a Polycentric City: The Case of Los Angeles, *Environment and Planning A*, 18, 161–173.

Griffith D. (1981) Modelling Urban Population Density in a Multi-Centred City, *Journal of Urban Economics*, 9, 298–310.

Henderson V. and Mitra A. (1998) The New Urban Landscape: Developers and Edge Cities, *Regional Science and Urban Economics*, 26 (6), 613–643.

Krakover S. and Adler E. (2009) Metropolitan Tel Aviv: From Monocentric to Dispersed Structure, pp. 315–337 in Kipnis B. (ed.) *Tel Aviv-Yafo: From a Garden Suburb to a World City*, Pardess Publications, Haifa.

Macauley M. (1985), Estimation and Recent Behaviour of Urban Population and Employment Density Gradients, *Journal of Urban Economics*, 18, 251–260.

Mahmassani H., Hadibaaj M. and Tong C.C. (1988) Characterisation and Evolution of Spatial Density Patterns, *Transportation*, 15, 232–1256.
McDonald J.F. and Bowman H.W. (1976) Some Tests of Alternative Population Density Functions, *Journal of Urban Economics*, 3, 242–252.
McDonald J.F. and Prather P.J. (1994) Suburban Employment Centres: The Case of Chicago, *Urban Studies*, 31, 201–218.
McMillen D.P. (2001) Polycentric Urban Structure: The Case of Milwaukee, *Economic Perspectives*, Federal Reserve Bank of Chicago, 2Q, 5–26.
McMillen D.P, and Lester T.W. (2003) Evolving Subcentres: Employment and Population Densities in Chicago, 1970–2020, *Journal of Housing Economics*, 12 (1), 60–81.
Muth C. (1969), *Cities and Housing*, University of Chicago Press Chicago, IL.
Newling B.E. (1969) The Spatial Variation of Urban Population Densities, *Geographical Review*, 59, 242–252.
Shachar A. (1975) Patterns of Population Density in the Tel Aviv Metropolitan Area, *Environment and Planning, A*, 7 (3), 279–291.
Small K.A. and Song S. (1994) Population and Employment Densities: Structure and Change *Journal of Urban Economics*, 36 (3), 292–313.
Stern D. (1994) Historical Path-Dependence of the Urban Population Density Gradient, *Annals of Regional Science*, 28 (2), 197–222.

3. The role of monetary and financial factors in regional development: an overview

Masagus M. Ridhwan, Peter Nijkamp, Piet Rietveld and Henri L.F. de Groot

INTRODUCTION

The re-emergence and fast development of regional science has been gaining momentum once more, especially in the last decade, after a period of "crisis" as diagnosed by Bailly and Coffey (1994). Considering regional economics in particular, despite this revival, there is still little attention devoted to the role of monetary policy in regional economic development.

An important reason why monetary policy tends to be largely ignored in economic theories of development is because of a traditional postulate that money is neutral (at least in the long run). Thereby, monetary variables are only considered as a medium of exchange with no implications for long-run economic development. Nonetheless, money is not necessarily neutral because of the potential relevance of market failures, such as asymmetric information problems which can result in, for example, spatial segmentation of capital markets.[1] In the presence of such imperfections, new-Keynesian economics has argued that indeed money may have a substantial impact on regional growth (see Samolyk, 1991, 1994; Greenwald et al., 1993; Faini et al., 1993). Money in this terminology is not limited to monetary policy conducted by a central bank but also embraces, more broadly, financial markets, such as banks, stock markets, venture capital, etc. As Levine (1997) pointed out, financial institutions have the important capacity to mobilize savings and reinvest them in a more efficient and productive way. In turn, they could facilitate output growth.

Meanwhile, in view of the growing trend among central banks around the world to confine monetary policy to the use of inflation targeting as their policy framework, a good understanding of regional economic structures and characteristics has now become even more important than it was previously, primarily because the national inflation rate originates from the aggregation of regional inflation rates within an economy. Therefore, understanding regional aspects is a fundamental pillar in meeting national targets.

The relevance of studying the complex relationship between finance and regional development has become gradually more apparent in view of the asymmetric shocks that have hit European economies since the introduction of the Euro in recent years. This also holds for less developed countries such as Indonesia, India and China, where large economic areas are divided into many (decentralized) regions, while monetary policy operates solely at the national level. Hence, a discussion of this issue is relevant for those countries too since it can contribute to finding optimal stabilization policies to counteract regional heterogeneous shocks.

The main objective of this study is to analyse and survey how monetary and, more broadly, financial factors may have an influence on regional development from both a theoretical and an empirical point of view. This will lead to an evaluation of the relevance of monetary policies and financial markets in facilitating regional growth.[2]

Furthermore, this study also attempts to make some refinements and extensions to the existing literature. More specifically, we aim to analyse the link between regional monetary impacts, optimum currency area theory and the role of interregional capital mobility as an adjustment mechanism to differential shocks. Finally, having argued that capital flows across regions may be imperfect, we could assert that there is a strong link between monetary/financial institutions and regional development.

The remainder of the chapter is organized as follows. The second section discusses briefly the theoretical aspect of regional monetary transmissions. It is divided into two subsections which focus, respectively, on the interest rate channel and the bank-lending channel. The third section surveys the empirical literature. Building on the optimum currency area literature, the fourth section critically discusses whether in reality capital across regions is as mobile as envisaged by the classical approach. Having illuminated by reference to several studies that the hypothesis may not hold, we then emphasize that spatial effects may play an important role in inhibiting the mobility of capital across regions or a monetary union. Finally, the fifth section concludes and ends with some policy lessons.

AGGREGATE MONETARY POLICY AND THE DISAGGREGATE REGIONAL ECONOMY

Theoretically, monetary policy may influence the real economy through different channels, such as the interest rate, bank-lending, the exchange rate, balance sheets, asset prices and expectation channels. In the remainder of this section we focus on the domestic transmission channels that are likely to be most relevant for regional economic development, namely the interest rate and bank-lending channels.

Regional Monetary Transmission Mechanism: A Brief Theoretical Approach

Interest rate channel

The interest rate affects the real economy through the cost of capital. Increased money supply reduces the cost of capital, which may positively affect investments and subsequent growth, provided that the increased money supply reduces the real cost of capital due to, for example, price stickiness, or alternatively that investments also in part respond to the nominal interest rate (e.g. Mishkin, 1995; Taylor, 1995; Mankiw, 1997).

The change in monetary policy could also affect consumption. First, an unexpected increase in the (real) interest rate will lower consumption. Second, households will become more pessimistic about future income and will reduce their consumption as compared with their planned consumption. Furthermore, anticipated increases of the (real) interest rate may have already resulted in lower consumption prior to the interest rate change through an expectations channel (Mahadeva and Sinclair, 2001).

At the regional level, the analysis of the transmission of monetary policies is slightly different, since every region has its typical economic structure and characteristics. Thereby a national monetary policy may affect different regions differently. A simple illustration is provided below in Figure 3.1, which describes how two economic regions in one country may be affected differently by a common interest rate.[3] The core region (A) typically has a positive perceived investment climate, and hence has a strong investment demand. Meanwhile the peripheral region (B) has characteristics of low investment perceptions and therefore experiences weak investment demand.

Monetary policy is typically targeted at the national economy, and hence the interest rate (namely the policy rate) is equal for all its regions (represented by a horizontal LM curve in Figure 3.1). As Owyang and Wall (2005) posit in this context: "the Fed holds the view that monetary policy cannot and should not be used to affect particular regions or states".[4]

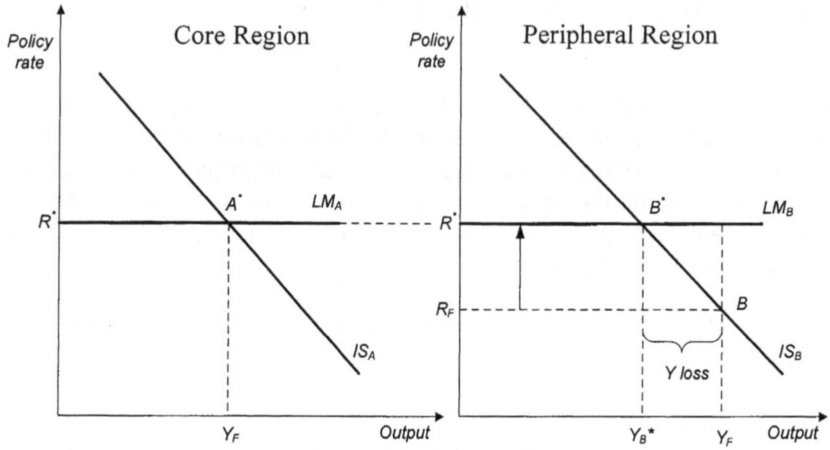

Figure 3.1 Regional monetary transmission through the interest rate channel: core region and peripheral region

Now suppose that, in order to achieve its national target (let us say maintaining price stability), the monetary authority has imposed a uniform interest-rate (R*). Because the regions have different economic structures, the common policy may have differential impacts.

This interest rate is favourable to region A since it is consistent with full employment or growth. In contrast, such an interest rate is unfavourable to region B which cannot maintain its natural growth at Y_F, but instead experiences an output loss equal to $Y_F - Y_{B*}$.

The asymmetric impact of the common policy across the two regions via the interest rate channel, as shown above, may only prevail in the short run, while in the long run the market will adjust towards the equilibrium situation (Y_F) following price changes. In the next subsection, we will consider the regional impact of monetary policy through the bank-lending channel.[5]

Bank-lending channel

In order to shed some further light on the effect of regional monetary policy through the bank-lending channel, we will use a similar approach as in the section on interest rate channel above. There are again two regions within a country (core and periphery), but with different economic structures. Assume that banks in the core region A have more access to credit from other sources of funding elsewhere, so they have a strong liquidity level. In contrast, banks in the peripheral region B face liquidity constraints. Thus demand for credit in region A is relatively high compared with region B. As previously discussed, the core region is perceived to have good investment opportunities, while the opposite holds for the peripheral region.

The better perceived investment climate following from the lower degree of uncertainty of region A compared with B implies that its (credit) interest rate is lower. Following best practice, the risk assessment conducted by banks will determine which part of a certain project will be funded, and in turn this could be mirrored by the magnitude of the lending rate. In other words, a riskier project will have a higher lending rate, since it is more costly[6] relative to a less risky one.

Following an approach by Dow and Montagnoli (2007), assume that there are two segmented regions. Each region is characterized by a demand curve for credit issued by the national banking system, represented by AR_A and AR_B in Figures 3.2a and 3.2b, respectively. Aggregate credit demand is given in Figure 3.2c (and equals to horizontal summation of credit demands in regions A and B). Borrowers in region A are assumed to have relatively easy access to alternative sources of finance whereas borrowers in region B have only few outside options. This is reflected in the relatively steep AR curve in region B (in other words: credit demand in region A is relatively elastic as compared to region B). The MR curve represents the marginal revenue of the banks, and the marginal cost of funds is represented by *MC*. The equilibrium supply of credit in the national economy is where marginal costs and revenues intersect. This determines the benchmark interest rate which is supplemented by a mark-up by local banks reflecting their liquidity preferences and a premium for perceived risk.

Now suppose region A faces macroeconomic instability due to inflation pressure, which potentially affects the entire economy. In response, the central bank increases its interest rate. This in turn, will induce an upward shift of the *MC* curve from MC_0 to MC_1 as is shown in third section of Figure 3.2 which represents the national economy. In the new equilibrium, the interest rate is higher and credit provision is lower.

Since region A has more elastic demand, the rise of the central bank's rate will result in a relatively strong increase of the lending rate (from R_0 to R_1), as well as a decrease of the supply of credits (from C_0 to C_1). Dow and Montagnoli (2007) summarize the effects as follows: "The cost of credit rises more in region A ...than in region B... in absolute terms, but proportionately more in region B relative to the fall in credit level". This is because region A has more access to alternative funding sources than region B, thereby reducing the adverse impact of the increased cost of credit.

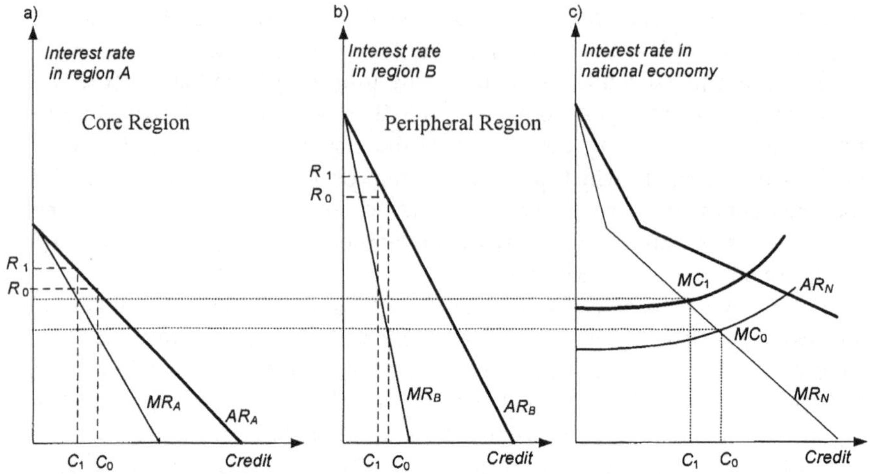

Figure 3.2 Regional bank-lending channel

EMPIRICAL EVIDENCE

Survey of Empirical Studies

Recent empirical studies on regional monetary transmission employ the vector autoregressions (VAR) approach based on the pioneering work of Sims (1980, 1986) and Bernanke (1986).[7] Relying upon this modern methodology, Carlino and DeFina (1999) studied the dynamic interaction between a national monetary policy and regional (state) economic development in USA. Unsurprisingly, such a common shock appeared to generate asymmetric responses across the states. The main reasons for these asymmetries are based on the variations in the interest-rate sensitivity of a region's industrial mix, the share of small firms in the local economy, and the share of small banks in the local economy. Mixes in the industrial sector are relevant because of the "interest rate channel", while the firms' and the banks' size have an effect through the "credit channel".[8]

Following the formation of the European Monetary Union (EMU), a robust understanding of how monetary policy affects regional economies became a "hot issue" for many European economists. Arnold's (1999) contribution focuses directly on the regional impact of monetary policy by simply measuring the correlation between regional GDP growth and interest rate changes across European countries. Even with the many caveats noted by the author (limited sample, annual frequency, non-homogeneity of the business cycle phase across regions), the results strongly suggest a negative relationship between interest rate increases and regional growth.

De Lucio and Izquierdo (2002) focus on the intensity of the reaction of Spanish regional economies to a common monetary shock using a vector autoregression (VAR) methodology. The endogenous variable used in the second step of their analysis is the long-run response in employment. The set of potential determinants of asymmetries is very similar to the one used by Carlino and DeFina (1999) for states in the USA.

Similarly, by using Spanish regional data, Ramos et al. (1999) have analysed the asymmetric response of regional economies to a national monetary shock. They provide support for the hypothesis that the distribution of industrial sectors such as manufacturing and construction across regions has been a main driving force behind asymmetric responses to the national monetary policy. Average firm size also played a key role in explaining regional differences. They also consider the importance of "firm size", since it could also be an indicator of credit constraints, since larger firms are more likely to use, for example, retained profits and find it easier to receive loans in the capital market.

Table 3.1 attempts to extract some underlying sources that explain regional variation of the impacts of monetary policy that have been studied empirically across countries. Regional differences in the industrial mix, the structure of the financial sector and firm size are most frequently cited as potential sources of this variation. In addition, some studies provide other sources of regional heterogeneity besides those mentioned here, such as: the degree of wage and price rigidity (Holmes, 2000); the regional housing market (Frantantoni and Schuh, 2003); the institutional features of the labour market (Arnold, 1999); and the share of state exports and fluctuations of the real exchange rate (Weber, 2006).

OPTIMUM CURRENCY AREAS AND THE ROLE OF CAPITAL MOBILITY

The most popular theory addressing economic interactions across regions or countries that share the same currency (or which have fixed exchange rates) under a monetary union is the theory of optimum currency areas (OCA). This theory is originating from the work of Mundell (1961), while further contributions were made by McKinnon (1963) and Kenen (1969).[9] In this line of thinking, economic differences among regions or participating economies do not matter so much for economic development and stabilisation. If some regions face a common shock, then adjustment mainly takes place through flexible factor markets (high factor mobility and adjusting factor prices), and hence will help to counteract economic divergence (see De Grauwe, 1994; Baldwin and Wyplosz, 2004).

Table 3.1 Taxonomy of differential impacts of regional monetary transmissions

Classification of Factors	Key Variables	Type of Region with respect to Monetary Policy Changes	
		More sensitive	Less sensitive
Compositional	Sectorial mix, Demand mix, Degree of openness, Firm size, Bank size	Manufacturing industries, Property High Investment, Consumption Low degree of export–import intensity Small–medium enterprises (SMEs), household businesses Small banks	Non-manufacturing industries (Agriculture, Services, etc.) Low investment, export–import oriented, Consumption demand High degree of export–import intensity Big companies /enterprises National banks, Foreign banks
Behavioural	Bank health Source of fund (for borrowers) Banking concentration	Poor/lack of soundness Highly bank-dependent Retail banking	Sound/healthy Availability of non-banks as alternative sources Corporate and international banking
Institutional	Labour structure Availability of financial market centre	Wage earners; income type/ Industrial labour Many and various types of financial institutions are available (bank, stock exchange, venture capital, etc.)	Profit/rent earners Limited alternatives in financial markets, mostly rural banks

Source: Extracted from various studies.

Having shed light on how regions respond to a uniform monetary policy, this section will look further into the role of factor mobility as an important adjustment mechanism. Specific attention is devoted to capital mobility across regions.[10] This issue is of particular interest for our review, since most relevant studies in this area focus on the international dimension instead of the regional dimension.

Is Capital Freely Mobile Across Regions?

As postulated by the classical approaches, differences in rates of return to capital among regions will not be sustained in the long run when financial markets are integrated. The mechanism behind this is illustrated in Figure 3.3. The figure assumes a peripheral region (left axis) which initially experiences a high interest rate (R_0), while the interest rate in the core region (right axis) is low (R_0^*). As a consequence, the rate of return on capital is relatively high in region 1 and it will attract capital in the presence of capital mobility, until equilibrium is achieved in point *E* (see Baldwin and Wyplosz, 2004).

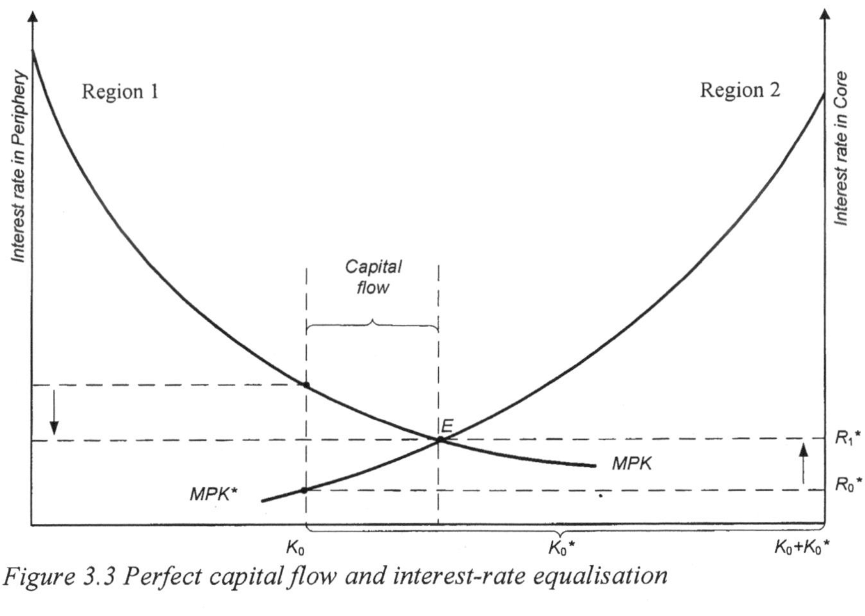

Figure 3.3 Perfect capital flow and interest-rate equalisation

The empirically relevant question is, of course, to what extent capital is truly mobile. The seminal contribution in this area of research is Feldstein and Horioka (1980). They showed that there was a high correlation between the national saving rates and domestic investment rates among OECD countries in the 1960s and early 1970s. They posit that savings and investments would be perfectly correlated in a closed economy, but should be unrelated in an open economy since savings would be directed to countries with the highest rate of return (Coakley et al., 1998).

Bayoumi and Rose (1993) use UK regional data from 1971 to 1985 to study intra-national capital mobility and found that capital in those regions is perfectly mobile (because saving and investment rates were not strongly correlated). Related studies also confirm their findings of perfect capital mobility at the sub-national level (for example, Dekle, 1996, based on Japanese prefectures; Bayoumi and Sterne, 1993, based on Canadian provinces; and Asdrubali et al., 1996, based on the US states).

Kalemli-Ozcan et al. (2006) use a different approach to study inter-state capital flows in the USA. Because of the lack of capital income flows data at the state level, they develop an approach using output/income ratios as a proxy for the relative magnitude of net inter-state capital income flows to a state. Using a simple neoclassical model, they found that, in the states where there are no barriers to movement of capital, capital should flow to states that experience a relative increase in capital productivity. In addition, they found evidence that net capital income flows between US states are consistent with the predictions of a simple "frictionless" neoclassical model.

Despite those conventional beliefs of perfect capital mobility at the subnational level, subsequent studies have arrived at different findings. The recent studies reviewed below have found that locational (spatial) effects should be incorporated in analysing mobility of capital.

Barriers to Capital Mobility and the Locational Effect

As mentioned above, the standard theory, which is supported by several empirical studies for developed countries, such as the USA, Canada, the UK and Japan, points out that intranational financial markets are fully integrated and that capital is thus perfectly mobile. However, some recent studies have contradicted these the results by showing that capital may not be freely mobile, not even at the subnational level, resulting in segmentation of markets.

One of the important gaps in the conventional view may be caused by ignorance or simplification of the role of spatial factors in shaping economic behaviour. This is, for example, illustrated in van Wincoop (2000) who simplifies the key barrier of capital mobility to: "asymmetric information that is closely related to language, cultural and regulatory barriers, [and which] … is likely to be less relevant for regions within a country." Later he asserts that intranational capital will be perfectly mobile.

In the remainder of this study, we will discuss the role of spatial factors in influencing economic (financial) behaviour. This sheds some interesting light on factors that may result in capital being imperfectly mobile, even at the subnational level.

Market imperfections and the financial sector: a spatial perspective

Despite the conventional belief that capital flows are perfectly mobile across regions within a single country, the information gaps and asymmetries[11] between investors and borrowers that are separated in different locations may be substantial. Apart from stressing the information problems, Keynesian approaches also shed light on other types of market imperfections such as "home-bias" problems that also render financial (equity) markets to function suboptimally as an intermediary.[12]

For the banking system in particular, a bank's reluctance to channel its credit to peripheral regions results in a concentration of its business in the core region.[13] In turn, such common information problems may result in an uneven distribution of financial capital across regions and persistent differences in rates of return with associated welfare losses.[14]

The theoretical literature has two underlying rationales for why distance should serve as a deterrent to lending. First, commercial loans to small businesses that are located close by (typically borrowers in regions, including households)[15] give lenders an advantage in terms of relatively low costs of screening and monitoring loans (see Stiglitz and Weiss, 1981). As argued by Petersen and Rajan (2000) and Berger et al. (2002), lenders lack the "hard" information provided by detailed public financial statements typically available for large firms. As a consequence, they have to rely on "soft" information which is informally collected through relationships between the lender and the borrower. The collection of this soft information is costly to the lender, as it may require multiple site visits by a loan officer to the small business, or specialised knowledge of the local market in which the firm operates.[16]

Second, in the traditional approach of spatial competition models, borrower travel costs are emphasized. Prospective borrowers must incur travel costs to do business with a lender, and pretty much the same also applies to depositors who want to deposit their money in one depository institution rather than others. Several studies have been conducted to understand this rationale (see, e.g. Chiappori et al., 1995; and Park and Pennacchi, 2003). Based on the two rationales above, the main implications of distance to lending disbursements are spatial price discrimination and limited credit supply faced by small firms, particularly those in remote regions.

However, it is crucial to differentiate between asymmetric information and transportation cost models of lending. In pure transportation-cost models, spatial discrimination only takes place through loan pricing because all borrowers who are deemed creditworthy obtain credit from the closest bank and never switch lenders. In an asymmetric-information setting, banks can strategically use proprietary information to create a threat of adverse selection for their rivals, thereby softening price competition. At the periphery of a bank's market, however, the informational advantage is smaller so that competitors attempt to poach customers more aggressively (see Agarwal and Hauswald, 2006; Porteous, 1995).

There are an increasing number of empirical studies that seek to explain the role of distance (information asymmetries) in lending decisions. A study by Degryse and Ongena (2005) established on the basis of Belgian data that distance is an important determinant of lending decisions and that banks actually engage in spatial price discrimination. Similarly, DeYoung et al. (2008) show that the probability of default on small-business loans increases with the distance between borrower and lender, although the adoption of credit-scoring techniques reduces this effect. Previously, Faini et al. (1993) also discussed the information asymmetries problem that has induced banks to discriminate between the Northern and Southern part of Italian regions in their pricing strategy.

Using a unique data set of loan applications by small businesses in US states, Agarwal and Hauswald (2006) have studied the determinants of lending decisions and, especially, the roles of private information and physical distance between a bank and its borrowers. They found that the likelihood of obtaining credit and the loan rate decrease with the bank–borrower distance and increase with the borrower–competitor distance, while controlling for a wide range of other aspects of the lending relationship.

This literature suggests that there are quite strong evidences of bank-lending problems and locational effects. However, the overall evidence is mixed. Contrasting the previously discussed evidence, there are studies such as Carling and Lundberg (2002) based on the Swedish data, who find no evidence that distance is a determinant of a loan's default probability. Also Petersen and Rajan (2000) have pointed out that the role of distance in the banking market tends to diminish due to technological factors, and that monitoring activities by banks tend to become more efficient in more developed local business environments.

In equity markets, several studies have shown that investors earn higher returns on investments in local companies than on investments in more distant companies. In other words, being located far from a company puts an investor at an information disadvantage that is clearly measurable. Other studies show that security analysts who are located closer to a company produce more accurate earnings forecasts than analysts who are located at a greater distance.

Analysing informational asymmetries as revealed by proprietary equity trading, Hau (2001), using data from Germany, finds that traders located near a company's headquarters outperform their competitors in intraday trading. Similarly, Coval and Moskowitz (1999, 2001) analysed a large sample of managed mutual funds in the USA and found that fund managers earn abnormal returns from investments made within geographical proximity. They use as their vicinity measure a cut-off distance of 100 kilometres of a fund's headquarters. In addition, Ivkovich and Weisbenner (2003) examine the stock investments of over 30,000 households in continental USA from 1991 to 1996. They find that the average household invests 31 percent of its portfolio in stocks located within 250 miles. Loughran and Schultz (2006) have also used location as a proxy for information asymmetries and found that information asymmetries between rural firms and investors are particularly large, so that firms avoid issuing equity in the presence of these asymmetries. Rural firms also wait significantly longer when they finally decide to go public.

The above mentioned studies are based on a seminal paper by Myers and Majluf (1984), who observed that information asymmetries between managers and outside investors can make it expensive to raise funds through equity offerings and may lead some financially constrained firms to forego valuable projects rather than sell stock. Myers (1984) takes this observation further and develops a pecking order theory of capital structure. In this theory, firms issue equity only as a last resort, and capital structure is determined in large part by the firms' ability to finance internally. Alternatively, wherever possible, it is argued that firms prefer to raise investment capital through debt financing rather than share issues. But then, as a result, as mentioned above, local or opaque firms may face an expensive source of financing. Again, greater distance implies a clear disadvantage in obtaining information.

Next, let us consider the effect of physical distance on an equity market that relates to information asymmetries and home-bias, which is defined simply as "a preference for investing locally".[17] Numerous studies have been done to identify home-bias effects. Grinblatt and Keloharju (2001) showed that investors tend to have more active and frequent transactions with stocks from Finnish firms that are located close to investors and also share the same language and culture rather than with foreign firms.

Similarly, Goetzmann and Kumar (2008) study the behaviour of portfolio investors over the period 1995–2000, and reveal that urban investors are less diversified (focus on certain stocks) in managing their portfolio, contrary to standard portfolio theory that suggests diversification. Indeed, their behaviour is biased toward urbanism compared with rural investors.[18] They furthermore collect some figures to explain the portfolio focus motivation, focusing on behavioural biases (bounded rationality and skewness preference), real and perceived informational advantage, local social competition and hedging of non-tradable risk.

In fact, the role of distance as an "informational barrier" in financial investment has been historically considered by the founding father of economics, Adam Smith (1976), who noted that, "in the home trade, his capital is never so long out of his sight as it frequently is in the foreign trade of consumption. He can know better the character and situation of the persons whom he trusts, and if he should happen to be deceived, he knows better the laws of the country from which he must seek redress."[19]

Spatial agglomeration in equity markets

Besides the aforementioned information problems that may deter interregional fund arbitrage from working perfectly, security markets especially also have an interesting feature, since investors like to trade in places where markets are thick. The reason is that the bigger the size of the market, the more liquid the market is, which in turn will guarantee more security and possibilities of portfolio diversification.

The important role of the "big caps" market can also be approached from the market price. As already pointed out by Samuelson (1965), financial markets are efficient because prices will follow a "martingale process", which results from the high level of transactions.[20]

With regard to the noticeable feature of the financial sector in that if displays a strong degree of spatial agglomeration in particular locations, usually large cities, the following observation by Glaeser (2005) is interesting:

> there are two major agglomeration economies at work. First, the role of the dense city as a centre for idea flows. The high value of knowledge meant that being in the city was particularly valuable. It may even be that New York's high density levels... helped New York finance continue to thrive because those high density levels are particularly conducive to chance meetings, regular exchanges of new ideas and the general flow of information.... [Second,] The costs of delivering manufactured goods depends only on transportation technology, but the cost of delivering services depends both on technology and on the value of the time involved by the participants in the transaction. Because services are by definition face-to-face, during an era of rising wages, there is an increased incentive to agglomerate these activities.[21]

A theoretical underpinning of such spatial agglomeration phenomena in the financial sector has been provided by Porteous (1995). He emphasizes the importance of path-dependent processes. These provide the reason why financial centres which had located early in a certain place may then well be sustained in the long run because of their early set-up, as happened in Amsterdam, New York City, London, Tokyo, Hong Kong and in other big cities around the world. Some sources of economies of scales that can be exploited by financial industry clusters located in urban areas are information spillovers (including tacit information), immediate access to the market (timing factor), the market for high-skilled labour, and good information and communication technology infrastructures.

Becker (2006) has supported this theory by providing evidence that the majority of the stock markets around the world are located in the largest cities, since much of the economic activity relating to the stock market takes place in these large cities. Therefore, countries with larger cities will have better developed stock markets because they can benefit from stronger agglomeration economies surrounding the stock market.

Where is Capital Flowing To?

The role of this subsection is to bring together the previous insights regarding information asymmetries and spatial agglomeration of capital that provide an important explanation for where capital tends to locate. However, because there are only a limited number of studies at the regional level, we need to further enhance our understanding of the destination of capital flows.

Lucas (1990) posits that eventually capital will flow from poor to rich countries, because of differences in human capital and capital market imperfections. This pioneering work has become known as the Lucas paradox. Tornell and Velasco (1992) also support the argument that capital flows from poor to rich countries. They found that both the lack of property rights (law enforcement) and institutions have been main driving forces behind this paradox, which they call "the Tragedy of the Commons". The argument is also supported by Krugman (1993), who asserts capital tends to flow asymmetrically from the poor to the rich countries.

Alfaro et al. (2005) emphasize that the Lucas paradox[22] challenges the standard neoclassical theory that predicts that capital should flow from rich to poor (countries) or from countries with a low return on capital to countries with a high return on capital. The sources of the paradox can be separated into two categories. First, there are fundamental aspects such as technological differences, lack of institutional structure, lacking factors of production and unreliable government policies.Second, there are capital market imperfections that stem from asymmetric information and sovereign risk. Previously, Gordon and Bovenberg (1996) also confirmed the lack of capital mobility internationally, because of the lack of international portfolio diversification, real interest differentials across countries, and the high correlation between domestic savings and investment.

If we were to project these studies that were carried out for the international level to the regional level, then we might end up with a similar conclusion – that capital tends to be imperfectly mobile, and that it may be asymmetrically mobile from the periphery to the core region. One crucial element to explain this phenomenon is that geographical proximity may play a role in affecting economic behaviour.

Moreover, the advent of the New Economic Geography that was developed by Krugman (1991) has pointed out the importance of the economies of scale in explaining the clustering of particular industries. In such industries (including the financial industry), the rate of return on capital is not necessarily a declining function of previous investments, but may well be an increasing function of the amount of capital which has been invested previously in a particular industry in a particular region. For this kind of industry, the integration of markets can lead to agglomeration and concentration in centres where the economies of large-scale operation can be best achieved.

All in all, this case may undermine the integration of regional financial markets. Combined with imperfections of capital mobility, this may have substantial effects on the growth prospects of regional economies, particularly in less developed countries. In other words, a large country with a core-periphery structure may well face problems of financial market segmentation.

CONCLUSIONS

We have discussed how monetary policy can have real impacts on regional development especially through the interest rate and the bank lending channels, by using both theoretical and empirical approaches. Consistent with our simple diagrammatic analysis, most empirical studies also report heterogeneity of regional responses with respect to a common monetary shock. The main driving forces behind the asymmetric effects are inherent differences in the structures of the economic and the financial markets.

Conventional studies have reported that, over time, regional financial markets especially in the USA have become more integrated, thus enhancing capital mobility. However, our own study argues that the generalization of such a hypothesis is problematic, especially if one takes into account some recent findings in Europe. Moreover, in huge less developed countries like Indonesia, India and China, where strong regional disparities are apparent, the extent of market segmentation could become more pronounced, resulting in limits to capital mobility. Barriers to mobility can be related to distance, culture, language and institutions, which tend to give rise to substantial information asymmetries.

Furthermore, even though technological advances such as electronic banking services seem to have reduced the importance of distance substantially, there is still much evidence in both the USA and Europe suggesting that local banks still have a strong position in controlling local financial markets because they have superior information about their local markets.

To conclude, this survey study of the literature has shown that traditional approaches that tend to neglect the role of money, monetary policy or, more broadly, financial markets in affecting growth, may be too simplistic. This argument is based on empirical evidence that capital is imperfectly mobile because of information asymmetries. We underlined the potentially important role of monetary policy in spurring regional development especially in emerging economies. As such, the ideas developed in this study may contribute to our understanding of covergence processes across nations and regions. The early work of Barro and Sala-i-Martin (1990), followed by Rey and Montouri (1999) and Shioji (2001) have shown economic convergence across the US states. Yet, a comprehensive survey by Magrini (2004) indicates that such conclusions are less clear-cut for European regions. Also the meta-analysis by Abreu et al. (2005) shown there are large variations in the estimates of convergence rate across countries (location) based on the particular study/author cited, model specification, methodology and data that are employed. Parts of these contrasting evidences may well be attributable to the functioning of financial and capital markets and deserve more attention in future research, especially at the regional level.[23]

NOTES

1. The way that money can affect output due to imperfect information was first studied by Lucas (1972) and Phelps (1970). The Lucas–Phelps model was used and further developed by, amongst others, Blanchard and Kiyotaki (1987) and Rotemberg (1987). It emphasizes that unanticipated monetary policy may affect real variables (in contrast to anticipated ones). Keynesian economics emphasizes price stickiness (including wages and interest rates) that arise in a market, where information does not flow properly among economic agents (for further discussion, see Romer, 1996).
2. Dow and Rodriguez-Fuentes (1997) and Rodriguez-Fuentes (2005) have undertaken literature surveys of how monetary policy may affect regional economies. They also discuss the differences between two main strands in economics, i.e. between monetarist (orthodox) and post-Keynesian theory in viewing the relationship between monetary policy (financial sector) and regional development. The former posits that regional monetary impact is exogenous to money, while the latter argues that it is endogenous (see Rodriguez-Fuentes and Dow, 2003).
3. Modified from McCann (2001).

4. Non-neutrality of money at the regional level has been long discussed in the Keynesian approach (see, for example, Dow, 1982; Cottrell, 1986; Tobin, 1991).
5. In our further analysis, we will use the notion of asymmetric effects of monetary policy only for analysis across regions, despite several studies which also use the notion for highlighting the differential effect of monetary policy over time (see Florio, 2004).
6. Banks are normally required to add more reserves (loss provision) for a riskier project, since this may be viewed as their cushion if the project quality is deteriorating. More provisions allocated for riskier project mean higher costs of fund for the bank since they have to pay interest to their depositors as well. Furthermore, banks are obliged to do this by the regulator.
7. Previously those studies employed "large" econometric models (see, for example, Fishkind, 1977; Miller, 1978, and Garrison and Chang, 1979). References for the VAR methodology can be found in, for example, Hamilton (1994) and Ender, (2004).
8. Quoted from McPherson and Waller (2000).
9. The OCA theory attempts to explain the desirability for a group of countries to unite within a monetary union, as they may expect that such a unification will enhance economic integration, through lowering transaction costs that stem from the use of the same currency and fixing their exchange rates. In turn, it will promote trade in goods and services, and subsequently foster growth and welfare amongst the participating economies.
10. Several studies highlight interregional labour mobility across regions in the USA and European countries (see, e.g., Barro and Sala-i-Martin, 1991; Barro et al., 1995; Nijkamp and Poot, 1998; Duczynski, 2000). Although conventional theory proposes labour mobility as a main adjustment factor, Beetsma and Vermeylen (2005) argue that such a mechanism may work at best imperfectly in the EMU, since there are still barriers to labour mobility in its member countries due to differences of language, culture and institutions (see also CPB and SCP, 2007). Given those limitations, another important type of adjustment mechanism may rely upon a well-functioning redistribution system (fiscal policy) that may be helpful for the regions/countries in accommodating temporary shocks (see, von Hagen, 1992, 1999; Obstfeld and Peri, 1998).
11. Standard theory postulates that asymmetric information is relevant when information between borrower and lender is unbalanced. The former party can observe closely what the outcomes of a project are, while the lender has to rely on the information submitted by borrowers and must incur costs to verify such information. Variations of information asymmetry problems in financial markets are adverse selection and moral hazard (see Mishkin, 2001).
12. Standard finance literature also documents other sources of imperfections in financial markets, such as agency problems, fundamental uncertainty and interdependence. However, these sources are beyond the scope of this study.
13. This notion is consistent with the previous ones, and thus we continue to use core- and periphery regions, city/urban/rich and rural/remote/poor area interchangeably in order to highlight differences in economic structures, and especially the role of geographical distance in the lending decision between those regions.
14. Myrdal (1957) has examined the "backwash" effect of polarizing regional developments, according to which banks potentially siphon the savings of people in the poor region and reinvest them in the rich region. Accordingly, he contends that continued growth in one region may occur at the expense of other regions or locations.
15. Credit to small businesses is an important driver for job creation and macroeconomic growth, both for regions and for the nation as a whole.
16. In addition, Berger et al. (2002) found that distance and the mode of interaction (preferably personal communication) between banks and borrowers may be to the advantage of small (local) banks that are better at using soft information than large banks that operate at a distance. Such a kind of information is very helpful in evaluating how to allocate their credits. This implicitly suggests that small local banks that operate in rural areas are better at disentangling such opaque information. In addition to that, even national banks have more sophisticated technology with on-line banking services and other type of modern services, "face-to-face" contact is inevitably needed especially before disbursing their

credits. This approach is very crucial to mitigate problems of asymmetric information due to geographical proximity (see Degryse and Ongena, 2004).

17. Even though there is a bulk of literature discussing the home-bias effect internationally either in the equity or the goods market, at the intranational level the literature is scarce. However, it is necessary to distinguish between information asymmetries and home-bias.
18. In other areas, such as the trade in goods, Wolf (2000) found that home bias is present within the US states, and the preferred explanation for such a trade barrier is a long-distance relationship between seller and buyer region. In commercial real-estate markets, Garmaise and Moskowitz (2004) find strong evidence in their analysis that buyers tend to be local when information asymmetries between the parties are severe, and more remote otherwise.
19. Quoted from Gordon and Bovenberg (1996).
20. A martingale process is one in which the price of a good varies randomly about a constant mean. Prices will conform to such a process if, on average, these remain stable. Since financial-asset prices include elements of risk and transactions costs, these will oscillate continually around a bid–ask price which becomes lower with larger volumes of transactions (Parr and Budd, 2000).
21. Quoted from Becker (2006).
22. The notion of Lucas paradox is related to a subject that is listed as one out of six major "puzzles" in international macroecoomics (Obstfeld and Rogoff, 2000).
23. See, for example, Barro et al. (1995) for a study on capital mobility in the context of a neoclassical growth model at the national level.

REFERENCES

Abreu, M., de Groot, H.L.F. and Florax, R.J.G.M.. (2005) A meta-analysis of β-convergence: the legendary 2%. *Journal of Economic Surveys*, 19, 389–420.

Agarwal, S. and Hauswald, R. (2006) Distance and information asymmetries in lending decisions. Mimeo, American University.

Alfaro L., Kalemli-Ozcan, S. and Volosovych, V. (2005) Why doesn't capital flow from rich to poor countries? National Bureau of Economic Research Working Paper, No. 11901.

Arnold, I.J.M. (1999) The regional effects of monetary policy in Europe. Mimeo, Universiteit Nyenrode.

Asdrubali, P., Sorensen, B.E. and Yosha, O. (1996) Channels of interstate risk-sharing: US 1963–1990. *Quarterly Journal of Economics*, 144, 1081–1110.

Bailly, A.S. and Coffey, W.J. (1994) Regional science in crisis: A plea for a more open and relevant approach. *Papers in Regional Science*, 73, 3–14.

Baldwin, R.E. and Wyplosz, C. (2004) *The Economics of European Integration*. New York: McGraw-Hill.

Barro, R.J., Mankiw, N.G. and Sala-i-Martin, X. (1995) Capital mobility in neoclassical models of growth. *American Economic Review*, 85, 103–115.

Barro, R.J. and Sala-i-Martin X. (1991) Convergence across states and regions, *Brookings Papers on Economic Activity*, 1991, 107–182.

Barro, R.J. and Sala-i-Martin, X. (1990) Economic growth and convergence across the United States. National Bureau of Economic Research Working Paper, No. 3419.

Bayoumi, T. and Rose, A. (1993) Domestic savings and intra-national capital flows. *European Economic Review*, 37, 1197–1202.

Bayoumi, T. and Sterne, G. (1993) Regional trading blocs, mobile capital and exchange rate co-ordination. Mimeo, Bank of England.

Becker, B. (2006) City size and financial development. Mimeo, Swedish Institute for Financial Research.

Beetsma, R and Vermeylen, K. (2005) Clustering, Optimum Currency Areas and Macroeconomic Policy. In: Brakman, S. and Garretsen, H. (eds), *Location and Competition. Studies in Global Competition*. New York: Routledge.

Berger, A., Miller, N., Petersen, M., Rajan, R. and Stein, J. (2002) Does function follow organisational form? Evidence from the lending practices of large and small banks. National Bureau of Economic Research Working Paper, No. 8752.

Bernanke, B.S. (1986) Altenative explanations of money-income correlation. *Carnegie-Rochester Conference Series on Public Policy*, 25, 49–100.

Blanchard, O.J. and Kiyotaki, N. (1987) Monopolistic competition and the effects of aggregate demand. *American Economic Review*, 77, 647–666.

Carling, K. and Lundberg, S. (2002) Bank lending, geographical distance, and credit risk: An empirical assessment of the church tower principle. Sveriges Riksbank Working Paper, No. 144.

Carlino, G. and DeFina, R.H. (1999) The differential regional effects of monetary policy: evidence from the U.S. states. *Journal of Regional Science*, 39, 339–358.

Chiappori, P.-A., Perez-Castrillo, D. and Verdier, T. (1995) Spatial competition in the banking system: Localisation, cross-subsidies, and the regulation of deposit rates. *European Economic Review*, 39, 889–918.

Cottrell, A. (1986) The endogeneity of money and money-income causality. *Scottish Journal of Political Economy*, 33, 2–27.

Coval, J. and Moskowitz, T. (1999) Home bias at home: local equity preference in domestic portfolios. *Journal of Finance*, 54, 2045–2073.

Coval, J.D. and Moskowitz, T.J. (2001) The geography of investment: informed trading and asset prices. *Journal of Political Economy*, 109, 811–841.

CPB and SCP (2007) *Diverse Europe: Public opinion on the European Union & Cultural diversity, economics and policy (European Outlook 4)*. The Hague: CPB Netherlands Bureau for Economic Policy Analysis.

Dekle, R. (1996) Savings–investment associations and capital mobility: on the evidence from regional data. *Journal of International Economics*, 41, 53–72.

De Grauwe, P. (1994) *The Economics of Monetary Integration*. Oxford: Oxford University Press.

DeYoung, R., Glennon, D. and Nigro, P. (2008) Borrower-lender distance, credit scoring, and loan performance: evidence from informational-opaque small business borrowers. *Journal of Financial Intermediation*, 17, 113–143.

Degryse, H. and Ongena, S. (2004) The impact of technology and regulation on the geographical scope of banking. *Oxford Review of Economic Policy*, 4, 571–590.

Degryse, H. and Ongena, S. (2005) Distance, lending relationships, and competition. *Journal of Finance*, 40, 231–266.

De Lucio, J.J. and Izquierdo, M. (2002) Local responses to a global monetary policy – the regional structure of financial system. *Journal of Economic Studies*, 29, 205–221.

Dow, S.C. (1982) The regional composition of the money multiplier process. *Scottish Journal of Political Economy*, 29, 22–44.

Dow, S.C. and Rodriguez-Fuentes, C.J. (1997) Regional finance: a survey. *Regional Studies*, 31, 903–920.

Dow, S.C. and Montagnoli, A. (2007) The regional transmission of UK monetary policy. *Regional Studies*, 41, 797–808.

Duczynski, P. (2000) Capital mobility in neoclassical models of growth: comment. *American Economic Review*, 90, 687–694.

Enders, W. (2004). *Applied Econometrics Time Series*. Hoboken, NJ: John Wiley and Sons.

Faini, R., Galli, G. and Giannini, C. (1993) Finance and Development: The Case of Southern Italy. In: Giovannini, A. (ed.), *Finance and Development: Issues and Experiences. London*: Centre for Economic Policy and Research.

Feldstein, M. and Horioka, C. (1980) Domestic saving and international capital flows. *Economic Journal*, 90, 314–329.

Fishkind, H.H. (1977) The regional impact of monetary policy: an economic simulation study of Indiana (1958–1973). *Journal of Regional Science*, 17, 77–88.

Florio, A. (2004) The asymmetric effects of monetary policy. *Journal of Economic Surveys*, 18, 409–425.

Frantantoni, M. and Schuh, S. (2003) Monetary policy, housing, and heterogeneous regional markets. *Journal of Money, Credit and Banking*, 35, 557–589.

Garmaise, M.J. and Moskowitz, T.J. (2004) Confronting information asymmetries: evidence from real estate markets. *Review of Financial Studies*, 17, 405–437.

Garrison, C.B. and Chang, H.S. (1979) The effect of monetary and fiscal policies on regional business cycles. *International Regional Science Review*, 4, 167–180.

Glaeser, E. (2005) Urban colossus: why is New York America's largest city? National Bureau of Economic Research Working Paper, No. 11398.

Goetzmann, W.N. and Kumar, A. (2008) Equity Portfolio Diversification. *Review of Finance*, 12, 433–463.

Grinblatt, M. and Keloharju, M. (2001) How distance, language, and culture influence stockholdings and trades. *Journal of Finance*, 56, 1053–1073.

Gordon, R.H. and Bovenberg, A.L. (1996) Why is capital so immobile internationally? Possible explanations and implications for capital income taxation. *American Economic Review*, 86, 1057–1075.

Greenwald, B.C., Levinson, A. and Stiglitz, J.E. (1993) Capital Market Imperfections and Regional Economic Development. In: Giovannini, A. (ed.), *Finance and Development: Issues and Experiences*. London: Centre for Economic Policy and Research.

Hamilton, J.D. (1994) *Time Series Analysis*. Princeton: Princeton University Press.

Hau, H. (2001) Location matters: an examination of trading profits. *Journal of Finance*, 56, 1959–1983.

Holmes, M.J. (2000) Monetary shocks, inflation and the asymmetric adjustment of UK industrial output. *Applied Economics Letters*, 7, 159–163.

Ivkovich, Z. and Weisbenner, S. (2003) Local does as local is: information content of the geography of individual investors' common stock investments. National Bureau of Economic Research Working Paper, No. 9685.

Kalemli-Ozcan, S., Reshef, A., Sorensen, B.E. and Yosha, O. (2006) Why does capital flow to rich states? Centre for Economic Policy and Research Discussion Paper, No. 5635.

Kenen, P. (1969) The theory of optimum currency areas: an eclectic view. In: Mundell, R. and Swoboda, A. (eds), *Monetary Problems of the International Economy*. Chicago: The University of Chicago Press.

Krugman, P. (1991) Increasing returns and economic geography. *Journal of Political Economy*, 99, 483–499.

Krugman, P. (1993) International finance and economic development. In: Giovannini, A. (ed.), Finance and Development: Issues and Experience. London: Centre for Economic Policy Research.

Levine, R. (1997) Financial development and economic growth: Views and agenda. *Journal of Economic Literature*, 35, 688–726.

Loughran, T. and Schultz, P. (2006) Asymmetric information, firm location, and equity issuance. Mimeo, University of Notre Dame.

Lucas R.E. (1972) Expectations and the neutrality of money. *Journal of Economic Theory*, 4, 103–124.

Lucas, R.E. (1990) Why doesn't capital flow from rich to poor countries? American Economic Review, 80, 92–96.

Magrini, S. (2004) Regional (Di)Convergence. In: Henderson, J.V. and Thisse, J.F. (eds), *Handbook of Regional and Urban Economics*. Elsevier, 4, 2741–2796.

Mahadeva, L. and Sinclair, P. (2001) The transmission mechanism of monetary policy. Mimeo, Centre of Central Banking Studies, Bank of England.

McCann, P. (2001) *Urban and Regional Economics*. Oxford: Oxford University Press.

McKinnon, R.I. (1963) Optimum currency areas. *American Economic Review*, 53, 717–724.

McPherson, S.H. and Waller, C.J. (2000) Do Local Banks Matter for the Local Economy? In Search of Regional Credit Channel. In: Hess, G. and van Wincoop, E. (eds), *Intranational Macroeconomics*. Cambridge: Cambridge University Press, 1–43.

Miller, R.J. (1978) *The Regional Impact of Monetary Policy in the United States*. Lexington, MA: Lexington Books.

Mishkin, F. (1995) Symposium on the Monetary Transmission Mechanism. *Journal of Economic Perspectives*, 9, 3–10.

Mishkin, F.S. (2001) *The Economics of Money, Banking, and Financial Markets*. Boston: Addison-Wesley Longman.

Mundell, R.A. (1961). A theory of optimum currency areas. *American Economic Review*, 51, 509–517.

Myers, S. (1984) The capital structure puzzle. *Journal of Finance*, 39, 575–592.

Myers, S. and Majluf, N. (1984) Corporate financing and investment decisions when firms have information that investors do not have. *Journal of Financial Economics*, 13, 187–221.

Myrdal, G. (1957) *Economic Theory and Underdeveloped Regions*. London: Duckworth,

Nijkamp, P. and Poot, J. (1998) Spatial perspectives on new theories of economic growth. *Annals of Regional Science*, 32, 7–37.

Obstfeld, M. and Peri, G. (1998) Regional non-adjustment and fiscal policy. *Economic Policy*, 26, 205-259.

Obstfeld, M. and Rogoff, K. (2000) The Six Major Puzzles in International Macroeconomics: Is there a Common Cause? In: Bernanke, B. and Rogoff, K. (eds), *NBER Macroeconomics Annual 2000*. Cambridge: MIT Press.

Owyang, M.T. and Wall, H.J. (2004) Structural breaks and regional disparities in the transmission of monetary policy. Federal Reserve Bank of St. Louis Working Paper, No. 2003–008C.

Parr, J.B. and Budd, L. (2000) Financial services and the urban system: an exploration. *Urban Studies*, 37, 593–610.

Park, K. and Pennacchi, G. (2003) Why does institution size matter for banking market competition? Mimeo, University of Illinois, Urbana.

Petersen, M.A. and Rajan, R. (2000) Does distance still matter? The information revolution in small business lending. National Bureau of Economic Research Working Paper, No. 7685.

Phelps, E.S. (1970) Introduction: The New Microeconomics in Employment and Inflation Theory. In: Phelps. E.S. et al. (eds), *Microeconomic Foundations of Employment and Inflation Theory*. New York: Norton.

Porteous, D.J. (1995) *The Geography of Finance*. Aldershot, Gower.

Ramos, R., Clar, M. and Surinach, J. (1999) EMU: Some Unanswered Questions. ERSA Conference Proceedings, Dublin.

Rey, S. and Montouri, B. (1999) US regional income convergence: a spatial econometric perspective. *Regional Studies*, 33, 143–156.

Rotemberg, J.J. (1987) *The New Keynesian Microfoundations*. Cambridge, MA: MIT Press.

Rodriguez-Fuentes, C.J. (2005) *Regional Monetary Policy*. London: Routledge.

Rodriguez-Fuentes, C.J. and Dow, S.C. (2003) EMU and the regional impact of monetary policy. *Regional Studies*, 37, 969–980.

Romer, D. (1996) *Advanced Macroeconomics*. New York: McGraw-Hill Companies-Inc.

Samolyk, K.A. (1991) A regional perspective on the credit view. *Federal Reserve Bank of Cleveland Economic Review*, 27, 27–38.

Samolyk, K.A. (1994) Banking conditions and regional economic performance: evidence of a regional credit channel. *Journal of Monetary Economics*, 34, 259–278.

Samuelson, P. (1965) Proof that properly anticipated prices fluctuate randomly. *Industrial Management Review*, 6, 41–49.

Shioji, E. (2001) Public capital and economic growth: a convergence approach. *Journal of Economic Growth*, 6, 205–227.

Sims, C.A. (1980) Macroeconomics and reality. *Econometrica*, 48, 1–48.

Sims, C.A. (1986) Are forecasting models usable for policy analysis? *Federal Reserve Bank of Minneapolis Quarterly Review* (Winter edition), 3–16.

Stiglitz, J.E. and Weiss, A. (1981) Credit rationing in markets with imperfect information. *American Economic Review*, 71, 393–410.

Taylor, J. (1995) The monetary transmission mechanism: an empirical framework. *Journal of Economic Perspectives*, 9, 11–26.

Tobin, J. (1991) On the Endogeneity of Money Supply. In: Nell, E. and Semmler, W. (eds), *Nicholas Kaldor and Mainstream Economics*. London: Macmillan.

Tornell, A. and Velasco, A. (1992) Why does capital flow from poor to rich countries? The tragedy of the commons and economic growth. *Journal of Political Economy*, 100, 1208–1231.

van Wincoop, E. (2000) Intranational Versus International Saving–Investment Co-movements. In: Hess, G. and van Wincoop, E. (eds), *Intranational Macroeconomics*. Cambridge, MA: Cambridge University Press.

von Hagen, J. (1992) Fiscal Arrangements in a Monetary Union – Some Evidence from the US. In: Fair, D. and de Boissieux, C. (eds), *Fiscal Policy, Taxes and the Financial System in an Increasingly Integrated Europe*. Deventer: Kluwer Academic Publishers.

von Hagen, J. (1999) A Fiscal Insurance for the EMU? In: Beetsma, R.M.W.J. and Oudshoorn, C. (eds). *Tools for Regional Stabilisation*. Dutch Ministry of Economic Affairs, Policy Discussion Paper, No. 9903.

Weber, E.J. (2006) Monetary policy in a heterogeneous monetary union: the Australian experience. *Applied Economics*, 38, 2487–2495.

Wolf, H.C. (2000) Intranational home bias in trade. *Review of Economics and Statistics*, 82, 555–563.

PART TWO

Industrial Innovation

4. Spatial mobility of firms

Piet H. Pellenbarg and Joris Knoben

INTRODUCTION[1]

In modern societies, firms are constantly adjusting to new situations due to changing consumer preferences, new regulations, new entrants, and (technological) innovation. This adjustment process very often also has a spatial dimension. Characteristics of the spatial environment of a firm (i.e. external factors) may change over time, but internal changes in firms may lead to other locational preferences as well. Firm relocation is a particular form of locational adjustment. Here, we define firm relocation as a move of a firm from location A to location B. This definition is best suited for single-site firms, but less so for multi-site firms. For multi-site firms locational adjustment usually involves the restructuring of the spatial layout of activities, which are spread out over multiple locations. These complex events can only partially be labelled as relocation. Often, the relocation component is but one element in a mix that also includes closing down, merging and splitting off of business units.[2] In this chapter we will limit ourselves to the discussion of the basic form of firm relocation: a firm moving from location A to B.

Research regarding the phenomenon of firm relocation started in the period around 1950. Since then, this field of research has seen different kinds of relocation (from long distance moves of the manufacturing industries to short distance moves of business services), different underlying policy interests (from steering firm relocations in order to reduce regional inequality to urban renewal), different theoretical perspectives (from neoclassical approaches to evolutionary approaches) and different levels of analysis (from analysing aggregated flows of firms between regions to micro-level models). The goal of this chapter is to provide an overview of the firm. As such, it adds to the literature by bringing together almost six decades of theoretical and empirical work and by identifying the most promising paths for future research on this topic.

To gain insight into the relevance and scope of firm relocation we start with an overview of theories which shed light on firm relocation and the underlying causal factors (the section on Relocation Theories). Next, a brief historical review of firm relocation research is presented in the third section. This review starts with the "classical studies" of the first post-war period, followed by a description of what can be called the golden era of the firm relocation studies: the 1960s and 1970s. This section ends with an overview of the firm relocation studies of the last decades. In the fourth section, we present an empirical (case) study that addresses relevant issues regarding firm relocation for the present era based on data for the Netherlands and Belgium. We conclude the chapter with a discussion of why and how the firm relocation research frontier can be pushed forward (the fifth section) and by drawing policy conclusions (in the last section).

RELOCATION THEORIES

Location theory has witnessed a proliferation of theories and approaches in the last three decades, none of which seems to dominate the field (Scott, 2000). Therefore, we will give an overview of several approaches which seem to be useful to gain more insight into firm relocation and the underlying (decision) processes. Here, we are not dealing with location theories per se, but with *re*location theories. Theories about firm relocation are rare. In general, they are treated as a special case of one of the existing location theories. Firm relocation differs from firm location because it explicitly takes into account that one location is substituted for another. The firm has a history, and this history is likely to have an influence on the outcome of the process, which is therefore conditional. The specific nature of these conditional effects is important for any theory of firm relocation. Another way to look at this is to separate the relocation process into two sequential steps: first, the decision to move, and second, conditional upon a move, the decision to relocate to a specific location. A similar distinction is between push and pull factors of relocation. Location theory focuses on the optimal location choice, which is about location factors determining the attractiveness of a site, or pull factors. Relocation theory also takes into account the first step, the push out of the present location. In this section, we will emphasise both elements of relocation. Broadly speaking five types of (re)location theories can be distinguished: a neoclassical, a behavioural, an institutional, an evolutionary and an organisational network approach.

The Neo-classical Approach

The neoclassical approach, which is derived from standard classical economic theory, focuses on cost-minimising or profit-maximising. In the neoclassical approach, the transportation costs of inputs and outputs determine a least transportation-cost surface (Weber, 1929). Other location factors, such as labour, determine similar least-cost surfaces. By aggregating the cost surfaces of all location factors, a total-cost surface is derived. In a similar vein, a spatial revenue surface may be calculated. The firm is able to make a profit in any location where total revenues exceed total costs. By subtracting the total cost surface from the revenue surface, the total area is divided into profitable and unprofitable areas. In this regard, the concept of the spatial margins to profitability for a firm may be defined (Rawstrom, 1958; Taylor, 1970; Smith, 1966, 1971; McDermott, 1973). These margins enclose the spatial area within which the firm is able to make a profit. Given the fact that, in the neoclassical approach, all firms are fully informed and behave completely rational, all firms will chose the optimal location. In an equilibrium situation, the optimal location for the firm is fixed, and relocation is irrelevant. However, both the firm (i.e. internal factors) and the environment of the firm (i.e. external factors) can change over time. These changes will lead to a changing shape of the cost and revenue surfaces and, hence, of the spatial margins to profitability of the firm. Assuming that location costs and revenues change over time, most existing firms will not occupy the "optimal" profit-maximising location.

Nakosteen and Zimmer (1987) provide a theoretical framework in which firms continuously monitor their profits relative to a fixed target threshold. As long as the firm exceeds this threshold profit rate, the firm will most likely stay at the present location and will not try to move to the 'optimal' location, because the costs of relocation are higher than the net present value of the increase in profit.[3] The other possible outcome of the monitoring may be that due to the changing shape of the cost and revenue surfaces, the profit drops below the aforementioned threshold level or even that current location is no longer inside the spatial margins to profitability. In these cases adjustments are necessary, otherwise the firm will go bankrupt. Relocation is one way to solve this problem.

One of the most common forms of internal change of the firm is growth, which often results in economies of scale. Within the neoclassical framework, these economies of scale imply that the level of the cost surface is higher for small firms than for large firms in all locations. Therefore, small firms cannot escape bankruptcy by relocating to another location, but must grow, possibly at another location in order to remain profitable. If the firm chooses to relocate, it is not driven by the traditional location factors, but by the need to adjust to internal dynamics. It is also possible that economies of scale can only be realised at particular locations (for instance urban areas with a large market).

The spatial adjustment process to firm growth in relation to the external environment is one of the key explanatory factors of firm relocation, which may be explained by the internal dynamics of the firm. Nevertheless, it has not received much attention in neoclassical location theory which has a strong focus on external location (pull) factors.

Summing up, it can be concluded that neoclassical relocation theory primarily focuses on location factors that are well covered in location theory, and could be denoted as locational pull factors, but also touches upon the factors triggering a relocation, the push factors. The spatial margins to profitability discriminate between profitable and unprofitable locations, and are therefore useful in determining where a firm should locate (pull). However, they usually span quite a large area within which firms may operate profitably. Changes in these boundaries are therefore not always sufficient in explaining why firms want to move (the push factors). In addition, firm internal processes need to be taken into account as well, of which firm growth as a result of economies of scale is the most common one.

The Behavioural Approach

Determined by its assumptions, neoclassical theory describes a radical simplification of real world conditions. Nevertheless it was and still is useful as a benchmark that defines the "optimal" behaviour of the firm in economic terms, under the assumptions of rationality and perfect information. However, it does not take into account the internal dynamics of firms in a context with imperfect information and uncertainty where profit-maximising behaviour is not the only or not even the ultimate goal. This motivated Simon (1955, 1957) and Cyert and March (1963) to develop a behavioural theory of the firm, which is based on more realistic notions of limited information and bounded rationality. Here, optimising behaviour is replaced by "satisfying" behaviour. The behavioural approach was also successfully introduced in location theory, primarily by Pred (1967, 1969). Apart from the general points of criticism towards neoclassical theory, the application of these behavioural ideas in location theory was also motivated by the limited discriminating power of regional economic conditions in determining the optimal location of industries (Benoit, 1995; Hayter 1997). If regional economic conditions show limited variation this leaves many profitable sites to choose from. As a result, firm-specific factors or non-economic factors may become of more importance for the explanation of firm relocation. The behavioural approach takes these factors explicitly into account.

The behavioural approach is especially geared towards firm relocation. We noted above that the key difference between location and relocation theory is that location theory is more concerned with pull factors, whereas relocation also deals with push factors: the trigger to moving. This distinction fits into a simplified description of the decision process of the firm that is based on the behavioural theory. In fact, more than two phases are distinguished in this model (Lloyd and Dicken; 1977, Hayter, 1978, 1997): (1) the decision whether to move or not; (2) the search for alternative locations; (3) the evaluation of alternative locations; and (4) the choice of the new location.[4]

There are four key elements in behavioural location theory: (1) the role of limited information; (2) the ability to use information; (3) perception and mental maps; and (4) uncertainty. These elements were combined by Pred (1967, 1969) into the behavioural matrix, where firms are classified along two dimensions: (1) the availability of information, and (2) the ability to use information. Firms with high information levels and a large ability to use it come close to the classical "Homo economicus", and may be expected to locate near optimal. Firms at the other end of the scale know little and cannot utilise the little information that they have, and thus are expected to locate at less profitable or even unprofitable locations. Despite its simplicity and popularity, Pred's behavioural matrix offers no more than a conceptual basis for constructing a behavioural location or relocation theory (McDermott, 1973). In behavioural theory it is the perception of reality, not only reality in itself, that matters. Mental maps (Abler et al., 1971; Cox, 1972), the perception of the geographic configuration, are what people use in their spatial decision making (Pellenbarg, 1985; Meester, 1999).

In this regard, limited information, limited ability, perception and uncertainty all lead to a large spatial bias in relocation decision making. First, more distant locations are less well known and therefore it is likely that closer locations are chosen more frequently. Second, distant locations are more difficult to imagine than closer places. Third, there is a strong distance decay in mental maps, which is of course partly related to the amount of information available, but also to the perceived attractiveness of the place. Finally, firms face uncertainty not only because they have a knowledge gap or are not able to digest the available information, but also because investment decisions are based on anticipated future situations, which are by definition uncertain. Anticipating the future for other locations that are not familiar adds to the uncertainty. The larger the relocation distance, the larger the amount of uncertainty about future points in time. Relocation to another geographical market may even be comparable to the inherent uncertainty of a start-up.

As a result of the behavioural approach to firm relocation, a large body of literature has developed that gives detailed descriptions of motives and reasons for moving, both on the push and the pull side of the process. Frequently given push reasons are both internal and external to the firm. Again, the main internal reason is related to firm growth: limited expansion space at the current location, or limited representativeness of the present location. External factors include limited accessibility, deterioration of the building, environmental considerations, limited labour supply, or high location costs. Pull factors are largely the opposite of the internal factors: enough space, accessibility to deliverers, suppliers, customers, the labour market, representativeness, low costs, and often also locational amenities.

In conclusion, the behavioural approach adds to the neo-classical view by exploring the many motives, both economic and otherwise, that are important in the decision-making process of the firm, and that leads to a particular location. The approach seeks to understand the actual behaviour of entrepreneurs, and focuses on the decision-making process that may lead to a relocation and also takes path dependency into account. On the one hand this is very valuable information, but at the same time it shows the weaknesses of the approach. Often based on questionnaires and detailed empirical work, it is largely descriptive and explorative and to a much lesser extent an explanatory model. Similar to the neoclassical framework, too much attention is given to locational factors as such, and the link with internal firm processes of production, investment and growth is weak (see also Sayer, 1982). Another drawback is that the behavioural approach focuses too much on sociological, psychological and other "soft" variables (Scott, 2000) often ignoring the "hard" (neoclassical) economic factors. Therefore, an eclectic combination of the behavioural approach and the neoclassical approach seems to be more fruitful. An even more promising combination can be found with the newer institutional and evolutionary approaches, which we will describe next. As we will see, the evolutionary approach in particular essentially recognises behavioural starting points in its reasoning.

Institutional Approaches

Although very different in their arguments the neoclassical and the behavioural approaches have one view in common: the firm as an active decision-making agent in a static environment (Hayter, 1997). In the 1980s this simplistic view of the locational behaviour of the firm was increasingly being questioned in a number of new research directions. These new developments, no matter how different they may work out, share one common belief that economic processes in space are mainly shaped by society's cultural institutions and value systems. In other words, we have to look not only at the behaviour of the firm, but also at the social and cultural context in which this behaviour is embedded. Institutional approaches have dominated the field since then (Martin, 1999). In institutional approaches, firms have to negotiate with deliverers and suppliers, local, regional or national governments, labour unions and other institutions, about prices, wages, taxes, subsidies, infrastructure, and other key factors in the production process of the firm. Locational behaviour is the result of the outcome of these negotiations. The implication of this view is that this approach is more suited for large corporations. Larger corporations have more negotiating power and are able to exert a substantial influence upon their environment, whereas small firms usually have to accept the restrictions and constraints imposed upon them by their environment (Hayter, 1997).

That regional systems can be important contexts for firm growth is illustrated by examples such as Silicon Valley or Emilia-Romagna. These regions have a particular favourable entrepreneurial culture, in which key resources such as venture capital and knowledge are shared through intensive networks. In this regard, new terms were introduced, such as incubator regions, new industrial spaces, learning regions, etc. (Scott, 1995, 2000). These terms may be important as general concepts that help to understand historical processes of regional growth, and the evolution of particular corporations, but they are less helpful in explaining the actual locational behaviour of the many small and medium-sized firms in the economy. Nevertheless, small and medium-sized firms also operate in an institutional environment, which has a significant impact on their locational behaviour. There are two types of institutions that are important for the (re)location behaviour of small and medium-sized enterprises: governments and the real estate market. Whereas in the 1960s and 1970s the role of the government in relocating economic activities was large, at least when viewed from many regional policy intentions in those days, nowadays it is more modestly seen as either facilitating or inhibiting the locational choices of firms. We will come back to these policy issues in the next section. The real estate market is likely to be an important institution for firm relocation because small firms are usually dependent on existing demand on the real estate market for commercial property. Therefore, the spatial characteristics of the supply of real estate (office space, industrial sites and commercial floor space) are of prime importance for understanding the locational choices of small firms. The role of commercial real estate developers should be taken into account as well. However, empirical studies dealing with the effects of this institution on firm relocation are scarce.

For a long time, the three theoretical approaches discussed above dominated firm relocation studies. Recently, two new theoretical approaches have emerged that, despite the fact that they strongly draw from earlier theoretical approaches, introduce two distinct perspectives on firm relocation. Both new approaches criticise the established ones for paying too little attention to the path dependency of relocation, as compared to location, decisions (Stam, 2007; Knoben and Oerlemans, 2008). The first new approach takes an evolutionary perspective and focuses on the impact of distinct phases of a firm's life cycle on its relocation decisions. The second new approach takes a relational perspective and focuses on the constraining effects of interactions with other (local) organisations on a firm's relocation decision. Both new approaches will be discussed briefly.

Evolutionary Approach

The evolutionary approach's main critique on the established theoretical approaches to firm relocation is that they fail to take into account the changing nature of firms and the developments that underpin it (Stam, 2007). This critique implies that the path dependency of location decisions is underemphasised and that life-course effects are neglected. In other words, the established location theories assume that the push and pull factors are similar for all firms, whereas the relevance of these factors is likely to depend on past choices as well as the phase of the life cycle of the firm.

The evolutionary approach to firm relocation has been developed in the context of entrepreneurial firms (young, fast-growing firms) (Stam, 2003a, 2007; Schutjens and Stam 2003) but its insights have been used in the context of more established firms and industries as well (Buenstorf and Guenther, 2007). The basis for this theoretical approach is the firm-level evolutionary theory by Penrose (1995). Firms are conceptualised as "historical entities" that accumulate capabilities through an endogenous learning process (Stam, 2003b), resulting in highly path-dependent behaviour.

The evolutionary approach distinguishes several phases in the life cycle of a firm, namely: (1) start up; (2) initial survival; (3) early growth; (4) growth syndrome; and (5) accumulation. In each of these phases, firms have different levels of willingness and ability to relocate, which are influenced by different factors depending on the phase of the life cycle in which a firm is. For example, the personal relations of an entrepreneur are extremely important during the early life course of a firm (Stam, 2007; Maurer and Ebers, 2006). Over time, however, other factors such as resource dependency and sunk costs increasingly influence a firm's location behaviour.

Empirically, it is found that "softer" factors derived from the behavioural approach are more important when a firm is young, whereas financial and economic factors derived from the neoclassical approach become increasingly important as a firm matures (Stam, 2007). An important empirical contribution of the evolutionary approach is the identification of "opportunity driven" relocations. In the established theoretical perspectives, firms are only expected to relocate when their current location is no longer optimal or satisfactory. By closely linking the relocation decision to the processes taking place within a firm, it is shown that entrepreneurial firms also relocate to take advantage of specific opportunities at other locations.

On the whole, the evolutionary perspective approach shows that distinguishing life-cycle phases of firms is a fruitful addition to the relocation research. In different phases, different kinds of factors determine the relocation behaviour of firms. Hereby, the evolutionary perspective incorporates elements from both the neoclassical approaches as well as the behavioural approaches.

Organisational Network Approach

The organisational network approach mainly criticises the established theoretical approaches to firm relocation for their lack of attention to the relations between a firm and other organisations in its environment (Knoben, 2008).[5] This critique implies that the relational context of relocation decisions is underemphasised and that firms are incorrectly treated as atomistic actors that make their relocation decisions in isolation.

The organisational network approach to firm relocation is strongly rooted in the sociological embeddedness literature (Romo and Schwartz, 1995). It starts from the notion that a large and growing share of firms does not operate in isolation but is engaged in inter-organisational relationships (IORs) (Schutjens and Stam, 2003; Knoben and Oerlemans, 2008). Even though such relations are generally found to enhance the performance of the participating firms (Ahuja, 2000), they are also assumed to constrain the behaviour of participating firms (Granovetter, 1985). The organisational network approach poses that spatial lock-in, the fact that firms would like to relocate from a cost perspective but are unable to do so, is such a constraining effect of high levels of embeddedness.

Two forms of embeddedness are distinguished in the organisational network approach, namely organisational and territorial embeddedness. Organisational embeddedness refers to the amount of IORs a firm maintains and the strength of these IORs. Territorial embeddedness refers to the geographical localisation of a firm's IORs (Knoben et al., 2008). The spatially constraining effect of organisational embeddedness is based on the fact that successful IORs require stable, intense and frequent interactions. Relocation is likely to have an impact on the degree of organisational stability (Knoben et al., 2006) and might therefore disrupt the functioning of a firm's IORs (Isabella, 1990). To prevent such a disruption, firms might refrain from relocation.

The spatially constraining effect of territorial embeddedness is based on the notion that the larger the spatial distance between actors the more difficult it is to successfully transfer (tacit) knowledge (Gertler, 2003). Depending on the direction and distance, a relocation can disrupt a firm's territorial embeddedness. Again, to prevent such a disruptive effect, firms can refrain from relocating.

The organisational network approach to firm relocation does not dismiss the findings from earlier research, but rather tries to add a relational dimension to existing theoretical perspectives. The empirical work based on this perspective has shown that, compared to a baseline model that includes internal, external and locational variables, adding a relational perspective enhances explanatory power of models that try to explain firm relocation (Knoben and Oerlemans, 2008). The outcomes show that firms with high levels of embeddedness are indeed less likely to relocate. This provides clear evidence of path dependency in firm (re)location decisions. Moreover, it is shown that firms that relocate despite having high levels of embeddedness experience negative performance effects on the short run. In the long run, however, most firms benefit from relocating (Knoben et al., 2008). In other words, breaking with path dependency imposes "costs" in the short run but often yields "benefits" in the longer run. This finding corresponds with one of the main ideas of the evolutionary approach discussed in the above.

On the whole, the available empirical work suggests that including a relational (i.e. network) dimension in research regarding firm relocation is fruitful. Among others, it provides a channel through which the path dependency and conditionality of relocation decisions, which is present in most theoretical perspectives yet lacking from a lot of the empirical research, can be incorporated in research regarding firm relocation.

The five above-mentioned schools of thought on location provide the theoretical background for studies of firm relocation. However, most empirical studies in the field of firm relocation were not initiated by scientists, driven by their wish to apply theoretical viewpoints on firm relocation data, but by policy makers who cast their glance on the firm relocation phenomenon because it was related to practical spatial planning problems or held a promise for economic or employment growth. Only in the last decade did this change. In the next section, a historical review of firm relocation studies is presented, to illustrate the varying backgrounds and results of firm migration research in the course of time.

A HISTORICAL REVIEW OF FIRM RELOCATION STUDIES

The 1950s and 1960s: Classical Studies on Firm Relocation

The earliest known study on firm relocation is beyond any doubt McLaughlin and Robock's (1949) study *Why Industry Moves South*. This book describes the mid-century shift of manufacturing industry in the United States from its original concentration area in the Northeast to the South-eastern states, where low cost labour was more abundantly available and trade unions were less active. McLaughlin and Robock in this "classical" study thus stressed the importance of firm external factors for firm relocation. Another publication in the US of this early post-war period (Garwood, 1953) concentrates on industrial moves to Utah and Colorado and highlights the influence of other external factors, namely markets and materials. Besides these US-based studies, several classical UK studies were published in the 1950s and 1960s as well (e.g. Luttrel, 1962; Keeble, 1968). Despite the fact that these studies paid some attention to firm internal relocation forces, especially the pressure resulting from growth in output, external relocation factors dominated these studies as well.

In this period, the impact of firm relocation on local and regional economic structures and the possibilities of influencing firm relocations through local and regional policy instruments are core interests of researchers. Redistribution of people and work became important topics in a time of rapid demographic and economic growth in the core regions of many countries. Cities expanded, suburbanisation increased the size of the urban agglomerations, and congestion became a part of everyday life in urbanised regions. At the same time, however, many more peripheral regions lagged behind this rapid growth. It could even be argued that the core regions developed at the cost of the peripheral regions, leading to a flow of people and resources from the periphery to the core. Two types of regional policies were implemented in many countries to cope with the effects of these developments. The first type of policy dealt with urban renewal and firm relocation to the outer zones of the cities and beyond (suburbanisation). The second type of policy aimed at redirecting growth from central to peripheral regions. Firm relocation was seen as a means to transfer work and prosperity to lagging regions, while at the same time easing the congestion, labour market, and space capacity problems in the core regions. Instead of reducing regional disparities by means of a "workers-to-the-work" policy, firm relocation can facilitate a "work-to-the-workers" type of policy.

In short, we can state that all early studies share a prime interest in the causes of relocation (push and pull) and the impact of relocation on local and regional economies, including the policy options to stimulate these economies. During these early years of relocation research, the focus lay on the outside world of the firm, more especially the external circumstances that push the firm from the old location and/or pull it to the new one. Neoclassical thinking provided a translation of the research findings in this period to cost-oriented policy measures like subsidies and tax reductions lowering the cost of land acquisition and factory building, and the reduction of transport cost levels for firms by improving public infrastructures. In several of the early studies such policy measures, geared to stimulate relocation to lagging regions, were evaluated with mixed results (e.g. Cameron and Clark 1966).

The 1970s: UK Dominance

In the next decade, the 1970s, most international publications on firm relocation were UK based. Some of the major and most cited studies from this period are: Townroe (1969, 1972, 1973), Keeble (1971, 1972, 1975, 1976), Lever (1972), Spooner (1972, 1974), North (1974), Sant (1975) and Rhodes and Moore (1976).[6] The profound interest of British researchers in the phenomenon of firm relocation is to be understood from the circumstance that UK regional policy during its heyday in the 1960s and 1970s relied heavily on steering manufacturing firms into assisted areas using the instruments of location controls, capital subsidies and labour subsidies. In fact, the British firm relocation studies are just part of the numerous studies in this period which attempted to estimate the effect of such policy instruments on the economy of the assisted areas (Armstrong and Taylor, 2000, p. 370). Given these interests, it is perhaps not surprising that the focus of the analyses in this period remained on external relocation factors.

Another factor that explains the dominance of British studies is the availability of complete and reliable data on firm relocation, on a national basis. On the basis of these statistics Keeble (1976) argues that earlier research has underestimated the importance of firm relocation. He refers to studies (e.g. Spooner 1972, 1974) which show that in peripheral regions in the UK more than a quarter of total employment in manufacturing is to be found in firms which moved there in the first post-war decades. For the UK as a whole it is interesting to see that in terms of employment, one third of all industrial movement (in the late 1960s) can be categorised as short distance spillover from the key conurbations and 50% as long distance "between regions" moves to the periphery (Keeble, 1976, p. 135). Furthermore, Keeble concludes that "regional policy was the single most important variable influencing the sub-regional pattern of migrant industries between 1966 and 1971" (1976, p. 145). This conclusion contradicts Cameron and Clark's (1966) more cautious conclusion about the same issue for the prior decade.

At the end of the 1970s, Townroe supplements the abundant information about industrial firm relocation in the UK with information about the US (Townroe, 1979). Perhaps the biggest contribution of his book is that he was the first to analyse firm relocation facts in a theoretical perspective – behaviourally oriented, as could be expected at the end of the 1970s – and he produces a first attempt to present a general (descriptive) model of the location decision-making process. Moreover, Townroe suggests that firm relocation research should be integrated with studies on the birth, growth, decline and death of firms to form a so-called components-of-change approach. As we now know, this suggestion has indeed been followed, and leads to what we now refer to as the demography of firms approach. We will follow up on this issue later.

A European Survey

In retrospect, Townroe's 1979 survey of research results concerning the US and the UK ends a period in which these two countries dominated the international literature on firm relocation research. Starting in the second half of the 1970s, researchers on firm relocation from other countries also started to publish in English. However, numerous publications from this period of researchers in the North, West and South of Europe in languages other than English exist. Impelled by the same curiosity concerning spatial-economic changes and the same wonder about the results of regional policies – which in the post-war period revived throughout Europe – researchers in many countries started to investigate the phenomenon of firm relocation. In hindsight this period of the 1970s may be considered as the florescence of classical firm relocation research. Aydalot et al. (1972) and Aydalot (1978) published results of firm relocation research in France, whereas Camagni (1976), Ortona and Santagata (1980), and Ortona et al. (1981) did the same for Italy. Christiansen (1978, 1979) published results in Denmark, whereas the Netherlands had an industrial relocation report by SISWO (1967) followed by a survey study by Pellenbarg (1976, 1977). Some information regarding Ireland, Belgium and Greece also became available. Klaassen and Molle (1983) made the content of all these original but not very accessible studies available to the international scientific audience with their book on *Industrial Mobility and Relocation in the European Community*. Through this book for the first time a European survey of information concerning firm relocation became available.

Klaassen and Molle's book presents information on firm relocation in all EU member countries of that time, with the exception of Luxemburg. An interesting outcome of the European comparison is that in all countries the same bi-partition of firm relocation is observable, i.e. the short distance "industrial suburbanisation" around the larger urban agglomerations on the one hand and the longer distance "industrial decentralisation" from the economic core areas to peripheral and/or development areas on the other hand.[7]

In most countries the research projects are concentrated in the manufacturing sector, apparently the most mobile sector of the 1960s and 1970s and at the same time in most countries the main target sector for regional development policies. All European researchers agree about the nature of the driving forces behind the firm relocation processes. Lack of space, transport-related issues, and labour market problems dominate the picture. The first two are the main causes for short distance moves in and around the urban core areas, while labour shortage is the dominant motive for long distance movements to development areas in national peripheries. So in line with the UK and US studies from the 1970s, external factors were found to be the main drivers of spatial firm mobility.

However, in the course of the 1970s the importance of the labour market motive decreased. Due to the economic recession unemployment went up sharply and labour shortages disappeared rapidly in almost all regions. After this, the magnitude of long distance relocation decreased. There are two reasons for this trend. First, firm mobility is related to firm growth, and tends to diminish in an economic recession. Firms are less likely to take fargoing investment decisions like a relocation when the future economic prospects are rather uncertain. The economic recession also had negative effects on redistribution policies. Redistribution of economic growth is a luxury problem. Without economic growth there is nothing to redistribute, and therefore these regional policies were abandoned in the recession period following the oil crisis of 1973. Second, the longer-term cumulative effects of redistribution of economic growth in the previous periods also led to a decrease in long distance relocation. For the Netherlands, Pellenbarg (1976, 1985) argued that the gradual rise of income and education levels in the 1960s and 1970s, especially in the more peripheral parts of the country, effectively diminished the potential of such regions to attract industrial branch plants, many of which are routine production facilities looking for cheap and low-skilled labour locations. Similar arguments can be applied to other European countries in the same period.

In the synopsis of their European survey Klaassen and Molle (1983) do not pay too much attention to the regional policy dimension. They conclude that with the change of the economic tide policy attention has shifted from influencing relocations to stimulating other components of economic change, especially new firms (birth) and growth. Despite this change of atmosphere, it cannot be denied that for many regions firm relocation has contributed considerably to employment growth. In the previous section we already referred to results for peripheral regions in the UK. Comparable analyses in the Netherlands for a number of regions varying in size and peripherality yield more modest but nevertheless sizeable outcomes. Parts of the provinces of Overijssel and North Brabant reach a percentage of 10% of industrial employment due to firm relocation over a post-war period of 25 years (Pellenbarg, 1985).

The 1980s: From the Regional to the Urban Level

In the 1980s, the number of international firm relocation studies dropped sharply. The studies of Erickson and Wasylenko (1981) and Ortona and Santagata (1983) stand out because they utilise the urban level, rather than the regional level, for their investigations. These publications stand out because they not only pay attention to government influence on firm relocation in the context of regional (development) policy, but also refer to other policy types such as environmental policy, urban renewal policy and local development policy (Pen, 1999). Just like regional policy, these other policy fields may be expected to result in firm relocations. For most of these new policies the urban agglomeration is the most relevant playing field, and local (and regional) governments are much more involved in policy formulation and its policy consequences than national governments. Despite the change in the spatial level of analysis, the focuses of the analyses from this period remain on external relocation factors.

The interest for the urban (i.e. local) level of analysis coincided with a strong increase of short distance firm relocation. While long distance relocation decreased for a number of reasons indicated earlier, the reverse was true for short distance relocation. The underlying causes can be related to the process of suburbanisation of firms, propelled by space shortage and increasing land prices in the large cities, parking problems, and growing congestion on city roads and beltways. No wonder that the urban firm exodus swelled, and that that city governments worried about the ensuing erosion of urban employment. The politicians' problems were even more complex because some of the government's own policies, especially environmental policy and urban renewal policy, in fact reinforce the pressure on firms to leave urban locations. In this context, it is easily understood that firm relocation studies from the 1980s onwards concentrate more heavily on urban agglomerations, and less on the balance between central and peripheral regions in a national context.

Looking back, we may conclude that for most modern countries urban renewal in its classical form (rebuilding and revitalising nineteenth century urban areas) has been a temporary activity, and consequently the same is true for the attached suburbanisation of inner city activities. In many cities, however, some large scale manufacturing sites have remained in the central parts of urban areas. The "outplacement" of such old industrial plants is an important task for the next decades, needed to create new space for urban development, and an interesting new research field for firm relocation researchers. At the same time, the majority of firms which leave the city are no longer industrial or wholesale firms, but business services. Hence it is not surprising to see that firm relocation research in the 1990s shifts its attention to this fast growing and most mobile sector.

The 1990s to Present: New Data and New Approaches

After 1990 some firm migration studies still follow the more classical path of monitoring regional or national (manufacturing) firm migration patterns and explaining them on the basis of external factor-oriented interview programmemes or regression analyses. Interesting examples from Europe and North America are the studies by Holl (2004) concerning firm migrations in Portugal (between 1986 and 1997) and a study concerning the American clothing industry (Van Dooren, 2003). The former demonstrated – on a national scale – the general significant influence of infrastructure on location choices, and the special influence of interregional road connections on relocations. The latter showed the interesting pattern of relocations in the North American garment industry, started by the 1994 NAFTA agreement (see Figure 4.1). This is, more than half a century after McLaughlin and Robock's classical study, still a fine example of "industry moves South"! In fact, many manufacturing firms in the US follow a stepwise migration path, looking for lower costs for land, labour, energy and so on, which takes them first from central urban agglomerations to urban fringe areas, then to cheaper states within the US, and finally abroad. In the case of the garment industry this leads first to Mexico, and later to other Central American states, or China (Verkoren, 2008).

Still, interesting as they may be, such studies no longer reflect the main trend in firm migration research in the 1990s and later. The more general picture is that the interest in firm relocation as a panacea for regional development has faded, and instead it is now felt that regions themselves should create the conditions for innovation and creation of new economic activities. Or, if they lack the resources to do this, effective regional subsidies should be given to them. Key concepts underlying the new policy agenda are regional agglomeration effects, endogenous growth, regional knowledge networks and learning regions. In this view, the focus is still on new economic activities, as in the 1980s, but there is also an emphasis on sustained economic growth. Only half of the new firms survive after ten years, and the conditions for survival of ideas and new firms are as important

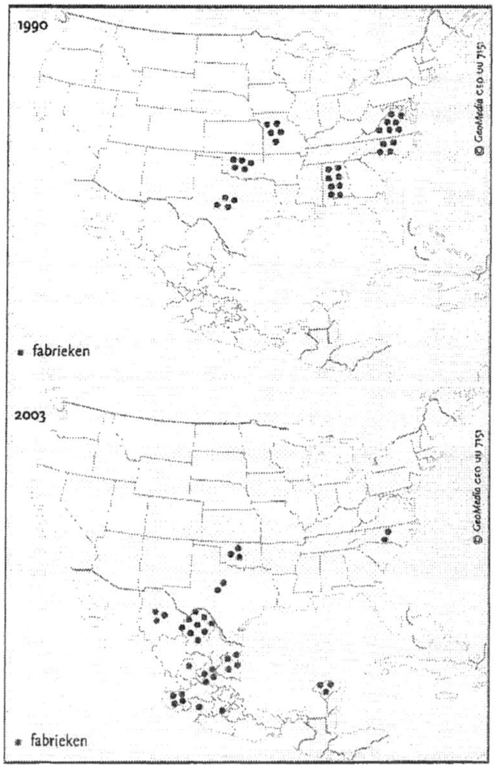

Source: Verkoren (2008)

Figure 4.1 Relocation of production locations of "branded jeanswear" in North America between 1993 and 2003

as those for the creation process itself. In this process of bringing new ideas to maturity or new and small firms to larger corporations, the spatial and regional conditions may be important as well. Firm relocation then may be considered as a locational strategy for removing restrictions to firm growth. This brings the issue of firm relocation studies to the fore again, but this time as the consequence of the coming of age of generations of new firms developing various locational strategies for coping with sustained growth. Consequently, firm relocation is no longer treated as a separate event, but incorporated in the aforementioned components-of-change approach. This is in fact a demographic approach to firm dynamics, which has gained popularity among regional scientists since the 1990s (Van Dijk and Pellenbarg, 2000a). This approach, which is studied both by geographers, sociologists and economists, is variously termed "industrial demography",

"demography of the firm or enterprise", or "economic demography" (Van Wissen, 2002). Some studies focus on changes at the macro-level of the population, as a result of the processes of firm birth, death and migration, as well as growth or decline and ageing of existing firms. Firm migration is a component of change of a regional or national population of firms. In this respect, the contribution of the migration component to the change in the size and composition of the population is studied. An example of such a macro-level study based on a micro-simulation model can be found in Van Wissen (2000).

The new angle on firm relocation as a specific locational strategy has resulted in two new streams of research. First, research has shifted from studying external causes of firm relocation to studying internal causes of firm relocation. Second, given the fact that relocating is only one of the possible strategies to remove restrictions to growth, research also delved into the issue of the relocation decision process. Both streams of research will be briefly discussed here, starting with the latter.

The new interest in the process of decision making in firm relocation situations is an important renewal in this research field. The identification of push, pull and keep factors only gives a superficial kind of explanation of firm relocation processes. In the 1990s, there is a desire to dig to a deeper level of explanation, which takes into consideration how decision processes regarding firm relocation develop. In the preceding discussion of (re)location theories we already indicated the interest in decision processes which springs from the behavioural approach. As early as the 1970s Townroe, representing this approach, developed a model with five successive decision stages: (1) stimulus, (2) problem definition, (3) search, (4) formulation and comparison of alternatives, and (5) choice and action. However, the application of this scheme in empirical research was scarce. The 1990s witnessed a revival of interest in this process aspect of firm relocation decisions. Louw (1996), for example, empirically studied the location choice behaviour of (relocating) large offices in the Netherlands. He divided the decision-making process into three phases: an orientation phase, a selection phase and a negotiation phase.[8] His findings indicate that spatial factors (i.e. geographical position, accessibility, parking possibilities, proximity of facilities and public transport, and quality of the spatial surroundings[9]) play an important part in the first two phases, whereas financial and contractual factors become more important in the third phase, when it comes to negotiating a result (see Table 4.1). Louw's process-oriented approach to firm relocation decisions was followed up by Pen (1999, 2000). In a study based on an extensive enquiry among over 1000 individual firms he considers relocation as one of six types of location strategies. He finds that the average length of the decision-making process is much longer than often assumed (on average two years) and shows that the number of stages in the decision process varies from a minimum of three to a maximum of seven.

Another important finding of Louw's study is that the average number of location factors mentioned per respondent varies considerably from one decision stage to the other. In the orientation phase the search is of a rather superficial nature. Only two location factors matter: the number of square metres available and the price for which the location/building is for sale or rent. In the selection phase, a much greater number of factors are considered. In this phase many spatial factors enter the scene as well. In the negotiation phase the focus lies again on one or two key issues, of which the price issue is always predominant.

Table 4.1 Factors in the location search process, mentioned by managers (percentage of all factors mentioned, per search phase)

Phase and Factors	Orientation	Selection	Negotiation	Total
Building factors	15.3	12.3	7.1	11.9
Functional factors	19.4	18.4	7.1	16.1
Technical factors	4.2	2.0	3.4	3.1
Financial factors	14.2	52.5	22.5	12.2
Location factors	43.9	36.0	12.1	32.3
Other factors	6.1	14.6	19.2	13.8
Total	100	100	100	100

Source: Louw (1996)

The other new research theme that emerged during the last decade is the focus on the role of firm internal characteristics as a cause of firm relocation. Most studies that examine this issue have done so by estimating micro level models that explain firm relocation by incorporating various firm internal, locational and firm external variables (Van Dijk and Pellenbarg, 2000b; Brouwer et al., 2004; Knoben and Oerlemans, 2008). Van Dijk and Pellenbarg (2000b) were the first to econometrically show with firm level data from the Netherlands that firm internal variables, such as firm size and age, are a better predictor of firm relocation than locational factors and external factors. Brouwer et al. (2004) performed a similar analysis but included more firm internal variables and utilised data from different countries. Their results also indicate that firm internal variables are the most important cause of firm relocation. Besides size and age they find that especially changes in the size of the firm (both positively and negatively) as well as mergers and takeovers are strongly associated with firm relocation. Knoben and Oerlemans (2008) add to the previous micro level analysis by explicitly taking into account the relations between a firm and other organisations in its environment (i.e. relational variables). They show that compared to a baseline model including internal, external and locational variables, adding relational variables significantly enhances the explanatory power of the models.

These micro level models have in common that they emphasise the role of firm internal variables and thereby deviate from many of the findings of earlier research. Perhaps the most important general finding of these studies is that relocation is not purely triggered by external factors and/or a shortage of expansion space. As such, these findings match-up nicely with the more recent (re)location theories that pose that relocations can also be opportunity driven (evolutionary approach) and/or relationally constrained (network approach). The new theoretical approaches to firm relocation have, in turn, created opportunities for new research. This new research frontier will be discussed in one of the upcoming sections. Before doing so, however, a case study of the Netherlands and Belgium will be presented. The choice of the Netherlands and Belgium for a case study is prompted by the availability, for these two countries, of recent, comprehensive and comparable data on firm relocation. This offers a rather rare possibility to, at the same time, identify recent tendencies and check the uniqueness of a number of key indicators on a national scale.

CASE STUDY: THE NETHERLANDS AND BELGIUM

Firm Migration in the Netherlands

Compared to most other countries, the Netherlands has been very intensively studied with respect to the firm migration phenomenon. The availability of data is certainly one of the explanatory factors here; another is the active interest of university research groups in economic geography, especially at the universities of Groningen and Utrecht that have witnessed the publication of a rather impressive series of PhD theses, most of them in the past decade.[10] If we add the publications of scientists, professional consultants and government institutions we count several hundreds of titles: books, reports, journal articles and papers. We will not try to sum up all the results of this magnitude of studies here, but extensive summaries can be found in Pellenbarg et al. (2002a, b). The most important milestones when it comes to the facts and figures of the firm migration process for the Netherlands as a whole until recently were the aforementioned SISWO report that documented industrial migration in the 1950s, and the firm migration datasets provided by the Dutch Chambers of Commerce (CoC) for the late 1980s and early 1990s (Pellenbarg and Kemper, 1999). For the 1960s and 1970s there are only fragmented pictures, for individual cities and regions, or sample-based studies (Pellenbarg, 1985). After 1995, the Chambers of Commerce stopped releasing the firm migration data for publication purposes because of a growing lack of confidence in the accuracy of the data. Although several tests showed that the data quality was better than expected (i.e. Inbo, 2003), the interruption of the CoC data series left us with the unanswered question whether or not the observed national trend of a gradual "fanning out" of firms from the Randstad area to adjoining regions was continuing or not. Figure 4.2 shows this process on maps, that have to be observed with care. The vast majority of firm migrations, taking place within provinces, is not visible. What is shown are the moves between provinces. Just like the internal moves, they concentrate in the Randstad area, but we also witness a deconcentration process to the adjoining provinces. The suburbanisation of firms around the major urban centres seems to increase in size and scale, surpasses provincial boundaries, and could be supposed to gradually reach the more peripheral

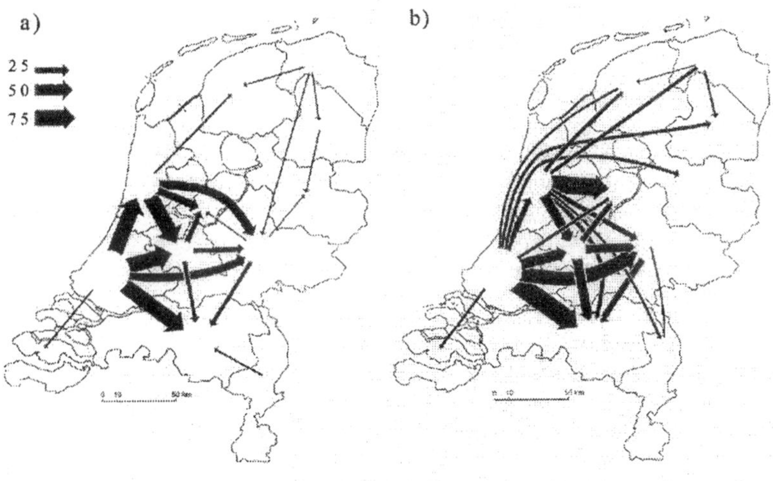

Source: Kemper and Pellenbarg (1997)

Figure 4.2 Interprovincial firm migrations (numbers of firms) in the Netherlands (in manufacturing, wholesale and commercial services; annual average)

a) 1990/91 *b) 1994/95*

provinces, if the data were indeed accurate enough and the tendency show persistency.

Recently, it has become clear that the expectations attached to the CoC firm migration data (i.e. Kemper and Pellenbarg, 1997) have not become true. In a project engaged by the Netherlands Institute for Spatial Research (Ruimtelijk Planbureau RPB) a totally new dataset concerning firm migrations in the period 1999–2006 was built up, using LISA instead of CoC data.[11] The analysis of the new dataset disclosed a fascinating new insight into the spatial tendencies of firm migration in the Netherlands (Van Oort et al., 2007). As Figure 4.3 clearly shows, the deconcentration from the Randstad regions has not grown in size and scale since the 1990s, and certainly does not yet reach the more peripheral provinces. On the contrary, it seems that the peripheries in the Netherlands in the North and South have become disconnected from the spatial dynamism in and around the Randstad area and are merely witnessing some internal firm migration activity. Also, the net outflow from the Randstad cities and provinces to the adjoining intermediate provinces of Flevoland, Gelderland and North Brabant that was observed earlier is much less pronounced, and counterbalanced by moves into the Randstad. One reason for this could be a growing share of business service firms in the migration streams, versus a further decrease of wholesale and manufacturing firms. Maybe the increase of available office locations in the Randstad area since the mid 1990s plays a role as well. The Randstad region appears to grow in space: parts of the adjoining provinces of North Brabant and Gelderland are now closely linked or even part of it. A regression analysis shows that the main features that determine whether a particular municipality is attractive to relocating companies are its accessibility by road, the economic climate and the proximity to other economic clusters (Van Oort et al., 2007).

Next to the necessary update of our insight into the spatial pattern of firm relocations, the 2007 RPB study offers a number of other important key figures concerning the actual state of affairs in firm migration in the Netherlands, and more especially its employment effects. First of all, the study reveals that each year some 18,000 organisations in both the public and the private sectors opt to relocate to new premises. This represents approximately 4% annually of the total number of firms in the Netherlands, and implies that over a period of 10 or 20 years considerable parts of the Dutch economy are being re-housed and relocated. With the firms move 200,000 staff, this is an annual 3% of all national employment.[12] Relocating firms enjoy a markedly higher employment growth than other firms. Around the time of the relocation, employment within migrant firms increases between one-and-a-half and two times more rapidly than in other firms. On the whole, the contribution of migrant firms to local (municipal) employment growth equals the contribution of newly established firms.

The majority of companies (94%) relocate within the same region and even (75%) within the same municipality. Therefore, the growth in local employment due to companies moving into municipalities from elsewhere, although important for some individual towns and cities, should generally not be overestimated. The RPB report's advice to municipalities is that they should focus their efforts on providing suitable accommodation for companies which are already established in the area but are now seeking larger premises, instead of competing with each other in attracting migrant firms. The "prisoner's dilemma" of no municipality feeling obliged to curtail its active marketing efforts can best be solved at the regional level.

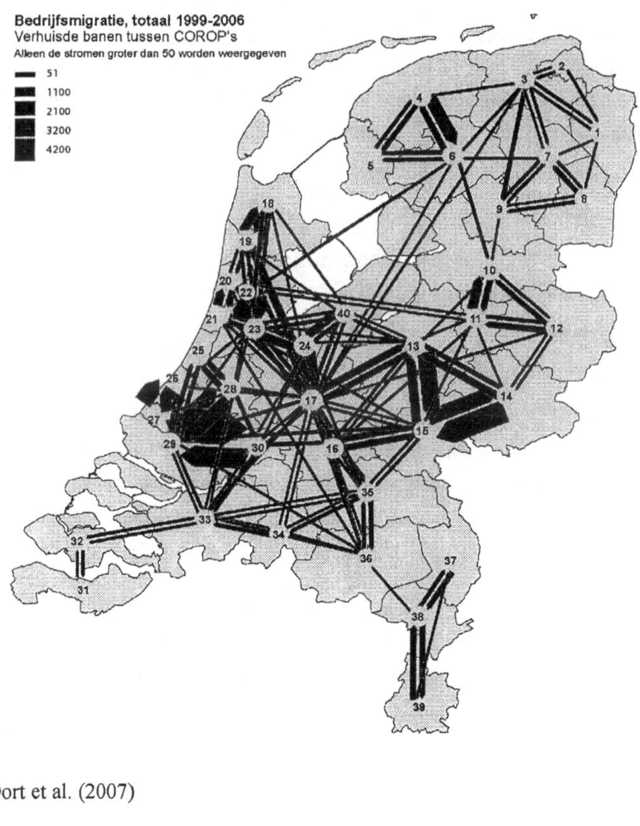

Source: Van Oort et al. (2007)

Figure 4.3 Interprovincial firm migrations in the Netherlands, 1999–2006 (numbers of jobs involved, only moves between COROP regions; only figures > 50 wp are pictured)

Evidentially, more and more municipalities are now making agreements about a more restricted regional supply of business areas; a strategy that fits quite well in the national goal of a more sustainable use of space and at the same time connects to the observation that intraregional firm moves tend to replace interregional migrations. A related observation is that the strong growth of "economic suburbanisation" which was observed earlier holds off. Here we find a stimulus to shift the focus of firm migration research to lower spatial levels, a theme that will be raised again in the next section, about possible new research frontiers.

Firm Migration in Belgium

Many firm demographic studies indicate that the components of employment change are characteristically different for different countries. This is certainly true for the main components of birth and death (Eurostat, 2008). But the ratio between the components of birth and death on the one hand, and migration on the other, also differs between countries. Within the Netherlands the components of new firm formation and firm migration are – measured on a local or regional scale and in terms of employment – of comparable magnitude. However, for a country like the United States it is reported that on the regional scale new firms are a much more important generator of new employment than migrant firms. Neumark et al. state that in the US interstate relocation has a "negligible effect" on state employment (Neumark et. al. 2006). But of course, the geographical scale of states in the US cannot be compared to regions (COROPs or provinces) in the Netherlands. Interstate relocation in the US should be compared to international relocations between countries within Europe. Although very little is known about such interstate moves in Europe, we may safely assume that they are much less frequent, just like interstate moves in the US. The conclusion may be that structural data on firm migration, as we just discussed for the Netherlands, should be considered in the context of its geographical scale. True comparisons with other countries require comparability of scale. Belgium offers a "safe" example here. The country, which neighbours the Netherlands to the South, is somewhat smaller in size and population, but the regions for which firm migration data are available (Flanders, Wallonia and the Brussels region) are somewhat bigger than the Dutch provinces, and its municipalities are generally comparable.

The firm migration data source which we can use for Belgium is a 2001 report by Cabus et al. Its main findings are based on a data set of Belgian firms that migrated in the period 1990–2001, i.e. slightly overlapping the period of the Dutch RPB study, which was 1999–2006. For the whole of Belgium the dataset includes 24,300 firm migrations, which represents approximately 2% annually of the total firm population. For the Netherlands this was 4%. Over the whole period of 1990–2001 no less than 25% of all firms in Belgium witnessed a move (Cabus et al., 2001). For the Netherlands such a figure for a ten-year period is unknown, but can assumed to be higher, as witnessed by the higher overall mobility figure of 4%. Incidentally, the height of the crude mobility figures for the Netherlands as compared with Belgium are to be considered with care because the data sources are not identical.[13] A more reliable comparison is found in relative indices concerning the age, size and sector composition of the mobile firm category and, most of all, the spatial trends in the firm migration process.

Unfortunately, the Belgian report does not show the firm migrations in an arrow map like the Dutch RPB study did, but the map showing mobility rates on a postal code basis (see Figure 4.4) nevertheless shows rather clearly that just like in the Netherlands, the highest mobility rates are found in the urban agglomerations, which in the Belgian case means many relocations in and around Brussels, Antwerp and Ghent. As Cabus et al. report, it is in these areas that we find both the highest degree of firm formation, especially in the service sector, and the highest level of "spatial pressure," i.e. shortage of locations and high land prices, prime causes for the movement of growing firms. The average migration distance is 19 kilometres, which indicates that a majority of the firms move within municipalities or to neighbouring municipalities, just like in the Netherlands. The intra-regional dimension of relocation far outranks the interregional dimension and within regions the same is true for intra-urban moves versus urban exits. No less than 40% of the migrant firms stay within the same central urban area and another 27% does not leave the urban agglomeration. Nineteen percent moves to suburbs. Municipal maps of employment gains and losses, available for both the Netherlands and Belgium (but not shown here) depict the big urban agglomerations in both countries as losers, and the surrounding urban fringe areas as the winners. As in the Netherlands, the Belgian peripheries, especially in the Southeast, are hardly involved in the national relocation process.

On a larger geographical scale, beyond the urban agglomerations, the regional impact of firm migrations in Belgium is unimpressive. Of all moves 83% are within the three major provinces of Flanders, Wallonia and Brussels (see Table 4.2). Half of the total mobility is within Flanders, the economic core area of Belgium. There is some interprovincial migration between Brussels and the two other provinces, with a negative balance for the Brussels (mainly urban) province. There is a remarkably low mobility (only 1% of all moves) between Flanders and Wallonia. The language barrier between these two provinces is a ready explanation for this.

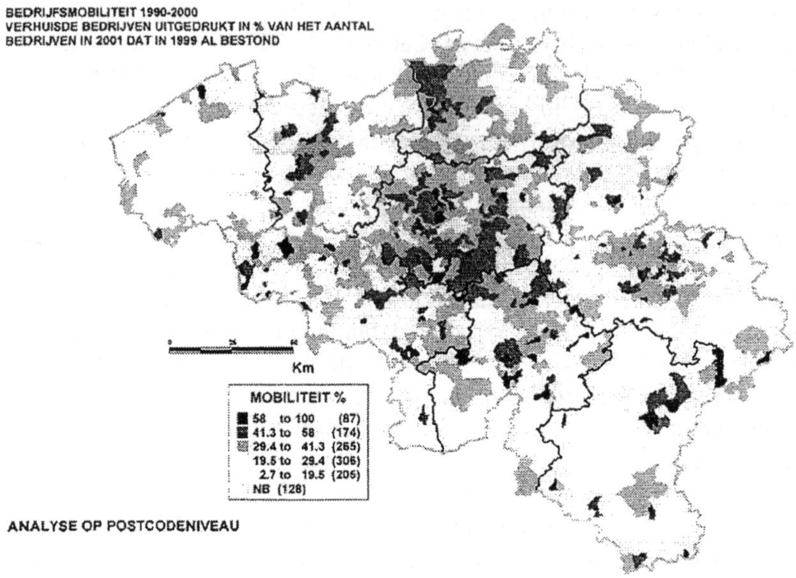

Source: Cabus et al. (2001)

Figure 4.4 Firm migration in Belgium, 1990–2001 (migrant firms as % of all firms, per postal code area)

Table 4.2 Firm migration in and between provinces in Belgium, 1990–2001 (numbers of firms)

From	To			
	Flanders	Wallonia	Brussels	Total
Flanders	12,138	292	881	13,311
Wallonia	225	3,815	550	4,590
Brussels	1,165	977	4,288	6,430
Total	13,528	5,084	5,719	24,331

Source: Cabus et Al. (2001)

The vast majority of the migrant firms in Belgium are service firms (59%) and most of them are small firms, with on average four employees, which is less than observed in the RPB study for the Netherlands but rather close to the earlier Dutch observation of three employees, based on CoC data

(Pellenbarg and Kemper, 1999). Relatively, most mobile firms are in the size category of 5–50 employees. A mobility peak is observable among firms that are nearing the age of five years after their start-up. On average, employment in mobile firms grows faster than in the firm population as a whole. These figures characterise the migration process as being dominated by young service firms, recently started up in urban incubation milieus and now looking for new locations elsewhere in the urban area that can accommodate their growth. These characteristics are exactly the same as observed in many reports for the Netherlands, including the most recent ones as discussed above, and also the indications for other countries, as we discussed in the literature review in the third section. It inspires a heavier concentration in firm migration research on lower spatial levels, i.e. the urban and even the neighbourhood level, to find out what determines the decisions to stay or go of firms, the relevance of their local networks, the relation to their performance and their feelings of local well-being. We will discuss these options for future research in greater detail in the next section.

PUSHING THE RESEARCH FRONTIER

Some of the issues on the firm migration research agenda of the past half century have been studied intensively enough to declare that they have had their day and only require more or less regular updating. This certainly applies to the main spatial and sectoral tendencies and the identification of external push and pull motives. But some of the newer research lines of the 1980s and 1990s are by no means exhausted and call for continued attention, i.e. the firm's internal motives for relocation, the phasing of the migration decision process and the possibilities of capturing this process and its main driving factors in a mathematical model, that should be able to forecast future migration patterns and the related size and spread of demand for business space (Van Wissen, 2000). At the same time, we perceive the appearance of several new and interesting research lines that can be expected to produce additional information for a better and deeper understanding of firm migration in the next period of time. We will discuss four of these new research frontiers.

Firm Migration and Firm Performance

Growth is a rather well known characteristic of firms that migrate. The growth-inspired search for space is push and pull motive number one for most firms. However, there are indications that the reverse of the relationship applies as well: firm migration is a cause of firm growth. This is partly because the larger size of the premises enables a more efficient production or service process, but this is just what is expected from the move, and only part of the story. An early firm migration study on the rise of 1970s publications in the UK from NEDO (1978) showed that many firms that had hesitated to invest in new premises in a new location experienced unexpected advantages on the new location afterwards. Besselink et al. (1988) found the same in a survey of migrant firms in the Netherlands and hypothesised the extra growth to be a consequence of the growth of trust of potential customers, who think the relocation of their supplier is a sign of the firm's doing well. Bouwers and Pellenbarg (1989) repeat this finding; in their survey a majority (70%) of firms do not expect the firm performance to improve in the new location, yet after the move there is also a majority of 70% who report that the firm performance did grow. In his recent study Knoben (2008) again explored the relation between firm migration and firm performance, which he finds insufficiently explored until now. Unfortunately, in most of the scarce literature sources that he finds, the performance variable is operationalised with stock prices and thus ignores innovative performance or growth of turnover, profit and employment. In his own research, focussing on the automation services sector and using a stepwise regression model approach, Knoben does include these other variables. As we mentioned briefly at the end of the second section, he finds that generally speaking most firms do indeed benefit from relocating, in the medium and long run. Still, the findings show that the effects of relocation on performance differ substantially between different performance indicators. Knoben's analysis also shows that the general positive performance effect of a relocation is influenced by the level of a firm's organisational embeddedness. In particular, the performance of firms with many and/or strong interorganisational relationships (IORs) seem to be hampered by a relocation. Interestingly, the liability caused by high levels of organisational embeddedness can be (partly) negated by high levels of organisational proximity! Knoben's findings may be regarded as a stimulus for further research along this line. Looking at some of the drawbacks he acknowledges in his own analysis (i.e. the focus on one particular branch of the service sector, and the fact that only movers, and specifically move survivors, were included in the survey) he pleads that such further research should be based on longitudinal and/or cross-sectional data (Knoben, 2008).

Firm Migration and Locational Satisfaction

It seems a logical series of steps in the research process: from the registration to the explanation of firm migrations first, then from the explanation of migrations to their effects in terms of firm performance, and finally from the firm's post-move performance to the degree of satisfaction (of the entrepreneur) following movement. However, until now the study of post-move locational satisfaction is even more exceptional than the study of migration and performance which we discussed just now, although there have been signals that it might be potentially interesting. A number of studies in the Netherlands and the UK in the 1970s and 1980s indicated that when measured within one or two years after the move, 20–30% of the firms (the percentage depending of the phrasing of the question) were not satisfied with their new location, meaning that such firms would have preferred another new location after all (Townroe, 1979; Pellenbarg, 1977, 1985). At first, one is inclined to think that such dissatisfaction may be caused by a lack of sufficient pre-move gathering of information concerning location alternatives. However, two recent tentative research projects of the university of Groningen's faculty of Spatial Sciences (Adema, 2003 and Schuilenburg, 2004) suggest that it is the other way around: intensive preparations more often lead to dissatisfaction, and many quite satisfied firms made more intuitive location choices. From Schuilenburg's thesis, which affected the population of manufacturing firms with ten or more employees[14] that moved to a new location within the Netherlands in 2001, we take Table 4.3 as an illustration. It shows that among the 2001 manufacturing industry relocaters the ratio of "regretters" against "satisfiers" was about 1 to 4, but more interestingly, the regret percentage was higher (in fact twice as high) among those who searched more seriously (Schuilenburg, 2004).

Table 4.3 Relocation and satisfaction (percentage of all firm: N=39)[15]

Satisfied with new location (%)	Considered alternative locations (%)		
	Yes	No	Total
Yes	73	88	82
No	27	13	18
Total	100	100	100

Source: Schuilenburg (2004)

In an attempt to achieve more clarity about this, a broader and more intensive new research project was set up, involving the questioning (by telephone) of a national representative sample of 200 firms with ten or more employees that relocated between municipalities in the Netherlands in 2005 and 2006 (Kobus, 2007), from which 154 firms (77%) responded. One third of the respondents (32.5%) answered that they would prefer another location if they could renew their choice. Following this, the firms were questioned about the intensity of their search process (i.e. the number of alternatives considered), the phasing of the search process, the role of objective and subjective search factors and external influences, external advice, firm profit and profit expectations, and possible new relocation plans. The relations between the variables were tested on significance using chi-square tests, and the influence of the various independent variables on the dependent variable, i.e. locational dissatisfaction, was tested in a regression analysis. The results can be summarised as follows:

- firms with more employees do not consider more alternatives in the search process;
- calling external advice does not correspond with considering more alternative locations or a higher locational satisfaction after the move;
- a process with more phases does not correspond with a higher level of satisfaction;
- having more alternatives does not correspond with experiencing external advice;
- firms stressing objective factors in the location search process are not significantly more content with their choice than firms stressing subjective factors;
- firms not expecting next year profits have more intentions to relocate in future;
- firms expecting a move (within 10 years) are less satisfied with their present location;
- the binary logistic regression showed that expected profits and subjective search behaviour were the main explanatory factors with respect to locational satisfaction;
- the regression analysis revealed no significant influence of having more location alternatives on the level of locational satisfaction, but the chi square test showed that there was a significant (negative) relationship between the number of location alternatives and location satisfaction as individual factors.[16]

The list as a whole is interesting, but the last two issues are especially relevant here. They confirm that subjective and less intensive search behaviour is positively correlated with the level of satisfaction with the new location, and vice versa that a more objective and intensive search is no guarantee for satisfaction afterwards. This result calls for discussion, especially if we want to consider it in the context of the various possible theoretical approaches we treated in the second section.

First of all, the results clearly do not fit in the neoclassical approach. In the neoclassical perspective, firm migration will always imply improvement of locational conditions, because a firm will only move when improvement is to be expected; and because the all-knowing decision maker always chooses the optimal new location, there will be no dissatisfaction afterwards. Generally speaking, in the neoclassical view any firm move, once decided for, will always be to the benefit of the firm and lead to satisfaction. In the institutional and organisational network approaches, things are totally different. These approaches attach great weight to the embeddedness of the firm in economic and social networks, that explain and sustain a firm's life and growth. Leaving the location endangers network relationships and thus generally speaking is not good but bad for the firm, and should when possible be avoided. Moving is risky and satisfaction cannot be taken for granted. The best fit with Kobus's results may be found in the behavioural approach. In behavioural theory, subjective expectancies with respect to a possible result of a location choice do play a role in the explanation of locational satisfaction. More intensive search behaviour may inspire higher expectations, followed more easily by dissatisfaction. To explore this line of thinking more seriously, it might be wise to revive an interest in the behavioural approach and look for inspiration to the many recent studies in behavioural economics that are inspired by the work of psychologist and 2002 Nobel Prize winner Kahneman and his fellow authors such as Tversky and Thaler, who have successfully integrated psychological insights into various field of economic analysis (Rabin, 2003).

Exploring the Local Perspective

From the historical survey of studies in the third section, firm migration research appeared to have a certain focus on the regional scale first, reflecting its relationship to regional policy objectives, and a subsequent shift to the urban level later, when short distance (service sector) migrations outgrew long distance (manufacturing industry) migrations, and the research interest shifted to urban development issues. Since then, the interest of spatial scientists in urban problems has increased, but it also tends to focus on even smaller scale entities, i.e. the individual urban neighbourhoods, where many of the most pregnant social problems of modern society become manifest. This is also reflected in a new explicitly local perspective in firm migration research, which can also be described as a new branch of the "demography of firms" approach focussing on the local and neighbourhood level.

The first good reason for a more intensive study of firm migrations at the local level is that although three quarters of all moves are local in nature, we know relatively little about the specificity of both firm internal and external migration factors. We may expect such specificity, for instance because on the local scale entrepreneurs will be better informed, less hampered by loss of staff or disruption of IORs, movement costs may be lower, and local governments may have more influence on the outcome of the migration decision process. Such suppositions have to be verified. But there is another and even more valid reason to push the firm migration research frontier to the local level, i.e. the increased policy interest in the social and economic well-being of urban neighbourhoods. Politicians increasingly argue for more economic activity in urban neighbourhoods, meaning more jobs and more prosperity, especially for such neighbourhoods that face a variety of problems with poverty of their inhabitants as a common denominator. In the Netherlands this is one of the three important goals of the so-called *Grotestedenbeleid* (Large Cities Policy) (Van der Burgh et al., 1999; Van Katwijk and van Eygenraam, 2005). One of the key elements in such a policy is to stimulate firms that have been set up in central city neighbourhoods not to leave these original locations when the necessity of a move (usually following success and growth) arises. This raises the question which keys of the entrepreneurial mindset have to be touched to bring this about. In the past, firm migration studies tended to concentrate on push and pull motives (which after all had caused migration) but have shown a certain degree of neglect for the keep motives, which keep entrepreneurs back from the decision to move.

An interesting example of a keep factor that thus far attracted relatively little attention in location choice studies is "place attachment" (Altman and Low, 1992). It seems to be comparable with the "attachment to the own region" factor identified by Meester and Pellenbarg in their first studies of the mental map of Dutch entrepreneurs (Meester and Pellenbarg, 1986). They attribute to this factor diverse historical and personal motives that makes firms stay where they are (Meester and Pellenbarg, 1986, pp. 75–76). Another intriguing concept that calls for analysis in the context of local firm migration studies is "location-specific capital". Da Vanzo uses this term to denote any or all of the factors that "tie" a person to a particular place (DaVanzo, 1981, p. 116). Like place attachment, location-specific capital is a reason for people not to move, and may be retraced with not-moving firms in city neighbourhoods. Finally, firm migrations exploring the local perspective should pay attention to neighbourhood characteristics that may be expected to influence firm moves at the neighbourhood level, such as population size and structure, the rates of employment, unemployment and self-employment, the sectoral and firm size structures, the housing and real estate markets and so on. We should not be surprised to find interesting differences between the importance of all these concepts and factors for different parts of urban spaces.

Integrating the Evolutionary and the Organisational Network Approach

Both the evolutionary approach and the organisational network approach are theoretical approaches in their infancy. The first theoretical and empirical steps have been taken, but both approaches show ample room for further development.

Even though the evolutionary approach to firm relocation has strong theoretical roots, the empirical research showing the validity of its claims is still rather modest. This can be attributed partly to the age of the approach, but the fact that empirical research based on this approach almost by definition requires longitudinal data reduces the ease of empirical research as well. As a result, the empirical evidence is limited to multiple case studies (Stam, 2007), which implies a limited extent of external validity of the available empirical work. To our knowledge, no large scale quantitative tests of the claims of the evolutionary approach to firm relocation have been conducted to date.

Another limitation to the extent to which the claims of this approach can be generalised lies in the fact that it focuses strongly on small entrepreneurial firms. This is not only an empirical limitation as the theoretical underpinning of the evolutionary approach is also strongly geared towards this type of firms. However, many firms that relocate do not fall under this header. As such, it seems useful to empirically explore if, and if so to what extent, the evolutionary approach can be extended to apply to incumbent firms as well.

Related to the above, as a result of the strong focus of the evolutionary approach on small entrepreneurial firms, social networks, rather than organisational networks, are given an important role in explaining firm (re)location behaviour. Interestingly, research has shown that especially new firms that are able to transform their social networks to more business oriented organisational networks are successful (Maurer and Ebers, 2006). This begs the question whether firms that are heavily influenced by their social networks in their (re)location decisions are worse off in the long run than firms that are able to "break the bonds" of their social ties. The evolutionary approach provides an excellent starting point to analyse such questions, even though more theoretical and empirical work is still required.

The main points of further development of the organisational network approach also lie in the realm of data collection and empirical testing. So far, the empirical work that draws upon this approach has been based on cross-sectional data, which is problematic for the establishment of the direction of causality. Therefore, the first challenge is to collect longitudinal data. However, collecting large scale longitudinal data with regard to (changes in) the geographical position of firms, (changes in) their inter-organisational relations and networks, and their performance is likely to lead to severe data collection problems. It seems more feasible to conduct (multiple) longitudinal case studies (e.g. Pen, 2002) and to use the opportunities offered by such a research approach to make fine grained analyses of, for example, to what extent different (embeddedness) variables are actually evaluated during different stages of the relocation decision. Studying the relocation decision and process over time and in detail could provide very valuable information about how managers deal with the trade-off between the need for geographical mobility and inter-organisational stability that is pivotal in this theoretical approach.

Second, many of the arguments in the organisational network approach are based on the structural network position of a firm. To fully assess a firm's structural network position, however, whole network data is required. To date, however, the empirical applications have been limited to ego-network data. Therefore, the second challenge is to test the impact dimensions of a firm's structural network position (e.g. eigenvector centrality or betweenness centrality) on its relocation behaviour and the subsequent outcomes. Ideally, this whole network data should also be collected longitudinally for reasons spelled out in the above. In order to limit the amount of work one could focus on a limited amount of networks (i.e. multiple longitudinal case studies) and collect data with regard to all nodes (firms) and ties in these networks. Subsequently, the impact of a firm's structural network position as well as the characteristics of its dyadic ties could be related to its spatial behaviour and the consequences thereof. One of the main problems with using such an approach is that the observations are not independent, and that, thereby, one of the main assumptions of many statistical procedures is violated. Recent methodological advances, however (e.g. the Sienna model of Snijders), are capable of dealing with dependent observations, which makes the gathering and analysis of longitudinal whole network data more feasible.

It is important to stress that future whole network research should not focus on the structural characteristics of a network only. Even though such data is easier to gather and analyse, the existing empirical work based on the organisational network approach has clearly shown that the combination of both structural and dyadic characteristics leads to the best explanations for the behaviour of actors and the subsequent consequences of that behaviour.

Perhaps one of the most fruitful paths for future firm relocation research is the integration of the evolutionary and the organisational network approach. Both approaches are different enough in terms of the stage of the life cycle to which they apply and the specific drivers of firm relocation that they identify, yet similar enough in terms of conceptualising the firm and the (re)location decision to be highly complementary. Both approaches view firms as embedded in (regional) networks and both attribute an important (constraining) role to this embeddedness. However, the evolutionary approach focuses on social networks of entrepreneurs, whereas the organisational network approach focuses on more formalised inter-organisational networks at the organisational level. As such, they can easily be combined in a life-cycle perspective in which different types of relations play an important role in different stages of development.

Furthermore, the inter-organisational network approach explicitly includes drivers of firm relocation derived from neo-classical insights (Knoben, 2008) as well as drivers derived from behavioral and relational theories. Thereby, it deals with the critique of Scott (2000) that behavioural approaches focus too much on "soft" factors. The evolutionary approach is explicitly longitudinal and focuses on life-cycle effects, whereas other approaches often treat all firms, despite differences in stage of development, as equal (Pen, 2000). By integrating both perspectives theoretically, and testing their validity empirically (see the points of improvements stressed above), a more complete approach to firm relocation than currently available can be developed. As such, we identify this as a fruitful endeavour to push the research frontier.

CONCLUSIONS

Since the publication of the first "classics" in the 1950s and 1960s, a research tradition concerning firm migration has developed that has been inspired by all major paradigms in economic geography. Both the paradigm shifts and the changes in public policies related to the theme produced a continuous renewal of research themes and questions. After half a century the basic features of firm migration as it developed in modern western societies are sufficiently known. The relocation of manufacturing industries from central urban areas to urban fringe zones and national peripheries characterised the 1950s and 1960s, but since then has decreased sharply. The migration of services (especially business services) within and between urban areas is now dominating the scene. Firm migrations crossing state boundaries are relatively scarce, although lack of research, especially in Europe (where no data are available for international firm migrations), makes it hard to draw definite conclusions. The disappearance of the Iron Curtain in 1989, the fading of the EU inner boundaries since 1992 and the accession to the EU of East European countries produced new circumstances which enable small and middle-sized firms to transfer (part of their) activities to relatively nearby cheap production locations. An eastward shift of activities has certainly occurred, but mostly in the form of offshoring and outsourcing. A clear insight into what has happened in the form of clear-cut firm migrations will have to wait until the national European business registers have found a way to couple their registration systems.

The mobility of firms in most countries is much higher than is generally assumed. In a decade, a quarter or a third of all firms in a regional or national population may move to a new location. Therefore, planning of new locations greatly profits from a sound knowledge of firm migration tendencies, and up-to-date information about the factors that dominate the location search process. Local and regional governments realise that migrant firms are growing firms, which makes them keen to attract them – and not to lose them! Still, we have to realise that the vast majority of moves is over (very) short distances, and therefore means little for the growth or decline of local and regional employment levels. Agreements between municipalities at the regional level to control the supply of business space (and prevent a race to the bottom of price levels) are generally advisable.

Many characteristics of migrant firms and their migration behaviour seem to be of a general nature, as our short excursion through the cases of Belgium and the Netherlands has demonstrated. Still, we have to be careful with such a conclusion, since not many reliable country-comparison studies have been done yet. To improve our knowledge about what is general in firm migration behaviour and what is not, comparative studies are needed, concerning more aspects, i.e. comparison between what happens to firm migration in different development stages of regions and countries, what happens on the scale of city neighbourhoods as opposed to suburbs and peripheries, what differences we observe between sectors, and between size and age categories, between firms that move and firms that stay put, between firms that are content with their new locations and firms that are not. The research agenda for firm migration is a long one. A special interest may be given to the questions addressed in the fifth section, especially those questions that are inspired by the old and new alternatives for the neoclassical approach: the behavioural view, the institutional and evolutionary theories, the network approach, and the possible links between them. If we relate the policy-related practical questions to the theory-related problems, we stand a good chance of deepening our understanding of the space economy and of improving our policy advice concerning space development at the same time.

NOTES

1. A large part of this chapter consists of an adapted and updated version of Pellenbarg et al. (2002a). We gratefully acknowledge L.J.G. van Wissen and J. van Dijk for their permission to use the original text as the basis for this chapter.
2. These events are certainly extremely interesting from a spatial point of view. All these events are taken into account in a demographic approach to firm dynamics that has gained

popularity in recent years (Van Dijk and Pellenbarg 2000a). This approach, which is studied both by geographers, sociologists and economists is called industrial demography, demography of the firm or enterprise, or economic demography (Van Wissen, 2002).

3. Three different costs can be distinguished. First, there may be significant relocation costs. Relocation costs may be direct costs of moving, as well as search and information costs of finding new markets, labour, suppliers and deliverers. However, these types of indirect costs are generally disregarded in the simple neo-classical framework, with its emphasis on full information and rational behaviour. Second, there may be a substantial amount on capital inertia (Auty, 1975). In many cases existing buildings and other equipment at the old location may already be written off, and still be operational at low costs. The firm is therefore able to make a profit at a suboptimal location where a new firm would not be able to make a profit. Third, the cost or revenue elasticity of any of the location factors is in general low, which means that the cost and revenue surfaces are rather flat. In other words, locational choice is often not a decisive factor in determining profit or loss. Only when the profits are much higher at another location may the firm decide to relocate in spite of the fact that also at the present location they make a profit.
4. A fifth stage might be added in which the implemented decision is assessed and evaluated.
5. Even though the institutional approach could incorporate such relationships it: (a) does not focus on the (characteristics of) relations between firms but instead assumes that in regions with particular "cultures" intensive networks exist; (b) has a rather specific view on the role of the environment (i.e. as a negotiation arena); and (c) is predominantly applicable to large and established firms.
6. There are of course some exceptions to this UK dominance. Struyk and James (1975), Jusenius and Ledebur (1977) and Schmenner (1978) deal with industrial mobility in the US, Söderman (1975) studies firm relocation cases in Sweden, Bade (1979) investigates firm relocation in Germany, and Molle (1977) and Kruyt (1979) describe firm relocation in the Netherlands, especially in and around the Amsterdam agglomeration. Where relevant, the insights from these studies have been incorporated in the other sections of this review.
7. Only Greece stands out as an exception, because it shows a reverse movement: firms move from peripheral to central areas (Athens and Thessaloniki).
8. Which roughly correspond to phases 3, 4 and 5 of Townroe (1973).
9. One interesting phenomenon is that Quality of Life (QOL) factors and more generally "soft" factors prove to be of growing importance in business relocation decisions, as shown by Love and Crompton (1999) and Salvesen and Renski (2003). This is especially true for firms employing a highly educated workforce.
10. This concerns the PhD theses of Ter Hart (1979), Pellenbarg (1985), Hessels (1992), Louw (1996), Ebels (1997), Van Wee (1997), Pen (2002), Stam (2003a), Van der Veen (2004), Mariotti (2005) and Knoben (2008).
11. The LISA dataset holds information about firms and employment, based on provincial employment registers. It registers firms as spatial establishments instead of juridical entities, thus counts lower numbers of firms than the CoC, but is generally considered more reliable for spatial-economic research. Firm migrations were identified by combining the LISA dataset with mutations in the CoC registration.
12. The less accurate registration of the Chambers of Commerce, including many "paper moves", used to count two or three times as many firms but, interestingly, about the same number of employees involved (i.e. Kemper and Pellenbarg, 1997).
13. The Dutch LISA data are assembled from provincial firm registers, the Belgian NBB dataset is based on accounts from the Belgian National Bank.
14. The sample included firms from the sectors manufacturing, wholesale, transport and business services.
15. The original population of relocating manufacturing firms consisted of 127 firms; the 39 firms included in this table are the ones that answered all the questions completely regarding both relocation and satisfaction.

16. In absolute numbers: of the 154 firms in the survey 104 were satisfied with their new location (i.e. would choose it again) and 82 of these satisfied firms had not had a choice between a greater number of alternative locations.

REFERENCES

Abler, R., R.S. Adams and P. Gould (1971) *Spatial Organisation: The Geographers View of the World.* Englewood Cliffs, NJ: Prentice Hall.

Adema, T. (2003) *Bedrijfsverplaatsingen: Een Beschrijvend Onderzoek naar de Tevredenheid na een Bedrijfsverplaatsing.* Groningen: faculteit Ruimtelijke Wetenschappen R.U.Groningen (unpublished MA thesis).

Ahuja, G. (2000), Collaboration Networks, Structural Holes, and Innovation: A Longitudinal Study, *Administrative Science Quarterly*, 45 (3), 425–457.

Altman, I. and S.M. Low (1992) *Place Attachment.* New York: Plenum Press.

Armstrong, H. and J. Taylor (2000) *Regional Economics and Policy*, 3rd ed. Oxford: Blackwell Publishers.

Auty, R.M. (1975) Scale economies and plant vintage; toward a factory classification. *Economic Geography*, 51: 150–162.

Aydalot, Ph. (1978) La mobilité des activités et de l'emploi, *Revue d'Economie régionale et Urbaine*, 4.

Aydalot, Ph., M. Noël and Cl. Pottier (1972) *La Mobilité des Activités Economiques.*, Paris: Gauthier-Villars.

Bade, F.J. (1979) *Die Mobilität von Industriebetrieben.* Schriftenreihe des Wissenschaftszentrums Berlin IIMV, Band 6. Berlin: Meisenheim.

Benoit, S. (1995) Local policies to attract mobile investment: a theoretical survey and an application to two sets of local organisations in France. In: P.C. Cheshire and I. R. Gordon (eds), *Territorial Competition in an Integrating Europe.* Aldershot: Avebury.

Besselink, E., P. Bijleveld, M. van Geene, R. Kerkhof, A. Rotink and A.R. Schaafsma (1988) *Veranderende Vestigingsplaatstendenzen.* Groningen, Faculteit der Ruimtelijke Wetenschappen R.U.Groningen

Bouwers, H.J. and P.H. Pellenbarg (1989) *Verhuizen en Bouwen.* Groningen, Faculteit der Ruimtelijke Wetenschappen R.U.Groningen.

Brouwer, A.E., I. Mariotti, and J.N. Van Ommeren (2004). The firm relocation decision: an empirical investigation, *Annals of Regional Science*, 38 (2), 335–347.

Buenstorf, G. and C. Guenther (2007) No place like home? Location choice and firm survival after forced relocation in the German machine tool industry, Jena Economic Research Papers, 2007-053.

Cabus, P., G. Allaert and W. Vanhaverbeke (2001) *Strategisch Plan Ruimtelijke Economie: de Ruimtelijk-Economische Hoofdstructuur van Vlaanderen.* Deel 3 Verhuisbewegingen van Bedrijven. Gent: Story-Scientia.

Camagni, R. (1976) La struttura settoriale della bilancia commerciale del Mezzogiorno: una indangine quantitativa. *Giornale degli Economist*, 7(8), 477–479.

Cameron, G.C. and B.D. Clark (1966) *Industrial Movement and the Regional Problem*, University of Glasgow Social and Economic Studies, Occasional Paper No. 5. Edinburgh: Oliver & Boyd.

Christiansen, U. (1978) Moves of firms 1961–1976 between Danish functional urban regions. IASA, Laxenburg (seminar paper).

Christiansen, U. (1979) Storbyerhvervenes mobilitet 1961–1976, *SBI-Byplanlaegning*, 35.

Cox, K.R. (1972) *Man, Location and Behavior*. New York: Wiley.

Cyert, R.M. and J.G. March (1963) *A Behavioral Theory of the Firm*. Englewood Cliffs, NJ: Prentice Hall.

DaVanzo, J. (1981) Microeconomic approaches to studying migration decisions. In: De Jong, G.F. and R.W. Gardner (eds), *Migration Decision Making. Multidisciplinary Approaches to Microlevel Studies in Developed and Developing Countries*, pp. 90–129. New York: Pergamon.

Ebels, H.J. (1997) Oudere stadsdelen en de ruimtelijke effecten van bedrijfsverplaatsingen. PhD thesis University of Amsterdam.

Erickson, R.A. and M.J. Wasylenko (1981) Relocation consideration by central city manufacturing firms. *Review of Regional Studies*, 11, 42–55.

Eurostat (2008) Eurostat database: http://epp.eurostat.ec.europa.eu/.

Garwood, J.D. (1953) An analysis of postwar industrial relocation to Utah and Colorado, *Economic Geography*, 29/1, 79–88.

Gertler, M.S. (2003), Tacit knowledge and the economic geography of context, or the undefinable tacitness of being (there), *Journal of Economic Geography*, 3 (1), 75–99.

Granovetter, M.S. (1985), Economic action and social structure: the problem of embeddedness, *The American Journal of Sociology*, 91 (3), 481–510.

Hayter, R. (1978) Locational decision making in a resource based manufacturing sector: case studies from the pulp and paper industry of British Columbia. *Professional Geographer*, 30, 240–249.

Hayter, R. (1997) The Dynamics of Industrial Location. The Factory, the Firm and the Production System. New York: Wiley.

Hessels, M. (1992) *Locational Dynamics of Business Services; An Intrametropolitan Study on the Randstad Holland*. Netherlands Geographical Studies 147. Amsterdam/Utrecht: Royal Dutch Geographical Society/Faculty of Spatial Sciences University of Utrecht.

Holl, A. (2004) Start-ups and relocation: manufacturing plant location in Portugal. *Papers in Regional Science*, 83 (4), 649–668.

Inbo (2003) *Kantoorgebruikers in beweging*. Woudenberg: Inbo Adviseurs.

Isabella, L.A. (1990), Evolving interpretations as a change unfolds: how managers construe key organisational events, *Academy of Management Journal*, 33 (1), 7–41.

Jusenius, C.L. and L.C. Ledebur (1977) *The Relocation of Firms and Workers in Ohio 1970–1975*. Washington, DC Academy for Contemporary Problems,

Keeble, D. (1968) Industrial decentralisation and the metropolis: the north-west London case. *Transactions, Institute of British Geographers*, 44, 1–54.

Keeble, D. (1971) Employment mobility in Britain. In: M. Chisholm and G. Manners (eds) *Spatial Policy Problems of the British Economy*. Cambridge: Cambridge University Press, pp. 24–68.

Keeble, D. (1972) Industrial movement and regional development in the United Kingdom. *Town Planning Review*, 43 (1), 3–25.

Keeble, D. (1975) Industrial mobility: in which industries has plant location changed most? A comment. *Regional Studies*, 9 (3), 297–299.

Keeble, D. (1976) *Industrial Location and Planning in the United Kingdom*. London: Methuen & Co.

Kemper, N.J. and P.H. Pellenbarg (1997) De Randstad een hogedrukpan. *Economisch Statistisch Berichten*, 82, 508–512.

Klaassen, L.H. and W.T.M. Molle (eds) (1983*) Industrial Mobility and Migration in the European Community*. Aldershot: Gower.
Knoben, J. (2008) *Firm Mobility and Organisational Networks: Innovation, Embeddedness and Economic Geography*. Cheltenham, UK and Northampton, MA, USA: Edward Elgar.
Knoben, J. and L.A.G. Oerlemans (2008), Ties that spatially bind? A relational account of the causes of spatial firm mobility. *Regional Studies*, 42 (3), 385–400.
Knoben, J., L.A.G. Oerlemans and R.P.J.H. Rutten (2006), Radical changes in inter-organisational network structure: the longitudinal gap? *Technological Forecasting and Social Change*, 73 (4), 390–404.
Knoben, J., L.A.G. Oerlemans and R.P.J.H. Rutten (2008), The effects of spatial mobility on firm performance, *Economic Geography*, 84 (2), 157–183.
Kobus, A. (2007) *Relocatie en Satisfactie: een onderzoek naar de tevredenheid na bedrijfsverplaatsingen*. Groningen: faculteit Ruimtelijke Wetenschappen R.U.Groningen (Unpublished MA thesis).
Kruyt, B. (1979) The changing spatial pattern of firms in Amsterdam: empirical evidence. *Tijdschrift voor Economische en Sociale Geografie*, 70 (3), 114–157.
Lever, W.F. (1972) Industrial movement, spatial association and functional linkages, *Regional Studies*, 6 (4), 371–184.
Lloyd, P.E. and P. Dicken (1977) *Location in Space. A Theoretical Approach to Economic Geography*. Second edition. London etc.: Harper & Row.
Louw, E. (1996) *Kantoorgebouw en vestigingsplaats*. Stedelijke en regionale verkenningen 12, Delftse Universitaire Pers (PhD thesis Technical University Delft).
Love, L. and J. Crompton (1999) The role of quality of life in business (re)location decisions. *Journal of Business Research*, 44, 211–222.
Luttrell, W.F. (1962) *Factory Location and Industrial Movement: A Study of Recent Experience in Great Britain*. Volumes 1 and 2. London: National Institute of Economic and Social Research.
Mariotti, I. (2005) Firm relocation and regional policy. Netherlands Geographical Studies 331. Utrecht/Groningen: Royal Dutch Geographical Society/faculty of Spatial Sciences University of Groningen.
Martin, R. (1999) The new "geographical turn" in economics: some critical reflections, *Cambridge Journal of Economics*, 23, 65–91.
Maurer, I. and M. Ebers, (2006), Dynamics of social capital and their performance implications: lessons from biotechnology start-ups, *Administrative Science Quarterly*, 51 (2), 262–292.
McDermott, P. (1973) Spatial margins and industrial location in New Zealand. *New Zealand Geographer*, 29, 64–74.
McLaughlin, G.E. and S. Robock (1949) *Why Industry Moves South: A Study of Factors Influencing the Recent Location of Manufacturing Plants in the South*. Kingsport Press, National Planning Association, Kingsport Tennessee.
Meester, W. (1999) *Subjectieve waardering van vestigingsplaatsen door ondernemers*. Netherlands Geographical Studies 261. KNAG/FRWRUG, Utrecht/Groningen (PhD thesis University of Groningen).
Meester, W.J. and P.H. Pellenbarg (1986) *Subjectieve waardering van bedrijfsvestigingsmilieus in Nederland*, Groningen: Geografisch Instituut Rijksuniversiteit Groningen.
Molle, W.T.M. (1977) Industrial mobility: a review of empirical studies and an analysis of the relocation of industry from the City of Amsterdam. *Regional Studies*, 11, 323–335.

Nakosteen, R.A. and M.A. Zimmer (1987) Determinants of regional relocation by manufacturing firms. *Economic Inquiry*, 50, 351–362.
NEDO (National Economic Development Office) (1978) *Construction From Industrial Recovery*. London: Crown, Milbank.
Neumark, D., J. Zhang and J. Kolko (2006) *Interstate Business Relocation: An Industry-level Analysis*. San Francisco: Public Policy Institute of California, Occasional Papers.
North, D.J. (1974) The process of locational change in different manufacturing organisations. In: F.E.I. Hamilton (ed.), *Spatial Perspectives on Industrial Organisation and Decision Making*. London: Wiley, pp. 213-244.
Ortona, G. and W. Santagata (1980) Mobilità industriale nell'area metropolitana torinese. Paper presented to the First Italian Congress of the Regional Science Association, Roma.
Ortona, G. and W. Santagata (1983) Industrial mobility in the Turin metropolitan area 1961–1977. *Urban Studies*, 20, 59–71.
Ortona, G., L. Parodi and W. Santagata (1981) Problemi della localizsatione dell'industria nel comprensirio di Torino. Torino: Ires.
Pellenbarg, P.H. (1976) *Bedrijfsmigratie in Nederland. Deel I Terreinverkenning.* Geografisch Instituut Rijksuniversiteit Groningen.
Pellenbarg, P.H. (1977) *Bedrijfsmigratie in Nederland. Deel II Onderzoeksresultaten.* Stichting Noord Holland-Noord, Industriecommissie Hollands Noorderkwartier, en Geografisch Instituut Rijksuniversiteit Groningen.
Pellenbarg, P.H. (1985) *Bedrijfsrelokatie en ruimtelijke kognitie* (Firm relocation and spatial cognition), Sociaal-geografische Reeks 33, FRW/RUG, Groningen (PhD thesis University of Groningen).
Pellenbarg, P.H. and N.J. Kemper (1999) *Industrial mobility in the Netherlands: patterns, causes and impacts for spatial policy*. SOM Research Report 99D34, University of Groningen.
Pellenbarg, P.H., L.J.G. van Wissen and J. van Dijk (2002a) Firm relocation. In: P. McCann (ed.), *Industrial Location Economics*, Cheltenham, UK and Northampton, MA, USA: Edward Elgar.
Pellenbarg, P.H., L.J.G. van Wissen and J. van Dijk (2002b) Firm relocation: state of the art and research prospects. Groningen: SOM Research report 02D31.
Pen, C.J. (1999) Improving behavioural location theory: preliminary results of a written questionaire about strategic decision making on firm relocations. Paper presented to the ERSA Congress, Dublin.
Pen, C.J. (2000) Actors, causes and phases in the decision-making process of relocated firms in the Netherlands. Faculty of Spatial Sciences, University of Groningen.
Pen, C.J. (2002) Wat beweegt bedrijven. Netherlands Geographical Studies 297. Utrecht/Groningen: Royal Dutch Geographical Society/Faculty of Spatial Sciences University of Groningen.
Penrose, E. (1995) *The Theory of the Growth of the Firm*, Oxford: Oxford University Press.
Pred, A.R. (1967) Behavior and location: foundations for a geographic and dynamic location theory: Part 1. University of Lund, Lund Studies in Geography B, no. 27.
Pred, A.R. (1969) Behavior and location: foundations for a geographic and dynamic location theory: Part 2. University of Lund, Lund Studies in Geography B, no. 28.
Rabin, M. (2003) The Nobel Memorial Prize for Daniel Kahneman. *Scandinavian Journal of Economics*, 105 (2) 157–180.

Rawstrom, E.M. (1958) Three principles of industrial location. *Transactions of the Institute of British Geographers*, 25, 132–142.
Rhodes, J. and B.C. Moore (1976) Regional economic policy and the movement of manufacturing firms to development areas. *Economica*, 43, 17–31.
Romo, F.P. and Schwartz, M. (1995), The structural embeddedness of business decisions: the relocation of manufacturing plants in New York State, 1960–1985, *American Sociological Review*, 60 (1), 874–907.
Salvesen, D. and H. Renski (2003) The importance of quality of life in the location decisions of new economy firms. Centre for Urban and Regional Studies. Chapel Hill: University of North Carolina.
Sant, M. (1975) *Industrial Movement and Regional Development: The British Case*. Oxford: Pergamon Press.
Sayer, A. (1982) Explanation in economic geography: abstraction versus generalisation. *Progress in Human Geography*, 6 (1) 68–88.
Schmenner, R.W. (1978) *The Manufacturing Location Decision: Evidence from Cincinatti and New England*. Economic Development Research Report, US Department of Commerce, Washington DC.
Schuilenburg, O. (2004) *Dynamiek in de industrie*. Groningen: faculteit Ruimtelijke Wetenschappen R.U.Groningen (Unpublished MA thesis).
Schutjens, V.A.J.M. and E. Stam (2003) The evolution and nature of young firm networks: a longitudinal perspective, *Small Business Economics*, 21 (2), 115–134.
Scott, A.J. (1995) The electric vehicle industry and local economic development: prospects and policies for southern California. *Environment and Planning A*, 27 (6), 863–875.
Scott, A.J. (2000) Economic geography: the great half-century. *Cambridge Journal of Economics*, 24 (4) 483–504.
Simon, H.A. (1955) A behavioural model of rational choice. *Quarterly Journal of Economics*, 69, 99–118.
Simon, H.A. (1957) *Models of Man: Social and Rational*. New York: Wiley.
SISWO (1967) *Verplaatsing industriële bedrijven*. Deelrapport 1, 2. Amsterdam: SISWO.
Smith, D.M. (1966) A theoretical framework for geographical studies of industrial location. *Economic Geography*, 42, 96–113.
Smith, D.M. (1971) *Industrial Location: An Economic Geographical Analysis*. New York: Wiley.
Söderman, S. (1975) *Industrial Location Planning*. Stockholm: Almqvist & Wiksell International.
Spooner, D.J. (1972) Industrial movement and the rural periphery: the case of Devon and Cornwall. *Regional Studies*, 6 (2), 197–215.
Spooner, D.J. (1974) Some qualitative aspects of industrial movement in a problem region in the United Kingdom, *Town Planning Review*, 45 (1), 63–83.
Stam, E. (2003a) *Why Butterflies Don't Leave: Locational Evolution of Evolving Enterprises*, Utrecht: Utrecht University.
Stam, E. (2003b) Development of entrepreneurial firms in time and space: an evolutionary perspective, Paper for EMAEE 2003, Augsburg.
Stam, E. (2007) Why butterflies don't leave: locational behaviour of entrepreneurial firms. *Economic Geography*, 83 (1), 27–50.
Struyk, R.J. and F.R. James (1975) *Intra Metropolitan Location*. Lexington, MA: Lexington Books.
Taylor, M.J. (1970) Location decisions of small firms, *Area*, 2, 51–54.
Ter Hart, H.W. (1979) *Vestigingsplaatsaspecten van top-management*. Amsterdam:

Universiteit van Amsterdam.

Townroe, P.M. (1969) Locational choice and the individual firm, *Regional Studies*, 3, 15–24.

Townroe, P.M. (1972) Some behavioural considerations in the industrial location decision, *Regional Studies*, 6, 261–272.

Townroe, P.M. (1973) The supply of mobile industry: a cross-sectional analysis, *Regional and Urban Economics*, 2 (4), 371–386.

Townroe, P.M. (1979) *Industrial Movement: Experience in the US and the UK*. Westmead: Saxon House.

Van der Burgh, Y., Y. Bernardt, R.P. uit Beijerse and C. Melchior (1999), *Ondernemerschap in de grote steden. De uitgangspositie van het Grote-Stedenbeleid*. Zoetermeer: EIM.

Van der Veen, H.C.J. (2004) *Het effect van milieubeleid op locatiebeslissingen van bedrijven*. Groningen: Interuniversity Center for Social Science Theory and Methodology (ICS).

Van Dijk, J. and P.H. Pellenbarg (2000a) The demography of firms: progress and problems in empirical research. In: J. van Dijk and P.H. Pellenbarg (eds), *Demography of Firms: Spatial Dynamics of Firm Behaviour*. Netherlands Geographical Studies 262, Utrecht/Groningen: KNAG/FRW RUG, pp. 325–337.

Van Dijk, J. and P.H. Pellenbarg (2000b) Firm relocation decisions in the Netherlands: an ordered logit approach, *Papers in Regional Science*, 79, 191–219.

Van Dooren, R.A.M. (2003) *Garments on the Move. The Local Dynamics of Export Networks in La Laguna, Mexico*. Amsterdam: Rozenberg Publishers.

Van Katwijk, J. and S. van Eygenraam (2005), *Kenniseconomie & de Stad. Naar een dynamische kennissamenleving in 2010*. Den Haag: Kenniscentrum Grote Steden.

Van Oort, F., R. Ponds, J. Van Vliet, H. Van Amsterdam, S. Declerck, J. Knoben, P. Pellenbarg and J. Weltevreden, (2007) *Verhuizingen van bedrijven en groei van werkgelegenheid*. Rotterdam/The Hague: NAi Publishers/RPB.

Van Wee, B. (1997) *Kantoor naar het Spoor. De Invloed van Bedrijfsverplaatsingen naar Openbaar-vervoerknooppunten op de Personenmobiliteit*. Amsterdam: University of Amsterdam.

Van Wissen, L. (2000) A micro-simulation model of firms: applications of concepts of the demography of firms, *Papers in Regional Science*, 79, 111–134.

Van Wissen, L. (2002) Demography of the firm: a useful metaphor? *European Journal of Population*, 18, 263–279.

Verkoren, O. (2008) De NAFTA, *Geografie*, 17 (4), 32–36.

Weber, A. (1929) *Alfred Weber's Theory of the Location of Industries*. Chicago: University of Chicago Press.

5. The location of knowledge economy and high-tech in Israel

Dafna Schwartz, Gil Avnimelech and Raphael Bar-El

INTRODUCTION

Knowledge-intensive and high-tech activities are known for their tendency to concentrate in central locations that exhibit specific qualities. As a general rule, such activities tend to be concentrated in metropolitan locations, and within these particularly in primary or capital cities (Capello, 2002; Cooke, 2004a, 2004b; Felsenstein & Ergas, 2002; Frenkel & Shefer, 2001; Schwartz, 2006; Schwartz & Bar-El, 2007). Such a concentration is explained mainly by the concept of learning regions (Florida, 1995; Morgan, 1997); this concept is based on processes such as regional knowledge accumulation, regional knowledge spillovers, generation of regional relational capital and various regional collective learning processes.[1]

Innovative activities performed by start-ups can certainly be classified as knowledge-intensive activity. In addition, small knowledge-intensive firms tend to be more innovative and more concentrated than large knowledge-intensive firms (Acs et al., 1994). Consequently, the various factors that lead to the geographical concentration of knowledge-intensive activities are expected to apply even more intensively for start-ups. VCs tend to invest mostly in innovative and technology-based start-ups. Therefore, start-ups backed by VC investors tend to be heavily involved in knowledge-intensive activities. Moreover, VCs tend to concentrate in highly innovative clusters (see Gompers & Lerner, 1999). Thus, VC-backed start-ups tend to be even more concentrated than non-VC-backed start-ups.[2]

This chapter follows on the article by Schwartz and Bar-El (2007) and analyses the location considerations of various knowledge-intensive and high-tech sectors on theoretical grounds, as well as from empirical data drawn from the Israeli experience. The second section is the literature review. The third section presents the context for our empirical work – a brief summary of the development of the Israeli high-tech sectors since 1969. The fourth section presents descriptive statistics on the geographical distribution of various categories of knowledge-intensive and high-tech sectors in Israel during the years 1995–2006. These statistics support our argument that as the activity became more knowledge-intensive, the geographical distribution of the activity became more concentrated in metropolitan locations and prime cities. Finally, the fifth section presents the results and the sixth section the conclusions.

LITERATURE REVIEW

The Knowledge Economy

The underlying idea of a knowledge economy refers to specific assets that comprise knowledge: "how to", "who to" and "what to", which are deployed to create value (Cooke et al., 2001). Knowledge economy activity expresses value through the scarcity of "knowledgeable" expertise (Cooke and Schwartz, 2008). Castell (1996) suggests that the key value-adding element in knowledge economy activities is knowledge. Knowledge takes three forms: exploration knowledge such as the kind that occurs in R&D; examination knowledge by which trials, testing and development or refinement are conducted; and exploitation knowledge, which takes the discovery or invention to market as a commercial innovation. Here scientific and technological skills interact intimately with legal, management, accounting, consultancy, financial and other relevant business skills, coming together for the project at hand (Cooke & Schwartz, 2008).

In the "applied knowledge" sectors, processes that may historically have been conducted by employees with tacit skills are replaced by codified skills controlled by technologies (Cooke et al., 2001). This implication is deeply embedded in knowledge economies. Cappellin (2007) suggests that "learning regions" have the ability to rapidly transform tacit knowledge into localized codified knowledge. Florida and Gates (2002) identified the rise of a class of knowledge economies that is highly spatially skewed towards large cities and is rarely present in rural settings.

High-Tech Concentration

Knowledge-based economic activities, particularly high-tech firms, have a strong tendency to cluster, primarily around metropolitan locations and in the satellite urban ring around a main metropolitan area (see, for example, Bar-El & Parr, 2003; Capello, 2002; Cooke & Schwartz, 2008; Frenkel, 2001; Frenkel & Shefer, 2001; Schwartz, 2006). As mentioned, the main factors behind such a concentration include local knowledge spillovers, tacit knowledge, spatial knowledge accumulation, relational capital, trust, collective learning, and innovation networks, as described briefly in this section (see also Bar-El & Schwartz, 2007; Cooke & Schwartz, 2008; Schwartz, 2006).

Knowledge spillovers (Quah, 2002) refer to the positive externalities firms receive in terms of knowledge from the environment in which they operate. Knowledge spillovers are geographically bounded (Baranes & Tropeano, 2003; Wallsten, 2001) and make an important contribution to absorptive capacity, which is a significant element of a learning region (Boschma, 2005; Cooke et al., 1997; Florida, 1995; Morgan, 1997).

Tacit knowledge, which refers to knowledge that is not coded (Howells, 2002; Maskell & Malmberg, 1999) is another characteristic of high-tech activity subject to spatial concentration, since it is primarily transferred through close, ongoing contacts. When such embedded knowledge is codified it can be more easily transferred, leading to a decrease in the importance of location, concentration, or proximity to other firms.

Proximity is an important factor in the process of knowledge acquisition from external sources to firms. There are various kinds of channels of knowledge acquisition, such as recruitment of personnel, external information networks of employees, and formal cooperation between the firm and other institutional agents. The type of knowledge required in each firm determines the attraction level of specific locations. A study on mechanisms of external learning by Almeida et al. (2003) reveals that smaller firms use more informal mechanisms for external learning, and therefore there is a higher expectation for them to locate closer to other related firms and in "knowledge-intensive" regions.

Proximity is also an important factor in the development of relational capital, as well as in the establishment of trust between economic agents within a region, which is an efficient functional economic tool for transactions (Gössling, 2004). Relational capital as describe by Capello and Faggian (2005) refers to the established relationship "that stems from a strong sense of belonging and a highly developed capacity of cooperation typical of culturally similar people and institutions" (p. 77). Proximity assists in the development of relational capital and in the establishment of trust between economic agents

Studies on the topic of regional innovation systems and the "learning region" (Boschma, 2005; Cooke et al., 1997; Florida, 1995; Morgan, 1997) highlight the importance of social interaction and networking for the diffusion of knowledge and innovation, and have emphasized the role of spatial proximity in this process. The impact of social capital on regional learning and innovation processes, as shown by Hauser et al. (2007), is highly significant and is comparable with the importance of the regional human capital. The absorptive capacity in a given region, with regard to enhancing knowledge-intensive activity, is an evolutionary process that is dependent upon regional knowledge accumulation and regional social capital development (Van Geenhuizen & Nijkamp, 2006).

A concentration of firms in a specific sector within a bounded geographical area, is not enough to create successful learning regions. Successful learning regions should include also high quality universities and other research institutions, developed financial institutions, and other business services and economic institutions. Therefore, the locations of knowledge-intensive and high-tech sector concentration are not random – they tend toward areas in close proximity to developed economic communities, such as metropolitan regions and primary cities. The basic conclusion derived from this brief survey is that knowledge economies, particularly high-tech activities, are highly spatially skewed towards metropolitan regions and the urban ring around them.

Hypothesis 1: Knowledge economy sectors are more concentrated than all other sectors in the economy.

Firm Size, Age and Spatial Concentration

Spatial concentrations assist firms to establish business networks which are important factors in a firm's innovativeness and business success (Cooke, 2004a, 2004b; Harmaakorpi & Melkas, 2005; Sternberg, 2000). This is particularly important for small firms, as it helps them overcome some of the disadvantages of their limited size (Havnes & Senneseth, 2001), as well as for young firms, as it help them overcome their immature formal learning routines. However the small size and young age of firms may limit their access to networks.

Based on the discussion above, it is anticipated that start-ups will be more geographically concentrated than other high-tech firms. Moreover, these start-up agglomerations will tend to grow near the metropolis and the primary cities. This notion is also supported by empirical evidence (Acs et al., 1994).

Hypothesis 2: High-tech start-ups are more concentrated than other knowledge-intensive sectors.

Main Factors in the Location of VC Investments

VCs invest in high-tech start-ups, and therefore they also have a tendency to concentrate (Mason & Harrison, 2002, 2003). VC investors act in the ventures they back as managers, business consultants and partners, in addition to their role as capital providers. Therefore, as presented by Schwartz and Bar-El (2007, pp. 627–628):

> The location of their investments depends on the one hand on factors that influence location of capital, and on the other hand on factors that influence location of entrepreneurship and management. Capital is relatively more mobile than other production factors, and is attracted to the location of high-tech activities with the highest potential return. Entrepreneurs, managers, and high quality employees are subject to a much lower level of mobility. Therefore, VCs tend to locate where most of the activities of the entrepreneurs are concentrated.

As already explained, it is advantageous for new, small high-tech firms to locate in central urban places, in proximity to other high-tech firms. However, in the case of VC-backed start-ups there is another factor to be considered: the proximity between the VCs and their portfolio companies is a significant factor in their success. Proximity is important due to the tacit nature embodied in the VCs' activities (see Powell et al., 2002; Zook, 2004) and because of day-to-day commitments. The transaction costs associated with frequent visits and intensive involvement are likely to be reduced if VC investors are close to the firms in their portfolios. Distance decreases the ability of the VC investors to be actively involved in their portfolio companies (Gompers & Lerner, 1999). Proximity is also important in reducing the asymmetrical information flow between the VCs and their portfolio companies (Allen, 2002). Since proximity to portfolio companies and to the other entities in the region is essential, VC investors prefer to invest close to their own location. Moreover, VCs activity has a significant interaction with the financial community (Gompers & Lerner, 1999).

Since syndication between several investors in each start-up is quite common, VCs tend to invest close to each other. The development process of a start-up is an ongoing process of rescue finance from various entities, and it is common that financing involves a "syndication" of two or more venture groups (Gompers & Lerner, 1999). The proximity between the investors enables these interactions to take place and facilitates the ability of the investors to be close to the firms which they back. Moreover, the proximity between VCs and relevant financial institutions such as investment banks and financial institutions (potential limited partners in the VC funds), is important for their successful operation. Therefore, we expect VCs to locate near developed financial communities.

The conclusion derived from the survey presented above is that the special management and consultancy role of VCs may lead to a higher level of concentration of their portfolio companies (as compared to other start-up activities), but at the same time this role may direct the VCs to very specific locations close to financial communities or to the locations of other investors.

Hypothesis 3: VC-backed start-ups are more concentrated than other start-ups.

THE ISRAELI CONTEXT – ISRAEL'S HIGH-TECH SECTORS' DEVELOPMENT

The meaningful starting point of the Israeli R&D-intensive sectors can be traced to the late 1960s with the creation of the OCS (Office of Chief Science) and the regular industrial R&D fund (see Avnimelech & Teubal, 2006; Trajtenberg, 2000). Since the early 1990s Israel has experienced a rapid growth in the high-tech sector, reaching a share of 9% of total employment in 2004 as compared with 6% in 1995. This is a higher share than that found in most European countries. This process has been characterized by a rapid increase in the number of start-ups, and was also accompanied by growth in the VC industry (Avnimelech & Teubal, 2004). The number of start-ups grew from approximately 300 in 1991 to over 2,500 in 2000. The annual VC investment increased from less than 50 million dollars in 1991 to more than 3 billion dollars in 2000. Between 2001 and 2008, the average annual VC investment was about 1.5 billion dollars and the average annual start-up creation rate was above 500 (Avnimelech & Schwartz, 2008).

Avnimelech and Teubal (2006) described the development of the Israeli high-tech start-up cluster as being divided into four phases: background conditions phase, pre-emergence phase, emergence phase and post-emergence phase. We will briefly describe the first three development phases.

In the background conditions phase (phase 1, 1969–1985) the cluster did not yet exist. During this phase the S&T fields in academia developed, as did a specialised labour force, and a segment of R&D-performing firms was created (enhanced by government R&D grants). In this phase a number of critical events took place, including the development of the military R&D industry and the development and diffusion of R&D/innovation capabilities in the business sector, the creation of a favourable environment for foreign investment, the involvement of business sector agents in high-tech activity financing, and the very first steps of technological entrepreneurship activity.

In the pre-emergence phase (phase 2, 1986–1992), the first signs of regional industrial concentration appeared. A central feature of this phase is the appearance of non-insignificant start-up activity and the gradual acceptance of technological entrepreneurship. Some start-up financing took place, however a VC industry did not yet exist. A critical mass of start-ups accumulated towards the end of this phase and, correspondingly, a measure of "demand" for VC services.

The emergence process (phase 3, 1993–2000) presented rapid quantitative growth of start-up and VC activity and the eventual emergence of both the entrepreneurial high-tech cluster and a VC industry. At some point in time the cluster reached a size which enabled it to sustain a large number of supporting services, such as local investment banks and underwriters, specialised attorneys and lawyers, and other start-up-oriented services. This induced entry of additional leading high-tech multinational companies and of additional domestic and foreign VCs, followed by rapid creation of start-ups and brisk growth in the acquisition activities of multinational companies. As long as external and internal conditions remained unchanged, the process of the creation of large numbers of start-ups continued and the cluster continued to grow at a fast pace.

The post-emergence process (phase 4, 2001–2008) started with the burst of the high-tech bubble in 2001. This global high-tech downturn had a significant impact on the Israeli high-tech sectors and VC industry. However in 2004 it was already clear that the Israeli entrepreneurial high-tech cluster had survived the downturn and it was entering its mature phase.

DATA AND METHODOLOGY

We now turn to a discussion of the methodology by which such a generalization of spatial economic patterns has been grounded in the Israeli empirical reality. In this chapter we use two types of databases concerning regional distribution of knowledge-intensive and high-tech activity in Israel: the IVC data on VC investment and start-up creation, and Israel's Central Bureau of Statistics (CBS) employment data, divided into economic sectors and geographical districts.

The Start-Up Creation and VC Investment Database

We analysed new start-up creation, VC investment and location data annually between 1995 and 2006, and in four groups for the period 1969–2008. We distinguished between start-ups backed by VC investors (Israeli VC, Foreign VC, or Corporate VC) and start-ups that were not backed by VCs. For each start-up we collected the following information: identification of the firm, year of creation, location of the firm (the district and the sub-district), technological sector, number of employees that were employed by the firm in August 2008, and type of investors that invested in the company.[3] The information used in the study for the start-ups is drawn from the Israel Venture Capital (IVC) online database. This is a comprehensive database of Israel's high-tech industry, conducted by the IVC Research Centre.

Definitions

In this study we define start-ups as young (up to five years) and small (up to 100 employees) high-tech firms that develop a highly advanced, innovative technology or device through their investment in extensive R&D activity. A firm is considered as Israeli, following the IVC definition, if it is registered in Israel or has at least one of the following characteristics: headquartered in Israel, has an R&D centre in Israel, has a senior Israeli management, or receives funding from an Israeli company.

The Employment Database

Employment data is based on the labour force surveys conducted by the CBS. Special district and sub-district analysis for the years 1980, 1985 and 1990, and each year from 1995 to 2006, were performed for this study by the Israel Social Science Data Centre of the Hebrew University of Jerusalem.

We collected data on employment in each district and sub-district: in the entire economic sectors (ET), in knowledge-economy sectors (KE), in knowledge-economy sectors excluding public services (KE.N), in high-tech sectors (HT), in high-tech manufacturing sectors (HTM), in high-tech services sectors (HTS), in high-tech services sectors excluding telecom services (HTS.N), and in start-up companies (SU) and VC-backed start-up companies (VSU).

Knowledge economy and high-tech definitions

The classification of knowledge economy sectors in this study is based on Schwartz (2006), which divided the knowledge economy sectors as identified by OECD (1999) and further elaborated upon by Cooke and Schwartz (2008), into groups of sectors with some common bases representing different aspects of the knowledge economy.

The knowledge economy is first divided into two broad categories: knowledge-based economy of services (KES) and knowledge-based economy of manufacturing (KEM). Next, the manufacturing category is divided into two groups and the services category is divided into three groups. The resulting six classes are the following: In the KEM category, the two classes are: (a) high-tech manufacturing (HTM), including all manufacturing that is defined as high-technology by the CBS; and (b) other knowledge-based manufacturing (OM), including all those manufacturing knowledge-intensive sectors that were not included in the previous group. In the KES category the three classes are categorized as follows. The first two classes are more business-oriented and their output is more tradable, and the third class focuses on less tradable outputs (the public services class includes mostly public sector activities) namely: (a) high-tech services (HTS), including services that are classified as high-tech by the CBS; (b) other business services (OBS) that include other services that are not classified as high-tech, but are closely related to the knowledge economy and are basically business-oriented; (c) public services (PS) that is, those public sector activities that support the growth of the knowledge economy and public services that are offered to the population as a whole.[4]

Geographic definition

Israel has a population of about seven million inhabitants within an area of about 20,000 km^2. According to the CBS, it is distributed across six statistical districts, which are divided into 15 sub-districts (the West Bank and Gaza are not included in Israeli statistics). Two of these districts are considered "centres": Tel Aviv and Central (actually the suburbs of Tel Aviv). Two are considered "other big city surroundings" (i.e., Jerusalem and Haifa). The last two districts (seven sub-districts) are considered to be part of the "periphery" (the Northern and Southern districts).

The main urban centre of Israel is the city of Tel Aviv which, together with a few smaller centres, forms the continuous urban area of Greater Tel Aviv (the Tel Aviv District). This has a population of 1.2m, concentrated in an area of 170 km^2. Thus, almost one-fifth of the population is concentrated in less than 1% of the area of the nation, representing a high density of 7,000 persons per square km. The Central District includes a wide area surrounding the Tel Aviv District, with four sub-districts. Its population is slightly greater than that of the Tel Aviv District (about 1.6m), but it is distributed over an area that is almost eight times larger. Most of the population is urban and is distributed among centres located within commuting distance of Tel Aviv, the largest city being Rishon Lezion in the sub-district of Rehovot, with a population of around 190,000. Compared to the Tel Aviv District, population density is relatively low (about 1,000 persons/km^2), although it is still greater than that of other districts with the exception of the Jerusalem District.

Other major urban concentrations in Israel are Haifa and Jerusalem. The Northern District surrounds Haifa on the landward side. The Jerusalem District includes the capital city and its surroundings. Finally, the Southern District, with two sub-districts which cover the majority of the nation's area and includes the Negev desert, has an extremely low population density. Most of its population lives in small towns, and the largest urban centre, Beer-Sheva, has a population of about 180,000.[5]

Methodology

After we present some general descriptive statistics on the dynamics of employment distributions across time, we calculate the share of employment in each district out of the total employment for each year during the main period investigated, 1995–2006, in knowledge economy (KE) in general and separately for each of the categories described above – high-tech manufacturing (HTM), High-tech Services (HTS), Knowledge-Economy without PS (KE.N), high-tech start-ups (SU) and VC-backed start-ups (VSU).

Two indices were calculated to measure the specific sectors activity in each region: first, the share of the employment in a specific industrial category in a district out of the total employment in that district (for example, $\%KE_i$ represent the share of knowledge-economy employment out of total employment in district i); second, using these basic distributions, we evaluate regional concentration levels by using location quotients[6] to compare between couples of distributions: for example:

$$LQ\,(HT\,/\,ET)_i = \%\,HT_i / \%\,ET_i \qquad (5.1)$$

This is the location quotient of high-tech employment in district i in relation to total employment in that same district. These scores create an index where the weighted average of all regions is equal to one. A value higher than one would indicate a district with a relatively higher level of concentration of high-tech employment than to its share of total employment.

RESULTS

Changes in the Distribution of Employment Throughout the Various Districts

Table 5.1 presents the geographical distribution of the entire employment between 1995 and 2006. The main observation from this table is that the distribution of employment around the country did not change significantly along the years of the study. In addition, we see that the two main districts (Tel Aviv and Central) captured about 50% of the employment, with minimal changes over the period.

Table 5.1 Geographical distribution of employment according to districts, 1995–2006 (in percent)

	1995	2000	2002	2004	2006
North	12.2	13.4	13.5	13.0	13.4
Haifa	14.0	13.6	13.3	13.1	12.7
Center	20.4	22.5	23.8	26.2	24.5
Tel Aviv	30.2	27.3	26.3	25.7	25.9
Jerusalem	11.1	10.9	10.7	11.0	10.7
South	12.1	12.3	12.4	13.0	12.8
Total	100.00	100.00	100.00	100.00	100.00

Concentration Levels Employment in Various High-Tech Categories

We begin by simply analysing the regional distribution of employment in various categories of high-tech along the years 1995–2006. Table 5.2 presents these employment distributions in all districts for six different industrial categories: total employment (ET), knowledge economy (KE), knowledge economy without public services (KE.N), high-tech industries (HT), high-tech start-ups (SU) and VC-backed high-tech start-ups (VSU). The figures for each category represent the share of employment in the category that is located in the relevant district. For example, 33.5% of high-tech employment in Israel is located in the Central District. The categories in the table become more R&D-intensive as we go to the right. Thus, the table presents a clear picture of more concentration in the two-core districts (Tel Aviv and Central) as the category becomes more R&D-intensive. These districts attract 49.8% of employment in the country, 53.4% of knowledge economy employment, 57.6% of the business sector's knowledge economy employment, 62.7% of high-tech employment, 68.9% of start-up employment and 76% of VC-backed start-up employment in the country.

The findings of Table 5.2 illustrate the well-known phenomenon of concentration of high-tech employment in and around a metropolitan region – in the Israeli case the Tel Aviv metropolis (Cooke & Schwartz, 2008). In general, the concentration of high-tech activity does not necessarily have to be in or around a metropolitan centre. This is, however, the case in Israel, where the city of Tel Aviv and its surroundings has developed as a cultural, social and economic centre (Bar-El & Parr, 2003). Other urban centres such as Haifa and Jerusalem play a relatively minor role in high-tech activity.

In Table 5.3, we present location quotients of the various industrial categories. In this table, we can see a more interesting phenomenon, which is the fact that the heaviest relative concentration of various categories of knowledge-economy activities is actually in the district of the Centre rather than in the district of Tel Aviv itself. For example, in the high-tech category we find the highest location quotient of 1.47 in the Central district, while in Tel Aviv the location quotient is only 1.08, showing a frequency of high-tech employment that does not differ significantly from the frequency of total employment. This perfectly fits the concept of Metropolitan Based Region as shown by Bar-El and Parr (2003), which explains the existence of a process of transition of economic activity from the core of Tel Aviv to the nearby region.[7] All other districts show a location quotient of less than 1 – in the urban centres of Jerusalem and Haifa as well as in the peripheral Northern and Southern districts.

Table 5.2 Employment distribution by districts of various economic sectors, 1995–2006

	%ET	%KE	%KE.N	%HT	%SU	%VSU
Jerusalem	10.9	10.8	10.2	7.4	7.7	6.2
Northern	13.4	11.7	9.7	9.5	8.7	7.5
Haifa	13.4	12.5	12.1	10.8	9.4	8.2
Central	22.8	25.5	26.1	33.5	32.7	38.0
Tel Aviv	27.0	27.9	31.5	29.2	36.2	38.0
Southern	12.4	11.6	10.3	9.6	5.3	2.1
Total	100.0	100.0	100.0	100.0	100.0	100.0

Notes: ET (total economy); KE (knowledge economy); KE.N (knowledge economy without public services); HT (high-intensive R&D industries (high-tech)); SU (high-tech start-ups); VCSU (VC-backed high-tech start-ups).

The next two columns in Table 5.3 – LQ (SU/ET) and LQ (VSU/ET) – support our thesis that the districts with the higher/lower relative specialization in high-tech activity (where the LQ (EHT/ET) value is higher/lower than 1) also enjoy a high/low concentration in start-up activity and VC-backed start-ups. Moreover, in most cases the concentration pattern is even stronger in the start-up categories (higher/lower level of concentration of start-ups than of high-tech in general when LQ (EHT/ET) value is higher/lower than 1). However we can see that in the start-up location quotients the differences between the Central District and Tel Aviv District are reduced, and both are significantly above 1.

Dynamic Spatial Patterns in the Knowledge Economy Sectors

The share of knowledge economy in the Israeli economy grew continuously over time in the last decade, 43.5% in 1995 to 51.9% in 2006. In the following section, we analyse the spatial pattern of the knowledge economy activity in Israel and the trends between 1995 and 2006. For each year we calculate: (1) the share of employment in knowledge economy activities out of total employment in each district in a given year, and (2) the location quotients of each district in knowledge economy activities in any given year (see Table 5.4).

Table 5.3 Employment location quotients by districts of various economic sectors, 1995–2006

	LQ (ET/ET)	LQ (KE/ET)	LQ (KE.N/ET)
Jerusalem	1.00	0.99	0.94
Northern	1.00	0.87	0.72
Haifa	1.00	0.93	0.90
Central	1.00	1.12	1.14
Tel Aviv	1.00	1.03	1.17
Southern	1.00	0.94	0.83
Total	1.00	1.00	1.00
Std.	0.00	0.07	0.15

	LQ (HT/ET)	LQ (SU/ET)	LQ (VSU/ET)
Jerusalem	0.68	0.71	0.57
Northern	0.71	0.65	0.56
Haifa	0.81	0.70	0.61
Central	1.47	1.43	1.67
Tel Aviv	1.08	1.34	1.41
Southern	0.77	0.43	0.17
Total	1.00	1.00	1.00
Std.	0.26	0.35	0.49

Note: * *LQ(x/y)* is the relative employment in sector *x* at a specific district divided by the relative employment in sector *y* at the same specific district.

Table 5.4 presents the following findings. First, all districts had an increase in the level of knowledge-economy activities during the 1995–2006 period, and second, the Tel Aviv District remained the leading knowledge-intensive location over the period. Its knowledge economy location quotients average was 1.09 (9% above the mean of all districts).

Table 5.4 Knowledge-economy employment by district (location quotients), 1995–2006

	Jerusalem	Northern	Haifa	Central	Tel Aviv	Southern	Average
1995	45.7% (1.05)	39.4% (0.90)	41.7% (0.96)	44.2% (1.02)	46.1% (1.06)	40.3% (0/92)	43.5% (1.00)
1996	46.4% (1.04)	37.2% (0.84)	42.8% (0.96)	44.9% (1.01)	48.8% (1.10)	41.3% (0.93)	44.4% (1.00)
1997	47.1% (1.05)	38.2% (0.85)	42.4% (0.94)	45.7% (1.02)	49.9% (1.11)	41.8% (0.93)	45.0% (1.00)
1998	49.6% (1.06)	41.1% (0.88)	45.5% (0.97)	47.1% (1.01)	50.7% (1.08)	43.1% (0.92)	46.8% (1.00)
1999	48.9% (1.04)	39.4% (0.84)	45.5% (0.97)	48.8% (1.04)	51.8% (1.10)	42.0% (0.89)	47.1% (1.00)
2000	48.7% (1.01)	41.7% (0.86)	46.8% (0.97)	50.0% (1.04)	53.1% (1.10)	42.3% (0.88)	48.2% (1.00)
2001	49.6% (1.00)	42.8% (0.86)	48.7% (0.99)	51.2% (1.04)	53.2% (1.08)	46.0% (0.93)	49.5% (1.00)
2002	49.2% (0.99)	42.1% (0.85)	49.6% (1.00)	50.8% (1.03)	54.4% (1.10)	44.3% (0.90)	49.4% (1.00)
2003	50.9% (1.01)	41.2% (0.82)	49.6% (0.99)	51.7% (1.03)	55.3% (1.10)	47.0% (0.94)	50.3% (1.00)
2004	51.3% (1.01)	42.8% (0.84)	49.7% (0.98)	52.2% (1.03)	56.2% (1.11)	46.8% (0.92)	50.9% (1.00)
2005	50.3% (0.99)	44.8% (0.88)	50.1% (0.98)	52.0% (1.02)	55.3% (1.09)	47.8% (0.94)	50.9% (1.00)
2006	52.3% (1.01)	45.4% (0.88)	51.1% (0.98)	54.0% (1.04)	55.7% (1.07)	47.3% (0.91)	51.9% (1.00)

Note: * It is worth mentioning that the changes in the share of knowledge-intensive employment in each of the districts can be a result of either the growth of knowledge-intensive activity or a decrease in non-knowledge-intensive activities.

Third, we notice the rise of the metropolitan "urban ring" (Central District) from third to second place during the period of the study (an increase in the knowledge economy employment share from 44.2% in 1995 to 54% in 2006). Correspondingly, the formerly second-place Jerusalem District slipped to the third rank over the period (the Central District passed the Jerusalem District in 2000). The fourth point to note is the relatively modest, fluctuating knowledge-economy profile of the periphery. Both the Northern and Southern districts remained behind the rising Israeli employment mean share, with less than 50% of knowledge economy employment. Moreover, in both the Northern and Southern districts the growth in knowledge-economy employment during the period of the study was below the national average growth in knowledge-economy employment.

Regarding the gap between the core of the country and the periphery, the findings show that inequality in the share of knowledge-intensive employment increased during the period of the study. We divided the country into three areas: core, other big cities and the periphery. The core area includes the Tel Aviv and Central districts. The other big cities areas include the districts of Haifa and Jerusalem. The periphery area includes the Northern and Southern districts.

For each area (core/big cities/periphery) we calculated the share of knowledge economy during the period of the study (see Table 5.5). The findings show that between 1995 and 2006 the share of knowledge economy in the core grew by 9.6% (from 45.3% to 54.9%) in the other big cities it grew by 8.3% (from 43.3% to 51.6%) and in the periphery it grew by 6.6% (from 39.8% to 46.4%). The result was an increasing inequality of knowledge-intensive activities between the core and the periphery (the gap grew from 5.5% in 1995 to 8.5% in 2006), and between the core and the other big cities (the gap grew from 1.9% in 1995 to 3.2% in 2006). These findings show, similar to other high-tech categories, that at least at this stage in the last few years, the knowledge-economy revolution is increasing regional gaps rather than decreasing them.[8]

Table 5.5 Knowledge-economy employment by area (location quotients), 1995 and 2006

	Core	Big Cities	Periphery	Total
1995	45.3% (1.04)	43.4% (1.00)	39.8% (0.91)	43.5% (1.00)
2006	54.9% (1.06)	51.6% (0.99)	46.4% (0.89)	51.9% (1.00)

Dynamic Spatial Patterns in the High-Tech Sectors

The share of high-tech industry employment in the Israeli economy continuously grew over time in the last decade; it grew from 6.0% in 1995 to 9.5% in 2006. In the following section, we analyse the spatial pattern of high-tech employment in Israel and its trends between 1995 and 2006. For each year we calculated the share of employment in high-tech industries out of total employment, and the location quotients in each district in a given year.

Table 5.6 presents the following findings. First, all districts showed an increase in the level of high-tech employment during the 1995–2006 period. Second, the Central District is by far the leading high-tech location. Its high-tech location quotients were higher than 1.33 during the entire investigated period, and its average was 1.50 (50% above the the mean of all districts). This led to an increase in the high-tech employment share in the Central District from 8.0% in 1995 to 15.3% in 2006. Third, the Tel Aviv District remained in second place during the 1995–2006 period, but its distance from first place increased significantly (the gap grew from 1.7% in 1995 to 6.4% in 2006). The fourth point to note is the very weak position of the Jerusalem

Table 5.6 High-tech employment by district (location quotients), 1995–2006

	Jerusalem	Northern	Haifa	Central	Tel Aviv	Southern	Average
1995	4.7% (0.79)	4.5% (0.75)	5.3% (0.88)	8.0% (.1.33)	6.3% (1.05)	5.5% (0.92)	6.0% (1.00)
1996	4.5% (0.73)	4.1% (0.66)	5.3% (0.86)	9.2% (1.49)	6.5% (1.05)	5.2% (0.84)	6.2% (1.00)
1997	4.2% (0.66)	4.6% (0.72)	4.7% (0.74)	9.6% (1.52)	7.0% (1.11)	4.9% (0.77)	6.3% (1.00)
1998	4.4% (0.64)	4.9% (0.71)	4.8% (0.69)	10.4% (1.50)	7.9% (1.14)	5.4% (0.78)	6.9% (1.00)
1999	4.8% (0.66)	4.7% (0.66)	5.7% (0.79)	10.7% (1.48)	7.9% (1.10)	6.0% (0.83)	7.2% (1.00)
2000	6.0% (0.70)	6.8% (0.78)	7.1% (0.82)	12.2% (1.41)	9.6% (1.10)	6.4% (0.74)	8.7% (1.00)
2001	6.5% (0.71)	6.6% (0.72)	7.2% (0.79)	13.3% (1.46)	9.9% (1.09)	6.7% (0.73)	9.1% (1.00)
2002	5.3% (0.63)	6.0% (0.70)	7.0% (0.83)	12.6% (1.48)	9.4% (1.10)	5.8% (0.68)	8.5% (1.00)
2003	5.1% (0.60)	5.7% (0.67)	7.2% (0.85)	12.2% (1.43)	9.4% (1.11)	6.9% (0.81)	8.5% (1.00)
2004	5.1% (0.61)	6.4% (0.76)	6.4% (0.76)	12.4% (1.47)	8.9% (1.06)	6.9% (0.82)	8.4% (1.00)
2005	4.5% (0.51)	7.1% (0.80)	7.3% (0.82)	14.1% (1.58)	8.4% (0.950	7.1% (0.80)	8.9% (1.00)
2006	5.3% (0.56)	7.2% (0.75)	7.5% (0.79)	15.3% (1.61)	8.9% (0.94)	7.7% (0.81)	9.5% (1.00)

district in high-tech industries. We believe that this district both did not advance enough to attract intensive R&D service sectors and in addition was too expensive and too crowded to attract high-tech manufacturing.[9]

For each area (core/big cities/periphery) we calculated the share of high-tech employment during the period of the study (see Table 5.7). The findings show that between 1995 and 2006 the share of high-tech employment in the core grew by 5.1% (from 6.9% to 12.0%), in the other big cities it grew by 1.5% (from 5.0% to 6.5%) and in the periphery it grew by 2.4% (from 5.0% to 7.4%). The result was increasing inequality in high-tech activity between the core and the periphery (the gap grew by 1.9% in 1995 to 4.6% in 2006) and between the core and the other big cities (the gap grew 1.9% in 1995 to 5.5% in 2006).

Table 5.7 High-tech employment by area (location quotients), 1995 and 2006

	Core	Big Cities	Periphery	Total
1995	6.9% (1.16)	5.0% (0.84)	5.0% (0.84)	6.0% (1.00)
2006	12.0% (1.26)	6.5% (0.68)	7.4% (0.78)	9.5% (1.00)

These findings reveal an interesting phenomenon where the periphery is stronger than the non-core big cities in high-tech. This may be explained by the fact that the high-tech industry comprises two different sub-sectors: the high-tech services sectors that are more advanced and are based mostly on tacit knowledge and highly skilled employees, and the high-tech services sectors that are based on codified advanced technologies and on less-skilled employees. Table 5.8, which presents these two sub-categories of high-tech industries, shows that the periphery is very strong in high-tech manufacturing sectors while it is very weak in the high-tech services sectors; the core is very strong in high-tech services sectors while it is around average in the high-tech

Table 5.8 High-tech services and manufacturing employment (LQ) by area, 1995 and 2006

	Core		Big Cities		Periphery		Average	
	HTM	HTS	HTM	HTS	HTM	HTS	HTM	HTS
1995	3.7% (1.11)	3.3% (1.22)	2.5% (0.77)	2.5% (0.92)	3.4% (1.02)	1.6% (0.61)	3.3% (1.00)	2.7% (1.00)
2006	3.8% (0.98)	8.2% (1.45)	2.6% (0.68)	3.9% (0.68)	5.1% (1.32)	2.3% (0.41)	3.9% (1.00)	5.7% (1.00)

manufacturing sectors; and the big cities area is weak with both sub-categories of high-tech industries.

Dynamic Spatial Patterns in the High-Tech Start-Ups

The share of new high-tech start-up employment in the Israeli economy grew continuously over time in the last decade, from 0.23% in 1995 to 0.51% in 2006. In the following section we analyse the spatial pattern of the high-tech start-up employment in Israel and its trends between 1995 and 2006. For each year we calculate the share of employment in high-tech industries out of total employment, and the location quotients in each district in a given year.

Table 5.9 High-tech start-ups employment by district (location quotients), 1995–2006

	Jerusalem	Northern	Haifa	Central	Tel Aviv	Southern	Average
1995	0.16% (0.75)	0.26% (1.20)	0.14% (0.18)	0.42% (1.93)	0.20% (0.90)	0.15% (0.69)	0.23% (1.00)
1996	0.20% (0.74)	0.25% (0.92)	0.17% (0.64)	0.48% (1.80)	0.26% (0.98)	0.12% (0.46)	0.27% (1.00)
1997	0.30% (1.03)	0.31% (1.09)	0.21% (0.72)	0.46% (1.58)	0.27% (0.94)	0.09% (0.30)	0.29% (1.00)
1998	0.27% (0.74)	0.23% (0.63)	0.28% (0.75)	0.59% (1.60)	0.42% (1.14)	0.20% (0.56)	0.37% (1.00)
1999	0.45% (0.74)	0.36% (0.58)	0.46% (0.74)	0.84% (1.37)	0.83% (1.36)	0.30% (0.49)	0.61% (1.00)
2000	0.53% (0.73)	0.40% (0.56)	0.37% (0.52)	1.22% (1.70)	0.90% (1.26)	0.28% (0.39)	0.72% (1.00)
2001	0.34% (0.85)	0.42% (1.07)	0.18% (0.45)	0.52% (1.30)	0.48% (1.21)	0.25% (0.63)	0.40% (1.00)
2002	0.23% (0.61)	0.26% (0.69)	0.32% (0.85)	0.51% (1.33)	0.50% (1.32)	0.20% (0.53)	0.38% (1.00)
2003	0.49% (1.10)	0.28% (0.64)	0.34% (0.77)	0.53% (1.21)	0.61% (1.39)	0.15% (0.34)	0.44% (1.00)
2004	0.35% (0.58)	0.26% (0.43)	0.60% (0.99)	0.77% (1.26)	0.94% (1.55)	0.21% (0.35)	0.61% (1.00)
2005	0.25% (0.47)	0.20% (0.38)	0.38% (0.72)	0.73% (1.39)	0.86% (1.61)	0.22% (0.41)	0.53% (1.00)
2006	0.24% (0.47)	0.26% (0.52)	0.39% (0.78)	0.61% (1.19)	0.88% (1.74)	0.15% (0.30)	0.51% (1.00)

Table 5.9 presents the following findings. First, all districts show an increase in the level of high-tech employment during the 1995–2006 period. Second, the Tel Aviv district presented a quantum leap in the share of start-up employees out of total employment, from third place in 1995 with only 0.20% to first place in 2006 with 0.88%. Its high-tech location quotient grew from 0.90 in 1995 to 1.74 in 2006. Third, both the Central and Haifa Districts significantly increased their share of start-up employment, achieving the second and third places in this category. The fourth finding is the very weak position of the Jerusalem District in start-up employment. The periphery area is the weakest one in start-up activity, with the Southern District closing the list.

For each area (core/big cities/periphery) we calculated the share of start-up employment during the period of the study (see Table 5.10). The findings show that between 1995 and 2006 the inequality between the core and the periphery grew significantly, while the gap with the other big cities did not change significantly (in relative terms – in 1995 the share of start-up employment in the core was 2.8 times more than in the big cities, and it decreased to 2.4 times that of the big cities in 2006).

Table 5.10 High-tech start-ups employment by area (location quotients), 1995 and 2006

	Core	Big Cities	Periphery	Total
1995	0.28% (1.31)	0.10% (0.46)	0.19% (0.89)	0.22% (1.00)
2006	0.75% (1.47)	0.32% (0.64)	0.21% (0.41)	0.51% (1.00)

CONCLUSIONS

The location decision of firms is naturally influenced by micro-economic or business considerations, with the ultimate objective being the maximization of profits. This does not contradict the fact that the spatial distribution of economic activities, and especially that of high-tech activities, has always been heavily influenced by regional macro-economic factors in a country – the availability of infrastructures of various kinds, the accessibility to research institutions, the supply of a highly-skilled labour force, the existence of agglomeration economies, etc. These and other factors actually lead to a pattern of a quite heavy concentration of high-tech activity as compared with other types of economic activities mostly in and around metropolitan centres. This same pattern has also been identified in the case of Israel (in all the various knowledge-intensive/high-tech categories that were analysed in this study).

Start-up activities are considered to be a leading engine for the future growth of high-tech activity. In this chapter, we have tried to initiate some preliminary evaluations of their location trends: do start-ups follow the same trend of heavy concentration; are they showing any indication of new trends? Since start-ups are themselves high-tech activities, they are also naturally subject to many of the macro-regional factors that lead to concentration.

The section "Concentration Levels Employment in Various High-Tech Categories" suggests that there seems to be significant regional advantages in the Tel Aviv District and the Central District. Moreover, it suggests that there are apparently two types of accumulation effects active in Israel's core regions – those that explain the concentration of high-tech activity and those that explain the concentration of start-ups and VC investments, as described in the literature review. The concentration of high-tech activity is explained by factors such as technological infrastructures and knowledge spillovers, but it is also significantly influenced by rent costs and other agglomeration costs. Therefore, these activities tend to concentrate in the Core Districts but within them in the cheaper locations (i.e. the Central District rather than the Tel Aviv District), while the concentration of start-up companies is influenced also by the location of financial services and VC investors. However the start-up companies seem to be less influenced by high rents, due to their small size, or high cost of labour relative to rent costs. Therefore, start-up companies tend to concentrate in the Tel Aviv District rather than in the Central District.

Such location behaviour may lead to changes in the geographical mapping of high-tech activity, and possibly to the increase of regional gaps. This is an important implication because of the suspicion that we may be experiencing a type of market failure in Israel. The fact that start-up creations are found in some focal places does not necessarily imply that other regions do not possess the necessary qualities for the long-term development of start-ups and high-tech activity. This may just be the result of the fact that, for example, VC funds are almost all concentrated in one region. Public involvement in the establishment of VC funds (or other start-up facilities such as incubators) in other regions may induce some change in that.

The sections "Dynamic Spatial Patterns in the Knowledge Economy Sectors" and "Dynamic Spatial Patterns in the High-tech Start-ups" suggest that in addition to the common patterns identified of knowledge-intensive and high-tech activities concentrating in the core districts in Israel, there are also differences in the concentration and location patterns of different sectors. For example, high-tech manufacturing sectors tend to concentrate in technological parks in the periphery, while high-tech services sectors tend to concentrate in technological parks in the core districts. Our conclusion is that while clearly there are some common location effects in all the categories of knowledge-intensive/high-tech activities, there are also some unique effects that are more dominant in some categories while less dominant in other categories. Specific identification of these unique regional effects requires additional research. However it is reasonable to argue that an advanced regional policy should take into consideration such unique regional effects. Using such an approach may lead to the identification of sectorial/technological areas where the specific location may have a competitive advantage, resulting in a specific targeted policy for those specific sectorial/technological areas.

An additional issue that arises in this research is the increasing regional gap in Israel between the core district and all other districts in terms of high-tech activity. We show that the gap between the two core districts in all categories of knowledge economy/high-tech activity increased. As the activity is more advanced (e.g. more technologically based and requiring a higher quality labour force), the initial regional gap is higher. Moreover, as the activity is more advanced the gap increases more rapidly. This seems to be a critical system failure that may lead to severe social problems in the following years.

In the high-tech industries we observed another problematic phenomenon, in which the districts of the big cities (big cities that are not in the core districts, e.g. Haifa and Jerusalem districts) declined significantly over time both relative to the core districts and relative to the periphery districts. This is due to the fact that they are "stuck in the middle" – they are not advanced enough to become leaders in the high-tech services sectors and do not have cost advantages that enable them to become leaders in the high-tech manufacturing sectors.

APPENDIX 5A VARIOUS KNOWLEDGE-INTENSIVE INDUSTRIAL CATEGORIES' CLASSIFICATION

Classification of Knowledge Economy by Categories – I.S.I.C. Definition (As presented in Schwartz. 2006, p. 35)

Knowledge economy manufacturing by categories

HTM

Manufacture of office, accounting and computing machinery (30), Manufacture of radio, television and communication equipment and apparatus (32), Manufacture of medical, precision and optical instruments, watches and clocks (33), Manufacture of aircraft and spacecrafts (353).

OM

Manufacture of chemicals and chemical products (24), Manufacture of machinery and equipment n.e.c. (29), Manufacture of electrical machinery and apparatus (31), Manufacture of motor vehicles (341), Manufacture of bodies (coachwork) for motor vehicles; Manufacture of trailers and semi-trailers (342), Manufacture of parts and accessories for motor vehicles and their engines (343), Building and repairing of ships and boats (351), Manufacture of railway and tramway locomotives and rolling stock (352), Manufacture of transport equipment n.e.c. (359).

Knowledge economy services by categories

HTS

Telecommunications (642), Computer and related activities (72), Research and development (73).

OBS

Water transport (61), Air transport (62), Post and courier activities (641), Financial intermediation, except insurance and pension funding (65), Insurance and pension funding, except compulsory social security (66), Activities auxiliary to financial intermediation (67), Real estate activities (70), Renting of machinery and equipment without operator and of personal and household goods (71), Other business activities (74).

PS

Higher education (803), Other education (809), Human health activities (851), Veterinary activities (852), Recreational, cultural and sporting activities (92), Primary education (801), Secondary education (802), Social work activities (853).

N.B.: There are slight differences in transforming the Israeli classification to the I.S.I.C. classification.

NOTES

1. It is also explained by traditional agglomeration effects. However such forces have a similar influence on both knowledge-intensive and traditional industries. Therefore, the traditional agglomeration effect cannot explain why knowledge-intensive industries tend to concentrate more than traditional industries.
2. It is important to mention here that there are many other considerations to the pattern of concentration of various economic activities that are not analysed here such as factors that influence VC investors, location investment (see Schwartz & Bar-El, 2007), or specific specialization of different locations due to historical events or natural resources.
3. In this specific study we used only part of the data.
4. The I.S.I.C. classifications for each category are presented in the Appendix.
5. There are seven universities in Israel (excluding the Open University): one in Tel Aviv, two in the centre, two in Haifa, one in Jerusalem, and one in the Southern district. In addition, there are 51 colleges and training centres, out of which 15 are located in the peripheral districts, 21 in Tel Aviv and the Central districts and 15 in the Haifa and Jerusalem districts. We do not discuss the impact on these institutions on the regional distribution of the knowledge-intensive industries due to the limited scope of the chapter.
6. Or Knowledge Economy Index (KEI), based on the OECD approach (Cooke et al., 2001)
7. According to Bar-El and Parr (2003) while the metropolis core has significant positive agglomeration effects it also has a significant agglomeration cost; on the other hand the nearby region enjoy most of the agglomeration benefits with much lower agglomeration costs.
8. It is worth mentioning two points regarding the regional gap: first, employment location is not identical to resident location – the actual gap could be either higher or lower than the gap captured by employment location; and second, for dealing with the social and financial gaps, many additional indicators must be considered.
9. However there could be another explanation to this result.

REFERENCES

Acs, Z.J., Audretsch, D.B. and Feldman, M.P. (1994) R&D, spillovers and recipient firm size, *The Review of Economics and Statistics*, 76 (2), pp. 336–340.

Allen, K.R. (2002) *Bringing New Technology to Market.* Pearson Education, Inc., Upper Saddle River, New Jersey: Prentice Hall.

Almeida, P., Dokko, G. and Rosenkopf, L. (2003) Startup size and the mechanisms of external learning: Increasing opportunity and decreasing ability? *Research Policy*, 32 (2), pp. 301-316.

Avnimelech, G. and Schwartz, D. (2009) Structural transformation of regional venture capital industry: Evidence from Israel, *Innovation: Management Policy & Practice*, 11 (1), pp. 60–73.

Avnimelech, G. and Teubal, M. (2004) Venture capital – start up co-evolution and the emergence and development of Israel's new high tech cluster, Part 1: Macroeconomic and Industry Analysis, *Economics of Innovation and New Technology*, 13 (1), pp. 33–60.

Avnimelech, G. and Teubal, M. (2006) Creating venture capital industries that co-evolve with high tech clusters: insights from an extended industry life cycle perspective of the Israeli experience, *Research Policy*, 35 (10), pp. 1477–1498.

Baranes, E. and Tropeano, J.P. (2003) Why are technological spillovers spatially bounded? A market orientated approach, *Regional Science and Urban Economics*, 33 (4), pp. 445–467.

Bar-El, R. and Parr, J. (2003) From metropolis to metropolis-based region: the case of Tel-Aviv, *Urban Studies*, 40 (1), pp. 113–125.

Boschma, R. (2005) Proximity and innovation: A critical assessment, *Regional Studies*, 39 (1), pp. 61–74.

Capello, R. (2002) Entrepreneurship and spatial externalities: Theory and measurement, *The Annals of Regional Science*, 36 (3), pp. 387–402.

Capello, R. and Faggian, A. (2005) Collective learning and relational capital in local innovation process, *Regional Studies*, 39 (1), pp. 75–87.

Cappellin, R. (2007) The territorial dimension of the knowledge economy: Collective learning, spatial changes, and regional and urban policies, *American Behavioral Scientist*, 50 (7), pp. 897–921.

Castell, M. (1996) *The Rise of the Network Society: The Information Age: Economy, Society and Culture*. Cambridge, MA, USA: Blackwell Publishers,.

Cooke, P. (2004a) Regional knowledge capabilities, embeddedness of firms and industry organisation: Bioscience megacentres and economic geography, *European Planning Studies*, 12 (5), pp. 625–642.

Cooke, P. (2004b) The role of research in regional innovation systems: new models meeting knowledge economy demands, *International Journal of Technology Management*, 28 (3–6), pp. 507–533.

Cooke, P. (2005) [not in text] Research, knowledge and open innovation: Spatial impacts upon organisation of knowledge-intensive industry clusters, Paper presented in a conference on: Regional Studies Association: "Regional Growth Agendas", Aalborg, 28th May to 31st May 2005.

Cooke, P. and Schwartz, D. (2008) Regional knowledge economies: An EU–UK and Israel perspective. *Journal of Economic and Social Geography* (TESG – Tijdschrift voor Economische en Sociale Geografie), 99 (2), pp. 178–192.

Cooke, P., Clifton, N. and Huggins, R. (2001) Competitiveness and the Knowledge Economy, Regional Industrial Research Report No. 30, Cardiff, Centre for Advanced Studies.

Cooke, P., Uranga, M.G. and Etxebarria, G. (1997), Regional innovation systems: Institutional and organisational dimensions, *Research Policy*, 26 (4–5), pp. 475–491.

Felsenstein, D. and Ergas, Y. (2002) Investing in an emerging node: foreign-owned companies in the Tel Aviv economy, *Geojournal*, 72, pp, 57–80.

Florida, R. (1995) Toward the learning region, *Futures*, 27 (5), pp. 527–536

Florida, R. and Gates, G. (2002) Technology and tolerance: The importance of diversity to high-technology growth, *Brookings Review*, 20 (1), pp. 32–36.

Frenkel, A. (2001) Why high-technology firms choose to locate in or near metropolitan areas, *Urban Studies*, 38 (7), pp. 1083–1101.

Frenkel, A. and Shefer, D. (2001) Firm characteristics, location and regional innovation: A comparison between Israeli and German industrial firms, *Regional Studies*, 35 (5), pp. 415–430.

Gompers, P.A. and Lerner, J. (1999) *The Venture Capital Cycle*. Cambridge, Massachusetts: The MIT Press.

Gössling, T. (2004) Proximity, trust and morality in networks, *European Planning Studies*, 12 (5), pp. 675–689.
Harmaakorpi, V. and Melkas, H. (2005) Knowledge management in regional innovation networks: The case of Lahti, Finland, *European Planning Studies*, 13 (5), pp. 641–659.
Hauser, M., Cushman, F., Young, L. and Mikhail, J. (2007) A dissociation between moral judgments and justifications, *Mind and Language*, 22 (1), pp. 1–21.
Havnes P.-A. and Senneseth, K. (2001) A panel study of firm growth among SMEs in networks, *Small Business Economics*, 16, pp. 293–302.
Howells, J.R.L. (2002) Tacit Knowledge, Innovation and Economic Geography, *Urban Studies*, 39 (5/6), pp. 871–884.
Maskell, P. and Malmberg, A. (1999) Localised learning and industrial competitiveness, *Cambridge Journal of Economics*, 23(2), 167–185.
Mason M.C. and Harrison, R.T. (2002) The geography of venture capital investments in the UK, *Transactions Institute of British Geographers*, 27, pp. 427–451.
Mason M.C. and Harrison R.T. (2003) Closing the regional equity gap? A critique of the Department of Trade and Industry's regional venture capital funds initiative, *Regional Studies*, 37 (8), pp. 855–870.
Morgan, K. (1997) The learning region: Institutions, innovation and regional renewal, *Journal of Regional Studies*, 31 (5), pp. 491–503.
OECD (1999) *Science, Technology and Industry Scoreboard 1999, Benchmarking Knowledge-Based Economies*, Paris: Organisation for Economic Cooperation & Development.
Powell, W.W., Koput, K.W., Bowie, J.I. and Smith-Doerr, L. (2002) The spatial clustering of science and capital: Accounting for biotech firm-venture capital relationships, *Regional Studies*, 36 (3), pp. 291-307.
Quah, D. (2002) Spatial agglomeration dynamics, *The American Economic Review*, 92 (2), pp. 247–252.
Schwartz, D. (2006) The regional location of knowledge based economy activities in Israel, *Journal of Technology Transfer*, 31 (1) pp. 31-44.
Schwartz, D. and Bar-El, R. (2007) Venture investments in Israel – a regional perspective, *European Planning Studies*, 15 (5), pp. 623–644.
Trajtenberg, M. (2000). R&D Policy in Israel: An Overview and Reassessment, Foerder Institute for Economic Research, WP No. 7-2000.
Van Geenhuizen, M. and Nijkamp, P. (2006) Learning regions in an evolutionary context: Policymaking for high technology firms, *International Journal of Entrepreneurship and Innovation Management,* 6 (3), pp. 265–282.
Wallsten, J.S. (2001) An empirical test of geographic knowledge spillovers using geographic information systems and firm-level data, *Regional Science and Urban Economics*, 31 (5), pp. 571–599.
Zook, M. (2004) The knowledge brokers: Venture capitalists, tacit knowledge and regional development, *International Journal of Urban and Regional Research*, 28 (3), pp. 621–641.

6. University–industry technology transfer: fostering and hindering factors and programs

Amnon Frenkel and Daniel Shefer

INTRODUCTION

Reciprocity between universities/research centres and industry is a high priority on the agenda of academic institutions.

The drastic changes around the world that are taking place in social, economic and organisational frameworks also influence academia. A manifestation of these continual changes can be found in the processes of privatisation and globalisation (free trade), which have exerted a direct effect, forcing academia to reorganise, expand and increase the scope of its outside financial resources partially through the commercialisation of knowledge created within these institutions. However, there is a feeling in academia of having missed opportunities to commercialise many of the inventions and innovations that have resulted from research carried out in the academic research framework.

Formal university policy, as to goals, faculty members' rights and obligations, academic freedom and publication, conflicts of interest, outside employment and consultancies, and so forth, is expressed in university guidelines. Among the wide range of topics dealt with is that concerning the distribution of profits that accrue from the commercialisation of patents.

In recent years, the vast changes have led to a new approach to commercialise universities' innovations the aim of which is to formulate better, more efficient means of making it possible to achieve the university's goals, on the one hand, and to provide answers to the new conditions that have evolved, on the other. Acknowledgement of the increased importance of innovation has been made, and the key role universities can play relates to the need to merge several different complex technologies to create a single successful innovation (e.g. nanotechnology, biotechnology, cognitive science and information science). This requires universities to integrate departments and faculties that previously did not collaborate. The potential gains for society as a whole and for the economy in particular through the strengthening of ties between academia and industry are profound. Furthermore, this reciprocity may offer an important source for funding research projects, as well as another source of income for university budgets, which are being depleted.

There is no doubt that "a proper balance" between sources of funding for basic research, from national foundations and from industry whose main aim is to apply the knowledge created in the universities, should be the main goal. Academic guidelines must preserve the character and goals of the university, on the one hand, and provide an incentive to implement and commercialise knowledge in the universities, on the other. Competitive public funding through national and international research foundations, without any bureaucratic influence or pressure to limit or restrict the research that is often exerted by industry, will ensure the continued realisation of the research curiosity that guides academics.

This chapter deals with factors and programmes that act to foster or to hinder the rate of technology transfer between academia and industry. We present the basic concepts of a technology-transfer production-function model in which human capital, investment capital and a technology transfer office (TTO) staff interact to produce innovations or patent registration.

A field study, carried out as part of the research reported here, involved six universities in the USA and six in Israel whose technology-transfer policies were thoroughly analysed. Additionally input from secondary sources (official data: IBC and AUTM database) concerning these universities was examined. Faculty members' opinions were reviewed and the extent of their links with industry examined through an evaluation of the data collated from the field survey, which was carried out at the Hebrew University, the Technion, and Tel-Aviv University. More than 150 faculty members participated in the survey. This additional viewpoint enabled us to better understand the existing links between faculty members and university institutions, on the one hand, and industry, on the other.

Next, government programmes designed to encourage research cooperation between academia and industry are briefly described. Inducing such cooperation could favourably influence the regional and national economy for the good of society. Finally, the last section summarises the chapter and draws some useful conclusions.

LITERATURE REVIEW

An avalanche of publications dealing with the issue of technology transfer from academia to industry has inundated academic journals in the past two decades. Numerous symposia and colloquia have also been held to discuss a variety of related aspects, such as the pattern of university patent registrations, licensing inventions, university policies on knowledge creation, the function of technology transfer offices, faculty attitudes, industry positions, and more (for example, Branscomb et al., 1999; *Journal of Technology Transfer*, 2001; *Management Science*, 2002; Bok, 2003; Mowery et al., 2004; Libecap, 2005).

Jaffe's (1989) seminal article, "Real Effects of Academic Research", illuminated the issue of university knowledge spillovers to commercial innovations. A statistical analysis led Jaffe to find "a significant effect of university research on corporate patents." He added: "University research appears to have an indirect effect on local innovation by inducing industrial R&D spending" (p. 957). Acs et al. (1991) also concluded that, "university spillover, relative to private-company R&D spending is considerably greater in the electronic sector" (p. 366).

Innovation and patent registration are vital to economic growth. Industrial innovation is recognised as an engine that fosters technological progress and economic growth (see Romer, 1986, 1990, 1994; Lucas, 1988; Dosi, 1988; Grossman and Helpman, 1990, 1991, 1994; Barro and Sala-i-Martin, 1995; Aghion and Howitt, 1998, and Jamison and Jansen, 2000). The endogenous economic growth models assume that investment in R&D yields new technologies. Thus, technological progress can explain continuous growth in output and, consequently, in income per capita or "standard of living" (Grossman and Helpman, 1991, 1994; Romer, 1994; Stokey, 1995). Innovation provides new and, in time, unique opportunities for creating new firms and expanding old ones. Consequently, market share is enlarged and, with increasing returns to scale, firms marketing their innovations enjoy greater production efficiency and a higher rate of economic growth. Greater production efficiency allows industries to expand their domestic market share through import substitution, to increase local consumption and, at the same time, to penetrate new foreign markets and increase their export share (Krugman, 1979, 1991, 1995; Porter, 1990).

Universities supply the most important players for the production and diffusion of knowledge and invention promoting economic growth. Universities constitute the major source of technological progress for industry (Henderson et al., 1998; Mowery and Shane, 2002; Bercovitz and Feldman, 2009).

The upsurge of patent registration and licensing invention by universities in the past two decades is attributed, among other reasons, to the now well-known Bayh-Dole Act of 1980 in the United States. This piece of legislation permitted universities to register patents for innovations resulting from federally funded research and subsequently to license them to industry (COGR, 1999). A large number of articles have been written on the impact of the Bayh-Dole Act (e.g., Mowery et al., 1999; Nelson, 2001; Mowery and Sampat, 2004; and Mowery, 2005).

Most of the authors agreed that Bayh-Dole gave considerable impetus to universities to establish technology transfer offices (TTO) and to pursue a more aggressive policy of patent registrations. At the same time, however, public financial support of universities gradually began to erode, forcing institutions to seek new sources of income for their budgets. A prime source toward the end became the commercialising of university innovation through patenting and the licensing of innovation derived from faculty research. Nevertheless, most authors could not trace any significant increase either in the rate of commercialisation or in income in consequence of the Bayh-Dole Act (Mowery et al., 1999, 2004; Nelson, 2001; Bok, 2003; Swamidass and Vulasa, 2009). In fact, some scholars maintain that the rush to register patents delays publications and thus restrains rapid dissemination of newly created knowledge.

Innovation and patent registration should respond to university policies dealing with intellectual property rights (ITR) and the incentives offered faculty members and researchers. Regarding these aspects, however, we can find diametrically opposing viewpoints in the literature. Lach and Schankerman (2003, 2004, 2008), who conducted a macro analysis of time series data from 102 American universities from 1991 to 1999, found strong statistical evidence that giving incentives to faculty significantly increased innovations, particularly in private universities. They concluded that incentives could attract new, able researchers and also increase the rate of commercialising new inventions.

To facilitate the registering of patents, the licensing of inventions, and the commercialising of these innovations, most research universities have established a TTO. The TTO staff also attempt to promote collaboration between university and industry. The slow rate of licensing and commercialising university inventions may be partially attributed to the inadequately trained staff budget shortages at the TTOs (Rogers et al., 2000; Allan, 2001; Swamidass and Vulasa, 2009).

Government-initiated programmes, such as the Advanced Technology Programme initiated in the US, are designed to promote industry–university research partnerships and are funded by federal money (Hall et al., 2003; Loof and Brostrom, 2008). The American programme is considered a success in that it has enabled industry to gain access to research conducted in university laboratories and provided a platform for engaging faculty members in industrial interests. On the other hand, it has also enabled universities to enhance their research budgets and thus increase their income.

In Israel, somewhat similar programmes were initiated by the Office of the Chief Scientist (OCS) at the Ministry of Industry, Trade and Labour under the names "Magnet" and "Magneton" (Barda, 2001; ENOSH, 2002; Gatz et al., 2009) (more on this later in the chapter).

Another way to promote university entrepreneurship and technology transfer is through the university incubator programme, which enables faculty to pursue their discoveries by establishing start-ups with financial assistance from government sources, such as from the OCS, or directly from the university's own resources, or from venture capital (Felsenstein, 1994; Allen and McCluskey, 1990; Etzkowitz, 2002; *Management Science*, 2002; Shefer and Frankel, 2003; Chukumba and Jensen, 2005; Boni and Emerson, 2005; Thursby, 2005; Bercovitz and Feldman, 2006). University entrepreneurship and technology transfer are critical issues that have attracted the attention of academia, as can be demonstrated by the special issue of *Management Science* (2002) and the colloquium on entrepreneurship, education and technology transfer held in Tucson, Arizona, in January 2005. A large number of the papers delivered at that colloquium have been published in Libecap (2005) see also D'Este and Patel (2007) and Trippl and Todtling (2008).

THEORETICAL CONSTRUCT

We will begin our theoretical construct analysis with a simple production-function model à la "knowledge production function", introduced by Zvi Griliches (1979) and subsequently modified by Jaffe (1989) (see also Arrow, 2000). The function relates major inputs to the desired output – patenting and licensing.

One way to describe the process of innovation or patent registration is through a production–function model in which patent registration is a function of human capital (faculty and graduate students), physical capital (research funds), equipment and laboratories, and technology transfer offices (TTO), size and quality of employees. This kind of relationship can be shown in the equation (Figure 6.1).

Thus we would expect that at universities with a large faculty and a large number of graduate students, the number of new innovations will be high as well, ceteris paribus. Similarly, universities that enjoy sizable research funds and maintain sophisticated equipment and laboratories will produce a large number of innovations. Finally, universities with a professionally diverse, skilled TTO staff are expected to produce larger number of innovations.

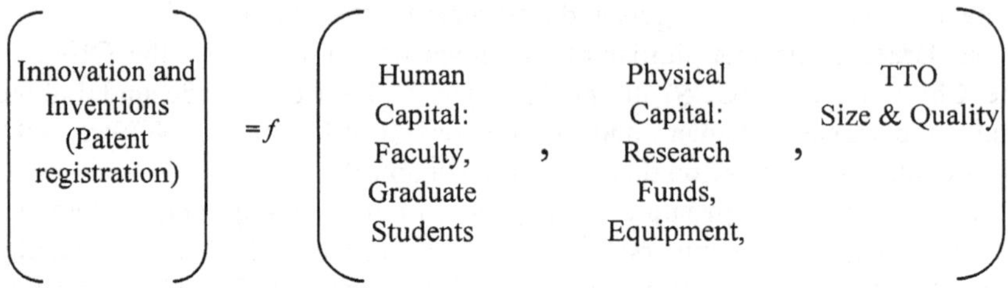

*Note: Patent Registration[1] = (Human Capital, Physical Capital, TTO)

Figure 6.1 Patents production–function model

We now proceed with a comparative analysis of data collected from selected leading research universities, six in Israel and six in the USA.

A COMPARISON OF SELECTED ISRAELI AND AMERICAN RESEARCH UNIVERSITIES

Data Analyses

Based on our theoretical construct, in the following section we will present a comparative analysis using basic statistics that show differences among leading research universities in the United States and Israel in order to verify our conceptual model. The comparison will point out some of the more important differences between the universities of the two countries and among universities within each country.

Table 6.1 presents the number of overall students and the number of graduate students in selected Israeli and US research universities. There are two outlying figures: one in Israel, the Weitzman Institute of Science, where there are only graduate students, and therefore the total number of students is relatively small – fewer than a thousand; the other in the US, University of Michigan, which is a significantly large university with approximately 40,000 students. However, when one looks at the number of graduate students in each university, the percentage at MIT, Stanford and Harvard is seen to be greater than 50% (at MIT, it reaches 60%).

As for Israeli universities, the percentage of graduate students at the Hebrew University and Tel-Aviv University is 45% and 41%, respectively, while at the Technion, it is only 30%. These significant differences should inadvertently impact the production of patent registrations.

Table 6.1 Number of students at selected research universities in Israel and the USA, 1999 and 2005

			Advanced-Degree Students*			
	Total Number of Students		Numbers		Percentage of Total	
Institution	1999/2000	2004/2005	1999/2000	2004/2005	1999/2000	2004/2005
Israeli Universities						
Weitzmann Institute of Science	790	960	790	960	100.0	100.0
Technion-Israel Institute of Technology	12,380	12,810	3,420	3,810	27.6	29.7
Ben-Gurion University of the Negev	16,020	18,640	4,740	5,090	29.6	27.3
Hebrew University of Jerusalem	21,220	21,985	8,510	9,890	40.1	45.0
Bar-Ilan University	21,030	25,025	5,230	7,250	24.9	29.0
Tel-Aviv University	26,120	28,740	10,800	11,815	41.3	41.1
US Universities						
MIT	9,972	10,253	5,672	6,126	56.9	59.7
Stanford	14,248	14,881	7,625	8,201	53.5	55.1
Cornell	19,021	19,447	4,095	4,354	21.5	22.4
Harvard	18,918	19,844	8,833	10,097	46.7	50.9
Columbia	22,068	24,417	6,482	6,087	29.4	24.9
Michigan	37,197	39,993	10,226	10,923	27.5	27.3

Note:* Refers to Graduate Students only.
Source: Israel – Council for Higher Education website; USA – individual university websites

When the overall faculty in each one of the selected research universities in the two countries are examined, one may observe the following (see Table 6.2). The number of senior faculty members, in general, correlates highly with the size of the university (in terms of the overall student body). Nevertheless, because the Weitzman Institute offers only graduate-student programmes, the ratio of students per faculty is very low (2.6 in 2004/5). This ratio, in contrast, is very high at two of the Israeli universities, Tel-Aviv University and Bar-Ilan University – 12.4 and 12.1, respectively; at the Hebrew University, the ratio is 8.8. What is alarming is the trend between 1999/2000 and 2004/5. Whereas these ratios remained constant or improved for the US universities, they rose appreciably at the Israeli institutions. The negative trend could adversely affect research output in terms of quality and quantity, as well as eventually also innovation output.

Table 6.2 Characteristics of selected research universities in Israel and the USA, 1999 and 2005

Institution	Number of Senior Faculty*		Students** Per Faculty Member: Ratio		Ranking	
	1999/2000	2004/2005	1999/2000	2004/2005	1999/2000	2004/2005
Israeli Universities						
Weitzmann Institute of Science	360	371	2.2	2.6	1	1
Technion-Israel Institute of Technology	658	638	5.2	6.0	2	2
Ben-Gurion University of the Negev	734	796	6.5	6.4	3	3
Bar-Ilan University	N/A	601	N/A	12.1	N/A	5
Hebrew University of Jerusalem	1,196	1,122	7.1	8.8	4	4
Tel-Aviv University	1,167	950	9.3	12.4	5	6
US Universities						
Cornell	1,520	1,594	2.7	2.7	1	1
Michigan	2,698	2,891	3.8	3.8	2	2
MIT	931	998	6.1	6.1	3	3
Stanford	1,213	1,280	6.3	6.4	4	4
Columbia	857	876	7.6	6.9	6	5/6
Harvard	1,216	1,471	7.3	6.9	5	5/6

Notes: *In Israel, senior faculty members are as defined in the data source. In the USA, the term refers to Full Professors, Associate Professors and Assistant Professors or Tenured Track appointments.
** Refers to Graduate Students only.
Source: Israel – Council of Higher Education, Planning & Budgeting Committee Report 2004/2005, Ch. 3, p. 128. USA – individual university websites

An examination of the research budgets of Israeli and US universities reveals striking differences. Unfortunately, the data from the Israeli institutions is incomplete: university authorities are reluctant to disclose this information and are not required by law to do so even though they are all state institutions. Despite the missing data, it is clear that Israeli universities are at a big disadvantage compared to the research budgets of US universities. In 2005, the MIT research budget amounted to $1.133 billion dollars in research grants, whereas the most for any Israeli university, Tel-Aviv University in this case, amounted to a meagre $64 million dollars. Stanford that year enjoyed a research budget of $930 million dollars. These huge disparities obviously affect the probability of producing innovations. What is interesting to note is that the trend at US universities was upward, meaning a significant increase in the research budget between 2000 and 2005, whereas Israeli universities (those with available data) showed only a small change during the amount budgeted per senior faculty member, which was particularly high at Stanford and MIT, $1.062 million and $885,000, respectively (see Table 6.3).

Table 6.3 Research budgets of selected research universities in Israel and the USA, 1995, 2000 and 2005

Institute	Total Research Budget (Million $)			Research Budget Per Faculty Member (000$)	
	1995	2000	2005	2000	2005
Israeli Universities					
Weitzmann Institute of Science	N/A	N/A	N/A	N/A	N/A
Technion-Israel Institute of Technology	31	30	24	45.6	N/A
Ben-Gurion University of the Negev	23	42	47	57.4	59.4
Bar-Ilan University	N/A	N/A	N/A	N/A	N/A
Hebrew University, Jerusalem	65*	78	N/A	65.0	N/A
Tel Aviv University	N/A	52	64	44.6	67.8
US Universities					
Cornell	310	400	561	148.3	194.1
Michigan	409	500	778	411.2	528.9
MIT	367	728	1,133	600.2	885.2
Stanford	373	444	930	518.1	1,061.6
Columbia	239	311	534	204.6	335.0
Harvard	344	431	623	462.9	624.2

Note: * The data refer to 1997.
Sources: Israel – individual university websites or data received following a request made to the universities' respective research authorities. USA – AUTM reports 2001, 2005; data sources for Stanford and Colombia for 2005 are the universities' respective websites

Table 6.4 presents the distribution of sources of research funding, basically government and industry. In the USA, industry funding of university research in 2005 was below 6% of the total academic research budget; in Israel, in contrast, the percentage of the research budget that universities (those with available data) received from industry was much higher, significantly above 10%. (There was one exception: in 2000, the Technion received 22.7% of its total research budget from industry.) What is particularly disheartening is that US government support of research at MIT reached 89% of that institution's research budget in 2005, whereas the Technion received only 40.5% of its total research budget in 2005 from governmental sources. Stanford and Columbia universities also enjoyed sizable governmental research support, 87% and 86%, respectively, of their research budgets.

Table 6.4 Research budgets of selected research universities in Israel and the USA by source of funding, 1995, 2000 and 2005(in percentage)

Institute	Financial Resources Percentage of Total[1]: 1995		2000		2005	
	Government	Industry	Government	Industry	Government	Industry
Israeli Universities						
Weitzmann Institute of Science	N/A	N/A	N/A	N/A	N/A	N/A
Technion-Israel Institute of Technology	48.2	19.2	42.5	22.7	40.5	13.0
Ben-Gurion University of the Negev	23.5	8.0	25.8	13.1	23.1	9.9
Bar-Ilan University	N/A	N/A	N/A	N/A	N/A	N/A
Hebrew University, Jerusalem	27.0*	9.0*	20.3	11.0	N/A	N/A
Tel Aviv University	N/A	N/A	32.2	12.0	N/A	N/A
US Universities						
Cornell	66.2	4.6	97.3	2.8	67.9	4.1
Michigan	68.0	7.1	68.4	7.2	73.5	4.5
MIT	74.7	16.6	82.0	10.2	89.1	5.7
Stanford	84.7	4.6	83.6	9.5	87.0	5.2
Columbia	85.4	4.2	88.1	3.5	85.8	3.2
Harvard	70.9	4.1	68.9	4.4	72.2	2.6

Notes: 1. The percentages do not add up to 100%, since there are other financial sources, such as public institutes, foundations, etc.
* The data refer to 1997.
Sources: Israel – University websites or data received following a request to the universities' research authorities. USA – AUTM reports 2001, 2005. Data sources for Stanford and Colombia for 2005 are the respective universities' websites

Unfortunately, the data concerning new patent registrations for the Israeli universities are incomplete. Nevertheless, as can be seen from Table 6.5, the number of new patents registered by US universities reached as high as 152 at MIT in 2000 and 133 in 2005. Stanford and Michigan universities also showed a high number of new patent registrations in 2005 – 84 and 80, respectively. For the Israeli universities, the number is much smaller although the Weitzman Institute registered 62 new patents in 2005. When the number of yielding patents (Table 6.5) is reviewed, MIT and Stanford University are seen to have a relatively high number: 436 and 428, respectively, in 2005. As far as patent revenue (royalties) is concerned, Table 6.6 shows that the amount earned at the US universities came to 384 million dollars at Stanford and 160 million dollars at Columbia in 2005. A great deal of data for the Israeli universities is missing, and what is available is very disappointing. The Technion reports royalty revenue of 8 million dollars in

2005, while Tel-Aviv and Ben Guri on universities had less than 1 million dollars each. The data on revenue from patents per senior faculty in each of the institutions revealed similar stories.

Table 6.5 New patent registrations and patent yields in selected research universities in Israel and the USA, 1995, 2000 and 2005

Institute	Number of Patents New in that Year			Yield Patent		
	1995	2000	2005	1995	2000	2005
Israeli Universities						
Weitzmann Institute of Science	N/A	N/A	62	N/A	N/A	N/A
Technion-Israel Institute of Technology	N/A	22	35	N/A	N/A	15*
Ben-Gurion University of the Negev	11	11	43	N/A	12	15
Bar-Ilan University	N/A	N/A	N/A	N/A	N/A	N/A
Hebrew University, Jerusalem	58**	59	N/A	N/A	N/A	N/A
Tel Aviv University	21	39	41	N/A	N/A	N/A
US Universities						
Cornell	N/A	68	51	N/A	188	196
Michigan	33	77	80	69	111	214
MIT	96	152	133	183	362	436
Stanford	70	98	84	220	378	428
Columbia	25	78	72	150	143	130***
Harvard	18	56	44	149	163	276

Notes:
* Data refer to 2007.
** Data refer to 1997.
*** Data refer to 2004.
[1] In 2007, the Technion reported 75 new patents.
Sources: Israel – the respective universities' websites or data received following request made to the universities' research authorities. USA – AUTM reports 2001, 2005. Data sources for Stanford and Colombia for 2005 are the respective universities' websites

Table 6.7 shows new start-ups in Israeli and US universities. Here, too, a significant amount of data concerning the Israeli institutions is missing. For the US universities, it may be observed that MIT established 29 start-ups in 2001 and twenty in 2005. Stanford and Columbia established 12 and 10 start-ups, in 2005 respectively. The figures for the Israeli universities are three each at the Weitzmann Institute and the Technion, and two at Ben-Gurion University in 2005.

Table 6.8 shows the total number of employees in Technology Transfer Offices at the US universities, along with the number of new patents, the number of yielding patents, and patent revenue. We ranked the numbers within each column. The very small number of observations precludes carrying out any meaningful statistical analysis. Therefore, we are able to make only some general comments. The ranking of the first and last column are exactly the s ame. That is, there is a perfect correlation between the rank order of number of TTO employees and patent revenues. Stanford and Columbia universities are prominent in this regard, with 384 million dollars and 160 million dollars, respectively.

Table 6.6 Patent revenue (royalties) earned by selected research universities in Israel and the USA, 1999, 2000 and 2005

Institute	Patents Revenue (Royalties) in millions of $			Royalties Per Faculty Member		
	1995	2000	2005	1995	2000	2005
Israeli Universities						
Weitzmann Institute of Science	N/A	N/A	N/A	N/A	N/A	N/A
Technion-Israel Institute of Technology	0.21	0.24	8.00	N/A	364	12,539
Ben-Gurion University of the Negev	N/A	0.21	0.61	N/A	286	766
Bar-Ilan University	N/A	N/A	N/A	N/A	N/A	N/A
Hebrew University, Jerusalem	N/A	N/A	N/A	N/A	N/A	N/A
Tel Aviv University[1]	N/A	0.61	0.89	N/A	523	937
US Universities						
Cornell	N/A	5.95	3.90	N/A	3,914	2,447
Michigan	1.20	3.97	17.00	445	1,471	5,880
MIT	4.80	31.48	40.00	5,156	33,813	40,080
Stanford	38.90	36.94	384.00	32,069	30,453	300,000
Columbia	34.20	149.00	160.00	39,895	173,862	182,648
Harvard	6.82	16.54	28.00	5,617	13,602	19,035

Note: [1] In 2007, the total revenue increased to $5.6 million because of new medical products.
Sources: Israel – the respective universities' websites or data received following a request to the universities' research authorities. USA – AUTM reports 2001, 2005. Data sources for Stanford and Colombia for 2005 are these universities' websites.

As far as the number of employees in a TTO is concerned, it is apparent that the size of the staff in US universities is significantly larger than in Israeli universities (see Figure 6.1). Among the US universities, Stanford and Columbia each enjoy a large staff of 35 employees, while Harvard and MIT have 29 employees each. In the Israeli universities, the TTO staffs are much smaller – 13 at the Hebrew University and only five at the Technion.

Table 6.7 Number of new start-ups in selected research universities in Israel and the USA, 2001–2005

Institute	Number of Start Up					
	2001	2002	2003	2004	2005	Average
Israeli Universities						
Weitzmann Institute of Science	N/A	N/A	N/A	N/A	3	
Technion -Israel Institute of Technology	N/A	N/A	N/A	N/A	3	
Ben-Gurion University of the Negev	2	2	3	2	2	2.2
Bar-Ilan University	N/A	N/A	N/A	N/A	N/A	
Hebrew University, Jerusalem	N/A	N/A	N/A	N/A	N/A	
Tel Aviv University	N/A	N/A	N/A	N/A	N/A	
US Universities						
Cornell	1	4	13	6	5	5.8
Michigan	12	5	9	13	7	9.2
MIT	29	23	15	20	20	21.4
Stanford	6	13	12	9	12	10.4
Columbia	7	8	9	9	10	8.6
Harvard	5	7	4	4	7	5.4

Note: [1] In 2007, there were six new start-ups.
Sources: Israel – the respective universities' websites or data received following a request to the universities' research authorities. USA – AUTM reports 2001-2005. Data sources for Stanford and Colombia for 2005 are these universities' websites

The Universities' Technology Transfer Policies

Extensive personal interviews were held with senior academic officials in Israel, such as VPs for research and heads of the universities' technology transfer offices (TTO). The information on university guidelines dealing with reciprocity between academia and industry that was collected in Israel and in the USA (received via mail and reviewed from websites) was sorted, categorised and analysed. The information included conflicts of interest, publications, inventions and innovations, patent registration, division profits and royalties, as well as the use of university facilities (laboratories, etc.).

Table 6.8 Total number of employees in technology transfer offices, new patents, patent yields, and patent revenue at the US universities (rating in parentheses)

Institution	Number of TTO Staff	New Patents	Patent Yields	Patent Revenue in Million Dollars
Stanford	35 (1)	84 (2)	428 (2)	384.0 (1)
Columbia	35 (2)	72 (4)	130 (6)	160.0 (2)
Harvard	29 (3)	44 (6)	276 (3)	28.0 (3)
MIT	29 (4)	133 (1)	436 (1)	40.0 (4)
Michigan	26 (5)	80 (3)	214 (4)	17.0 (5)
Cornell	26 (6)	51 (5)	196 (5)	3.9 (6)

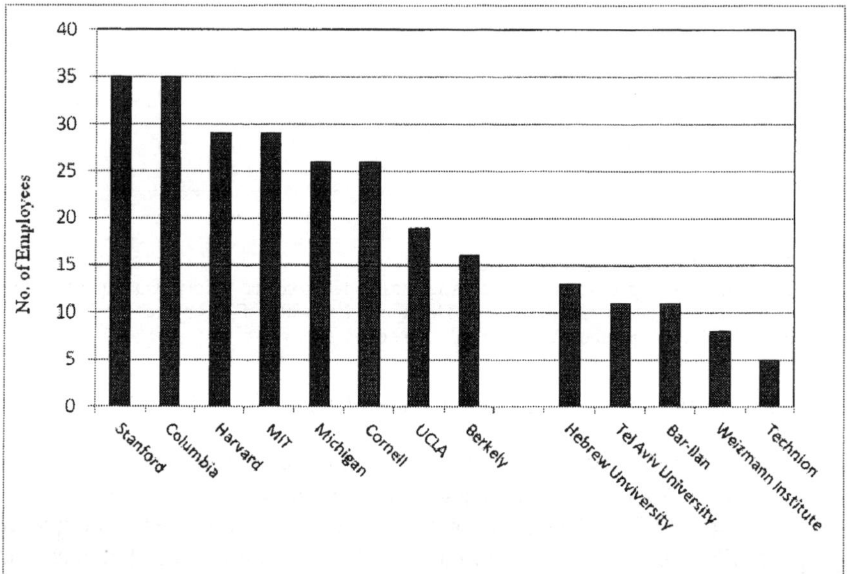

Figure 6.2 Number of employees in technology transfer offices in universities in the US and Israel, 2006

Furthermore, the extent of research funding, sources of funding for research, and patent registration and commercialisation in universities were examined.

Most Israeli and US universities cherish the academic freedom of all faculty members, valuing it over any economic benefit that may ensue. It is for this very reason that these institutions protect the faculty's right to publish. Three of the universities did impose some restrictions on publication: the University of Michigan in the USA and the Weitzmann Institute and Ben-Gurion University of the Negev in Israel. Most often, these restrictions required faculty to report to, and receive clearing from, the institution's administration prior to the publication of a new discovery.

The funding of research at a university by commercial enterprises in which a member of the faculty has a personal or an economic interest could create a conflict of interest or commitment, having to choose between one's responsibilities to the university or to the economic enterprise. Thus, MIT forbids receiving research funding from a company in which the researcher has a vested interest (shares).

In Israel, the Weitzmann Institute, Ben-Gurion University and the Hebrew University of Jerusalem do not object to such funding, provided that the institution's authorities approve it beforehand. It is conceivable that this lenient attitude derives from a desperate desire by Israeli institutions to enlarge their financial base, as well as an assumption that the university's control over such permission acts as a barrier to any possible conflict of interest and/or commitment.

Gaining ownership of discoveries made by researchers at a university is one of the most loaded issues in academia. One contentious issue relates to discoveries that emanate from research that was funded by government or public sources: Who has ownership, the university or the funding agency? In the United States, this delicate issue was resolved with the Bayh-Dole Act of 1984, which permits US universities to register patents that are derived from research funded by government offices. As a consequence of this new act, the number of registered patents climbed appreciably. It is not clear, though, whether the legislation significantly increased the rate of technology transfer between universities and industry.

At Israeli universities, inventions derived from public funding are registered as patents of the university. Nevertheless, there are no clear, unequivocal rules governing this particular issue.

Another sensitive matter concerns patent registration and intellectual property rights (IPR). This dispute here between university authorities and faculty members is generated by the dissatisfaction of researchers, who strongly believe that the invention or discovery is their sole intellectual property right, albeit created while conducting research at the university.

All the universities surveyed maintain a formula by which royalties received from commercialised patents and new discoveries are proportioned between the institution and the faculty member. The most generous university in Israel is Ben-Gurion University, which rewards the inventor with 60% of the benefit received. The Technion and the University of Michigan split the royalties with the researcher, while the Hebrew University,

MIT, the University of California and Stanford University allot only 33%–35% of the benefits received.

In four of the six Israelis universities surveyed, some funds were transferred to the TTO over and above the amounts needed to cover the expenditures associated with patent registration, etc. As mentioned, the allocation of royalties between the university and the faculty member has emerged as a central policy issue in the academic discourse in Israel concerning new initiatives to promote technology transfer in particular and collaboration in general between academia and industry.

Faculty Attitudes

In order to identify faculty attitudes toward collaborative research with industry, a survey of faculty members was undertaken at three of the Israeli universities: Technion, Hebrew University and Tel-Aviv University. Faculty attitudes were assessed by means of a questionnaire designed specifically for this purpose. A total of 159 researchers participated (76% of the interviewed personnel were at the rank of full or associate professor). Most interviews were conducted face-to-face, and a few were completed through a mailed-in survey (Shefer and Frenkel, 2003b).

The survey included a group of researchers engaged in collaborative research with industry (53%) and a control group of faculty involved only in basic research and who do not collaborate with industry (47%). In this way it was possible to identify both obstacles to and motivations for collaborating with industry. Based on survey answers, we were able to learn whether non-collaboration reflects researchers' preference or whether such a choice is the result of administrative barriers, technological difficulties, and the like.

SURVEY FINDINGS

Willingness to Collaborate with Industry

A total of 57% of the faculty members interviewed noted that they were interested in collaborating with industry; 14% more answered that they had no objections in principle to collaborating. However, they conditioned the extent of collaboration with industry on the type of research, its scope, contribution and connection with other research projects in which they were presently engaged. Researchers at the Technion, it seems, demonstrate a greater willingness to collaborate with industry than do those at any other Israeli institution. Among the researchers interested in applied research, more than half motioned that they would not initiate a connection, but would respond positively if approached by industry. This finding indicates that a relatively large unexploited potential exists that could be developed if the universities, through their applied research departments, took steps toward identifying relevant, interested industries. Reliance on faculty members to initiate contact appears to contribute to the failure in collaboration with industry and to a loss of more than a third of the existing potential among researchers currently engaged in basic research, but willing to conduct applied research.

Analysis of the answers of faculty members who are not interested in conducting applied research for the benefit of industry reveals that their primary objection stems either from concern that a connection with industry would change the character of their research (Technion) or from the fact that their academic interest is solely in basic research (Hebrew and Tel-Aviv universities).

University and its Institutional Support of Collaboration with Industry

Based on the survey, it was found that 90% of all Israeli faculty members interviewed believed that it was the role of the academic institution to encourage technology transfer from academia to industry. Many went on to say that the transfer of knowledge and collaboration constituted an integral part of the university's role and societal objectives. In spite of this belief, however, many faculty members believe that the academic institution at which they work does not encourage collaboration with industry.

Despite the consensus that the university should encourage collaboration with industry, half of the faculty members think there is no room for formal university recognition of a researcher's collaborative work with industry as part of the process of academic promotion (although this kind of recognition already partially exists). In addition, a considerable number of faculty members who answered that this type of activity should be taken into account, conditioned their statement on the type of activity and the field of research.

Importance of the Contribution of Collaboration

There is no doubt that collaboration between academia and industry contributes positively to all parties involved. The survey tested the importance of the contribution of collaboration in several subject areas.

The exposure of researchers to commercial market activities, product development and applied processes were found to be of the greatest importance. The significance of these subjects for faculty members was evident in answers given to additional questions on whether a connection with industry changes the character of academic research as a result of exposure to new ideas and new areas or from a faculty member's interest in receiving funding. About half of those interviewed noted that this latter attitude existed and that they were acquainted with faculty members who behaved in this manner. Many of them answered that they, too, had engaged in such activities. The data here concur with findings by Siegel et al. (1999) showing that about 65% of the academic faculty members they interviewed admitted that a connection with industry influenced the character of their basic research.

These answers show that academic researchers are open to commercial market activities and are willing to adopt new ideas. Many are aware that such collaboration could bring about a conflict of interest in the event that academic researchers felt compelled to change the direction and character of their research in order to obtain funding from industry. In the extreme case, research conducted for the benefit of industry is even thought to contradict the principle of research undertaken for the sake of research.

Expected Contributions of a Variety of Methods Intended to Strengthen the Relationship Between Academia and Industry

Collaboration between academia and industry contributes to a wide variety of subjects. To enable the preservation and expansion of this contribution, it is necessary to actively strengthen the connection between faculty and industry. In order to test this hypothesis, the faculty members interviewed were presented with a series of different methods that could help strengthen ties with industry and asked to rank their importance. An analysis of the answers found that most of the methods presented were categorised as making from an average to a large contribution to strengthening ties. The method considered to make the greatest contribution was extending the flexibility of university policies. The next most important was awarding a greater percentage of royalties to the researcher. Granting copyright and ownership rights of patents solely to the researcher and the free use of university facilities received the lowest grades.

The distribution of answers among researchers engaged in basic research, compared with those involved in applied research as well as for the benefit to industry, showed significant statistical differences. These differences relate to the expected contribution of strengthening ties between academia and industry as a result of increasing the flexibility of university policies in regard to collaboration with industry and to a wider scope of activities within the partnership.

One of industry's primary motivations for collaboration with academia is financial saving on research and necessary infrastructure. Therefore, it is reasonable to assume that most of the initiatives for this type of connection would come from industry. This assumption is supported by the survey results, which showed that in most cases industry plays a full or partial role in initiating the research. Close to 40% of the 133 research projects that were acknowledged in the survey were initiated exclusively by industry, and an additional 20% by both academic researchers and industry. Only about one third of the initiatives came from academia.

Most research that was either initiated by industry or conducted for the benefit of industry has clearly applied objectives, such as product development, process development, and the like. Many of the researchers who conduct research for industry integrate this activity with their current research interests, such that it often combines the applied side with the basic. The survey shows that three of the primary objectives of research conducted for the benefit of industry are basic research (47%), product development (39%) and process improvement (28%). These objectives were indicated as primary at all three Israeli universities surveyed.

Collaborative research between academia and industry is aimed especially at assisting industry in problem-solving. From the answers received on the questionnaire, it was found that, in addition to problem-solving, many of the research projects included basic research at some level. On average, about 50% of the research conducted for the benefit of the companies for which we received data culminated in a scientific article, and an additional 30% concluded with the development of a new process. Only 6% of the research projects led to the establishment of a start-up company.

GOVERNMENT PROGRAMMES IN ISRAEL

The Magnet Programme

The mandate of the office of the Chief Scientist of Israel's Ministry of Industry, Trade and Labour is to implement public policy and support industrial R&D. The Chief Scientist's objective to develop Israel's technology is a means of encouraging technological innovation and realising the country's scientific potential, increasing the knowledge-base of its industry, promoting high-level value-added R&D, and encouraging cooperation in R&D both at the national and international levels. The Law for Encouragement of Industrial Research and Development (1984) is the primary mechanism for providing government assistance to industrial R&D. Its objective is the advancement of innovative scientific industrial development through the utilisation and expansion of existing technological and academic infrastructure in order to increase production and exports. The activities of the Office of the Chief Scientist include direct funding of industrial R&D through the R&D foundation and support of generic R&D through special programmes, among them the Programme for the

Development of Technological Infrastructure for Industry – the Magnet programme.

The purpose of the Magnet programme, which began in 1994, is to promote the development of long-term competitive advantages for civilian industries through the establishment of consortiums of corporate businesses and research institutes in fields that are important in the global market and in which Israel already has a comparative advantage. The programme is based on collaboration among industrial companies, as well as between them and research institutes engaged in technological R&D that has yet to reach the competitive market. Government funding consists of up to a 66% exemption from payments of royalties and recognising many more expenditures as grant eligible. The university partner's expenses are fully covered.

In 2002, the Magnet programme's administration conducted an evaluative survey of its activities (ENOSH, 2002) that included companies that had partnered in the consortiums and concluded their activities up to 2001, as well as some of the academic researchers. The primary conclusion was that Magnet fulfilled its objectives with great success; new generic technologies were developed and technologies were purchased from abroad. Some companies, however, did not contribute at all or did not contribute enough through their participation in the programme. An evaluation showed that collaboration between companies was good, but with academia it was limited. Although it may not have satisfied all expectations, Magnet was still deemed beneficial. Statistical data collected from the companies that had filled out questionnaires showed that research groups established as a result of a consortium employed more than 225 people, published more than 63 articles, submitted more than 30 patent applications, developed more than 88 products, along with new processes and services at various levels, and established or were planning more than 21 new companies.

Survey responses regarding collaboration with academia were very mixed. Very productive collaboration was reported, sometimes after initial struggles, in some of the consortiums. In others (especially in the communications field), however, industrial personnel made it clear that collaboration results were not good. The roles filled by academic personnel were very diverse. In some cases, they transferred a great deal of advanced, innovative knowledge to industry and served as prime technological catalysts. In other cases, they served as a kind of "sub-contractor" who performed specific tasks for the consortium. In extreme cases, they accepted funding and did as they wished. In general, the evaluation determined that collaboration with academia, including cases where there was no concrete collaboration within the consortium, had many positive outcomes. The inclusion of academia in Magnet consortiums achieved better personal acquaintance between industrial and academic personnel, leading to immediate assistance and hopes for further assistance in the future. In addition, the engagement of academic personnel with industry helped in the establishment of start-up companies and the training of engineers and graduate students in fields new to academia – or what might be called entrepreneurship education (Shefer and Frenkel, 2003b).

The Magneton (Mini-Magnet) Programme

The "Magneton" programme began operating in 2000 as part of the expansion of the Magnet programme. Its purpose is to form a research and development plan that will confirm the ability of the consortium to convert scientific ideas into products or technologies for industry. "Magneton" is designed to encourage the undertaking of activities that do not occur under other different programmes and to bring about wider utilisation of the technological capabilities present in academic research institutes in Israel for the benefit of Israeli industry. Designed to augment the benefits of academic research by advancing its application, the programme promotes cooperation between Israeli industrial corporations and research groups or research institutes – in order to validate technological feasibilities. This stage concludes with the industrial company making a commitment of entering the process leading to the development of a new industrial product. The "Magneton" programme was meant to expand Israeli industry's access to scientific research attainments in research institutes in Israel for the benefit of the Israeli market. The programme's overall purpose is to encourage a process that will contribute to technological innovation, which is the primary growth engine of industry and of the Israeli economy as a whole.

The "Magneton" programme received 280 funding applications for projects up to the year 2000. Of those, 150 projects were approved and 130 refused. In August 2008, there were 24 active projects in the programme and 16 additional projects that had been approved but not begun operating. Recently the Samuel Neaman Institute of the Technion-Israel Institute of Technology conducted an evaluative study of the Magneton project in which top researchers in academia and managers of approved projects were interviewed (Getz et al., 2009). Primary findings showed that the "Magneton" programme was very successful, highly valued, and greatly appreciated by both academic and industrial participants (see also Barda, 2001). The programme succeeds in creating a supportive environment and structure for advancing cooperation between academia and industry; projects make significant contributions to both academia and industry, and importantly to the process of transferring technologies between them. Most "Magneton" projects are considered highly to very highly innovative, and the knowledge developed within them can be considered cutting-edge technologies, whether as new knowledge or as improving on existing knowledge.

Most project participants both from academia and from industry assessed the "Magneton" projects in which they participated to be successful. The highest level of successes was found in projects with a technological classification of electronics, optoelectronics and communications that together made up the highest percentage of projects in the programme (31%). The direct results of the projects, as indicated by research participants, were primarily as follows: technological development, new processes or products, development of prototypes, publication of articles, patent registration, participation in seminars/conferences and continued research. In addition, most of the academic researchers indicated that the collaboration resulted in research work for graduate students.

University Incubator

Shefer and Frenkel (2003a) evaluated the Israeli Government Technological Incubator Programme ten years after its establishment. Among the 24 public incubators that were in operation at the time, three were regarded as university incubators: the Technion Entrepreneurial Incubator Co. Ltd., connected to the Technion-Israel Institute of Technology; the Weitzmann Entrepreneurial Incubator Co., connected to the Weitzmann Institute in Rehovot; and the more recently established biotechnology incubator of the Hebrew University of Jerusalem (created in 2004). All three are classified as specialised incubators, since more than 75% of their projects belong to no more than three fields of activity. The Technion incubator specialises in medical equipment and chemicals and raw materials; the Weitzmann incubator in the fields of drugs, chemical and raw materials and in mechanical engineering (Shefer and Frenkel, 2003a); and The Hebrew University incubator in biotechnology products, such as pharmaceuticals (for more on university incubators see: Hannon and Chaplin, 2003; Soetanto and van Geenhuizen, 2007).

CONCLUSION

Discoveries and inventions constitute a *sine qua non* for the existence of research universities. Such discoveries provide vivid testimony of the creation of knowledge and, eventually, of the transfer of knowledge that could be commercialised.

Closer integration of the universities in the national innovation system is in the interest of universities, since such a move enhances teaching and research. It also demonstrates the universities' contribution to the economic welfare of the state, thus justifying continuous public support for these institutions.

The primary goal of this chapter was to compare parameters and programmes that are likely to affect innovation in selected research universities in Israel and the USA. We investigated inputs, such as the number of graduate students, number of senior faculty, sources of research funding and number of TTO staff. Policies maintained in these areas by universities are very likely to affect the academic staff's desire to engage in applied research.

Our findings show significant differences between universities in the USA and Israel, as well as noticeable variations among universities in the USA and in Israel in regard to such policies.

A comparison of the policies and the rules and regulations of selected USA and Israeli universities that were surveyed revealed no significant differences between the two countries in regard to the specific subjects investigated. The differences that do exist are university-specific and relate to the issue of how a university treats a certain subject. In general, it can be concluded that all universities deal with similar subjects and try to resolve problems that arise from a conflict of interest or a conflict of commitment. Additionally, all universities are concerned with issues relating to intellectual property rights (IPR), registered patents, distribution of royalties, and the use of a university's infrastructure, such as its laboratories, etc.

The universities in Israel and in the USA are cognisant of the "conflict of interest" that is likely to emerge owing to the collaboration between researchers from academia and those from industry and other commercial enterprises. Nevertheless, all universities, without exception, encourage such collaboration. Future research should also focus on the new rise of Asian giants, and particularly on the university–industry cooperation in China and India.

As a result of the vast changes in recent years, we are witness to a new approach that aims to formulate better, more efficient means by which to achieve the university's goals, on the one hand, and provide an answer to the new conditions that have evolved, on the other.

The potential gains for society in general and for the economy in particular through strengthening the ties between academia and industry are vast. Furthermore, this reciprocity may offer an important source for funding research projects, as well as another source of income for university budgets, which are being depleted.

The opinion of Israeli faculty members and the extent of their links with industry were examined through an evaluation of data collated in a field survey carried out at the Hebrew University, the Technion, and Tel-Aviv University. More than 150 faculty members participated in this survey.

The answers show that academic researchers are open to commercial market activities and are willing to adopt new ideas from it. Furthermore many researchers believe that the transfer of knowledge and collaboration are an integral part of the university's role and social responsibility. Despite the consensus of most interviewed faculty that the university should encourage collaboration with industry, half of the faculty members think that there is no room for formal recognition of a researcher's collaborative work with industry as part of the process of academic advancement. In addition, a considerable number of faculty members who believe that it is necessary to account for this type of activity, stated that it must be conditioned on the type of activity and on the field of research. Many of the faculty are aware that such collaboration could bring a conflict of interest in the event that academic researchers feel compelled a change the direction and character of their research in order to obtain budgetary allowances from industry.

From analysis of the answers provided it is found that the greatest contribution of strengthening ties between academia and industry could be gained by extending the flexibility of university policies. The next most important issue is awarding a greater percent of royalties to the researcher.

These viewpoints stemming from this survey enable us to better understand the existing links among faculty members, university institutions and industry, and to design new ways and policies that will enhance future collaboration.

NOTE

1. Patent registration is not the best measure of commercialisation success. It is well known that not all patents result in commercialisation. In addition many commercialised innovations are not registered as patents. Nevertheless, patent registration is the most commonly used measure of innovation success frequently employed by innovation researchers. In the current study we use patent registration as a surrogate for research and development outputs

REFERENCES

Acs, Z.J., Audretsch, D.B., Feldman, M.P. (1991), Real Effect of Academic Research: Comment, *The American Economic Review*, 81(1), 363–367.

Aghion, P. and Howitt, P. (1998), *Endogenous Growth Theory*, Cambridge, MA: The MIT Press.

Allan, M.F. (2001), A Review of Best Practices in University Technology Licensing Offices, *Journal of the AUTM*, 13, 57–69.

Allen D.N. and McCluskey R. (1990), Structure, Policy, Services, and Performance in the Business Incubator Industry, *Entrepreneurship Theory and Practice*, (Winter), 61–77.

Arrow J.K. (2000), Knowledge as a Factor of Production, Annual World Bank Conference on Economic Development 1999. Washington, DC: World Bank.

Barda, S. (2001), "Two for Tango" The Magneton Project as a Case-study of Academy–Industry Collaboration. Shaine Working Papers no.7, Faculty of Social Sciences, The Hebrew University (In Hebrew).

Barro, R.J. and Sala-i-Martin, X. (1995), *Economic Growth*, New York, McGraw Hill.

Bercovitz, J. and Feldman, M. (2006), Entreprenerial Universities and Technology Transfer: A Conceptual Framework for Understanding Knowledge-Based Economic Development, *Journal of Technology Transfer*, 31, 175–188.

Bok, D. (2003) *Universities in the Workplace: Commercialisation of Higher Education*, Princeton, NJ: Princeton University Press.

Boni, A. and Emerson, T. (2005), An Integrated Model of University Technology Commercialization and Entrepreneurship Education, University Entrepreneurship and Technology Transfer. Edited by G. Libecap. Elsevier Press, 16, 241–274.

Branscomb, L.M., Kodama, F. and Florida, R. (1999), *Industrializing Knowledge: University–Industry Linkages in Japan and the United States*, Cambridge, MA: The MIT Press.

Chukumba, C. and Jensen R., (2005), University Invention Entrepreneurship and Start-ups, NBER WP 11475. Cambridge, MA. http://www.nber.org/papers/w11475

COGR (1999), Council on Governmental Relations- The Bayh-Dole Act, http://www.cogr.edu/

D'Este P. and Patel P. (2007), University–Industry Linkages in the UK: What are the Factors Underlying the Variety of interactions with Industries? *Research Policy*, 36, 1925–1313.

Dosi, G. (1988), Sources, Procedures and Microeconomic effects of Innovation, *Journal of Economic Literature*, 26(1), 120–171.

ENOSH Systems (2002), *Evaluation of the Magnet Program*, final report submitted to the Magnet administration in the office for industrial research and development, Israel Ministry of Industry and Trade (unpublished report in Hebrew).

Etzkowitz, H. (2002), Incubation of incubators: innovation as a triple helix of university–industry–government networks, *Science and Public Policy*, 29(2), 115–128.

Feldman, M.P. and Link A.N. (2001), *Innovation Policy in the Knowledge-Based Economy,* Boston: Kluwer Academic Publishers.

Feller, I. (1999), The American University System as a Performer of Basic and Applied Research, in Branscomb, L.M., Kodama, F. and Florida R. (eds.), *Industrializing Technology: University–Industry Linkages in Japan and the United States*, Cambridge, MA: The MIT Press, pp. 66–101.

Felsenstein, D. (1994), University-Related Science Parks - 'Seedbeds' or 'Enclaves' of Innovation? *Technovation*, 14(2), 93–110.

Florida, R. and Cohen, W.M. (1999), Engine or Infrastructure? The University Role in Economic Development, in Branscomb, L.M., Kodama, F. and Florida, R. (eds), *Industrializing Technology: University–Industry Linkages in Japan and the United States*, Cambridge, MA: The MIT Press, pp. 589–610.

Getz D., Segal, V. and Leck, E. (2009), *Evaluation of the Magneton Program*, The S. Neaman, Technion, Haifa, Israel (unpublished report in Hebrew).

Griliches, Z. (1979), Issues in Assessing the Contribution of Research and Development to Productivity Growth. *Bell Journal of Economics*, 10(1), 92–116.

Grossman, G.M. and Helpman, E., (1990) Trade, Innovation and Growth, *American Economic Review*, 80(2), 86–91.

Grossman, G.M. and Helpman, E. (1991), *Innovation and Growth in the Global Economy*, Cambridge, MA: MIT Press.

Grossman, G.M. and Helpman, E. (1994), Endogenous Innovation in the Theory of Growth, *Journal of Economic Perspectives*, 8(1), 23–44.

Hall, B.H., Link, A.N. and Scott, J.T. (2003), Universities as Research Partners, *The Review of Economics and Statistics*, 85(2) 485–491.

Hannon, D.P., and Chaplin, P. (2003), Are Incubators Good for Business? Understanding Incubation Practice–The Challenges for Policy, *Environment and Planning C: Government and Policy*, 21, 861–881.

Henderson, R., Jaffe, A.B. and Trajtenberg, M. (1998), Universities as a Source of Commercial Technology: A Detailed Analysis of University Patenting, 1965–1988, *The Review of Economics and Statistics*, 80, 119–127.

Jaffe, A. (1989), Real Affects of Academic Research, *American Economic Review*, 79(5), 957–970.

Jamison, D.W. and Jansen, C. (2000), Technology Transfer and Economic Growth, *The Journal of the Association of University Technology Managers*, 12, 23–46.

Journal of Technology Transfer (2001), Symposium on Technology Organizational Issues in University-Industry Technology Transfer, 26(1).

Krugman, P. (1979), A Model of Innovation, Technology Transfer and Trade. *Journal of Political Economy*, 87(April), 253–266.

Krugman, P. (1991), *Geography and Trade*, Cambridge, MA: MIT Press,.

Krugman, P. (1995), *Development, Geography and Economic Theory*, Cambridge, MA: MIT Press.

Lach, S. and Schankerman, M. (2003), Incentives and Invention in Universities, NBER WP 9727, Cambridge, MA. http://www.nber.org/papers/w9727.

Lach, S. and Schankerman, M. (2004), Royalty Sharing and Technology Licensing in Universities, *Journal of European Economic Association*, 2(2–3), 252–264.

Lach, S. and Schankerman, M. (2008), Incentives and Invention in Universities, *Rand Journal of Economics*, 39(2), 403–433.

Libecap, G.D. (ed.) (2005), *University Entrepreneurship and Technology Transfer*, (Volume 16), Stamford: Elsevier Science/JAI Press.

Loof, H. and Broström, A. (2008). Does Knowledge Diffusion between University and Industry Increase Innovativeness?, The Journal of Technology Transfer, 33(1), 73–90.

Lucas, R.E. (1988), On the Mechanics of Economic Development, Journal of Monetary Economics, Vol. 22, 3–42.

Management Science (2002) Special Issue on: University Entrepreneurship and Technology Transfer, 48(1).

Mowery, D.C. and Shane, S. (2002), Introduction to the Special Issue on University Entrepreneurship and Technology Transfer, *Management Science*, 48(1), v–ix.

Mowery, D.C. and Sampat, B.N. (2004), The Bayh-Dole Act of 1980 and University–Industry Technology Transfer: A Model for Other OECD Governments? *Journal of Technology Transfer*, 30(1–2), 115–127.

Mowery, D.C. (2005), The Bayh-Dole Act and Entrepreneurship in the US Universities: Chicken, Egg or Something Else? *Advances in the Study of Entrepreneurship, Innovation & Economic Growth*, 16, 39–68.

Mowery, D.C., Nelson, R.R., Sampat, B.N. and Ziedonis, A.A. (1999), The Effects of the Bayh-Dole Act on U.S. University Research and Technology Transfer, in

Branscomb, L.M., Kodama, F. and Florida, R. (eds), *Industrializing Knowledge: University–Industry Linkages in Japan and the United States*, Cambridge, MA: MIT Press, pp. 269–306.

Mowery, D.C., Nelson, R.R., Sampat, B.N. and Ziedonis, A.A. (2004), *Ivory Tower and Industrial Innovation: University–Industry Technology Transfer Before and After the Bayh-Dole Act in the United States*, Stanford: Stanford Business Books.

Nelson, R.R. (2001), Observations on the Post-Bayh-Dole Rise in University Patenting, *Innovation Policy in the Knowledge-Based Economy*, Boston: Kluwer Academic Publishers, pp. 165–170.

Roberts, B.E. (1991), *Entrepreneurs in High Technology, Lessons from MIT and Beyond*, New York: Oxford University Press.

Rogers, E.M., Yin, J. and Hoffmann, J. (2000), Assessing the Effectiveness of Technology Transfer Offices at US Research Universities, *The Journal of the Association of University Technology Managers*, 12, 47–80.

Romer, P.M. (1986), Increasing Returns and Long-run Growth, *Journal of Political Economy*, 94(5), 1002–1037.

Romer, P.M. (1990), Endogenous Technological Change, *Journal of Political Economy*, 98(2), S71–S102.

Romer, P.M. (1994), The Origins of Endogenous Growth, *Journal of Economic Perspective*, 8(1), 3–22.

Shefer, D., Frenkel, A. (2003). *An Evaluation of the Israeli Technological Incubators Programme and Its Projects, Final Report*, The S. Neaman, Technion, Haifa, Israel.

Shefer, D. and Frenkel, A. (2006), *Analysis and Formulation of Policies for the Transfer of New Technologies from Universities to Industry*, The S. Neaman Institute, Technion, Haifa, Israel (in Hebrew).

Siegel, D., Waldman, D. and Link, A. (1999), Assessing the Impact of Organisational Practices on the Productivity of University Technology Transfer Offices: An Exploratory Study. WP 7256, Cambridge, MA: NBER.

Soetanto P.D. and van Geenhuizen, M. (2007), Technology Incubators and Knowledge Networks: A Rough Set Approach in Comparative Project Analysis, *Environment and Planning B*, 34, 1011–1029.

Stokey, N.L. (1995), R&D and Economic Growth, *Review of Economic Studies*, 62, 469–489.

Swamidass, P. and Vulasa, V. (2009), Why University Inventions Rarely Produce Pncome? Bottlenecks in University Technology Transfer, *The Journal of Technology Transfer*. 34(4), 343–363.

Thursby, J.G. and Thursby C.M., (2005), Pros and Cons of Faculty Participation in Licensing, *Advances in the Study of Entrepreneurship, Innovation & Economic Growth*, 16, 187–210.

Thursby, C.M. (2005), Introducing Technology Entrepreneurship to Graduate Education: An Integrative Approach, University Entrepreneurship and Technology Transfer, G. Libecap, Elsevier Press, 16, 211–240.

Trippl, M. and Todtling, F. (2008), From the Ivory Tower to the Marketplace: Knowledge Organisations in the Development of Biotechnology Clusters, *The Journal of Regional Analysis Policy*, 38(2), 159–175.

PART THREE

Human Capital, Migration and Labour Force

7. The effect of asset and credit constraint on interregional labour migration

Patricio Aroca, Geoffrey J.D. Hewings and Michael Sonis

INTRODUCTION

In the last two decades, many developing countries have implemented the free market model to promote growth and development. In Latin America, the Chilean case is probably the first and the most successful among the countries that decided to let the market allocate the productive factors to produce goods and services. It was with this country in mind that the model proposed in this chapter was developed.

In this context, the mechanism to reallocate workers among the different regional labour markets is migration. In contrast with the 1960s and 1970s, where the focus of migration research in developing countries was concentrated on rural to urban migration (see the seminal work of Harris and Todaro, 1970), our chapter is devoted to study what we refer to as a second stage of migration, the movements of the labour force among regional labour markets, since today the urban population in Latin America averages close to 90 per cent.

In a simple free market model, it is easy to show that under suitable conditions such as perfect competition, perfect information, free mobility of the factors, the market signals will reallocate people efficiently among regions. However, some typical characteristics of developing countries suggest that they have a financial market that is far less developed than those in developed countries. The effect of this limitation is that an important share of the workers is credit constrained by different reasons such as unemployment and lack of wealth to be used as collateral, and so forth. Therefore, in order to show the effect of credit constraints and lack of assets on the efficiency of labour migration to reallocate workers among regional labour market, we need to develop a more general model than those currently in use.

For any time period, we can separate the labour force between those who want to stay where they work and those who want to migrate. Among the workers who decide to migrate, the following distinction can be made:

- Workers who migrate as soon as they make the choice because they can afford the moving cost (MC);
- Workers who have to save money until they can afford the moving cost (MC);
- Workers who never can migrate because they can never afford the moving cost (MC).

These three different types of worker are the objects of this research. Figure 7.1 summarizes this process that distinguishes between the decision time and the realization time of migration.

In this chapter, we will clarify the role of assets and credit constraints in the decision to migrate from an efficiency point of view; we will focus on the realization time of the workers who made the decision to migrate (Movers). We understand the lack of assets and the credit constraint as intervening obstacles (Lee, 1966) in conditioning the market efficiency of migration in developing countries.

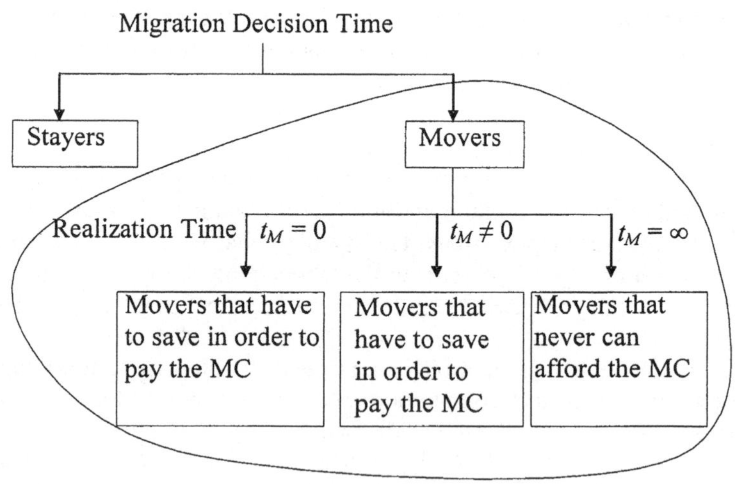

Figure 7.1 The migration decision-making process

In addition, different types of migration will be considered (e.g., speculative and contracted, following Molho, 1986) through analysis of the behaviour of workers facing different employment status, and also different levels of wealth in an imperfect capital market context. While these factors are not unique to developing economies, their manifestation and interaction present particularly important explanatory mechanisms in the determination of migration decision-making. The model will offer insights to explain why unemployment rate differentials among regions are not the equilibrium ones. It will also show that an inefficient migration process could make poor people worse off.

In brief, the model results in a series of assumptions. First, it is assumed that workers make decisions under certainty. Therefore, when they decide to move from region *i* to region *j* it is because they have found a job in region *j* (contracted migration). In addition, they know the salary they will earn. This information allows them to make an evaluation about the profitability of moving. Secondly, the difference between an employed worker and an unemployed worker is that the former has access to credit in the capital market while the latter cannot because of his or her employment status. Finally, when workers make the decision to migrate, they must be able to afford the moving costs. These can be covered by borrowing, by drawing down assets, or by saving an amount each period until they accumulate enough money to pay the moving costs.

On the basis of these assumptions, it will be possible to show that the migration mechanism could be inefficient if the worker cannot afford the moving costs. As a result, selective migration may result, based on wealth or assets, with the more endowed leaving the less prosperous regions, further exacerbating the status of the economy of these regions; in the destination regions, the impacts may be positive since along with the additional wealth, the in-migrants may bring embodied occupational capital that further enhances the competitive position of these regions.

A further extension of the model introduces uncertainty in the job search process in the destination region (speculative migration); as a result, migration for less affluent migrants becomes more difficult than under conditions of perfect information.

This chapter is organized as follows; in the next section, a brief review of some of the received theory is provided. The third section addresses the perspective of the unemployed worker. The employed worker's decision process occupies the fourth section; the role of uncertainty is introduced in the fifth section and a concluding section reviews the findings and offers some summary perspectives.

PREVIOUS ANALYSIS

One of the major differences between developed and developing economy migration would appear to be rooted in the financial ability of migrants to move on the one hand, and the level, quality and quantity of information that circulates about the relative advantages of two or more locations in the context of the choice set. Notwithstanding these differences, the volume of migration in developing economies has continued to rise with a dominating rural–urban flow (see, for example, Becker and Mills, 1986). One of the more important explanatory models was provided by Harris and Todaro (1970); their explanatory mechanism for migration flows was derived from differences in expected earnings between rural and urban areas, with the unemployment rate serving to act as an equilibrium mechanism. This approach has formed the basis of migration decision-making in many computable general equilibrium models. Fields (1975) extended the formulation by considering several other important factors. In particular, he offered a more generalized approach to the job search process; further, he explicitly considered the role of the informal sector as an intermediary between rural and urban employment and he allowed for varying degrees of job turnover. A more recent review by Bhattacharya (1993) has been critical of many of the probabilistic approaches on the one hand, and the treatment of the informal sector on the other hand. In particular, the notion of the informal sector as an intermediate staging post between rural and formal urban employment has been questioned on the basis of empirical evidence pointing to long-term attachment to the informal sector.

One of the problems with a two-sector (agriculture–manufacturing), two-region (urban–rural) approach is that it necessarily simplifies the migration decision-making process by assuming away any impact associated with distance – such as the costs of movement, information flows, the increasing role of uncertainty and the impact outmigration may have on source regions. This phenomenon has been illustrated in the Canadian context by a series of studies by Vanderkamp (1970). In particular, he was able to show the way in which outmigration further undermined the fragility of many less prosperous regional economies as the reduction in circulation of income from out-migrating residents further decreased employment levels and made out-migration even more attractive to the remaining residents. In one estimate for the Atlantic provinces of Canada, Vanderkamp (1970) found that for every five unemployed residents leaving the region a further two employed persons became unemployed.

For developing economies, one of the major issues centres on the ability of residents of rural or less prosperous regions to assemble the necessary resources to migrate to regions perceived to offer more favourable economic prospects. Aroca and Maloney (2005) find for Mexico that relative wage capture substitution effects, and the freestanding initial wage at the origin region (to capture credit constraint effects) are both significant in their estimation. They conclude that the overall strong performance of this variable suggests that the poor results for the wage at origin region variable in many previous studies arise because they capture two contradictory tendencies, with the credit constraint effect especially important in developing countries. In the next section, a formal model is proposed to explore some of these issues in a more explicit fashion.

UNEMPLOYED WORKER DECISION-MAKING

When an unemployed worker makes the decision to move from region *i* to region *j*, we assume that she or he has found a job. The worker has to pay the moving costs (MC), which are mainly transportation costs and housing price differentials between region *i* and region *j*. These costs can be paid with savings from the past or other accumulated assets ($a(0) = a_0$). Otherwise, if the worker does not have enough money to pay the moving costs, then he or she has to save until the necessary amount can be accumulated. We assume that assets and savings earn the same rate of return (*r*); therefore, we will use both as synonymous from now on.

Traditional neoclassical theory assumes that moving costs can be covered by obtaining a loan from the perfect capital market, an obligation that would be retired with proceeds from future income. However, in practice, it is very difficult to find a financial institution that is willing to loan money to an unemployed worker. In this model, the main difference between an employed and an unemployed worker is indicated by their capability to borrow from the capital market. Evidently, if an unemployed worker has the ability to borrow from the credit market, then this person would move as soon as she or he finds a job. Nevertheless, if the worker does not have access to the credit market, then he or she has to spend his or her assets to pay the moving costs. If we assume that the assets are liquid or there is an easy, efficient way to make them liquid, and they are large enough to pay the moving costs, then there is no inefficiency in the working of the market.

The inefficiency arises when the unemployed worker does not have enough saving or assets to pay the moving costs and has to wait until he or she can save the moving cost amount. The inefficiency becomes worse the poorer is the unemployed worker. If the unemployed worker's income is too low, it could be the case that he or she can never move, even though she or he had found a job in other region. It is assumed that there is not an informal capital market from which the unemployed worker could obtain credit.

To illustrate the above argument, we set the migration decision problem of an unemployed worker as a maximization of the intertemporal utility function subject to a constraint where the assets have to be greater or equal to zero. This means that he or she does not have access to the credit market. In addition, the assets are liquid and they increase (or decrease) by the interest paid by the financial institution and also by the differences between current income and current consumption. At this point, we should make clear that we are talking about income instead of just the wage. The reason for this distinction is that we are assuming that the worker could have other sources of income such as transfers from the government. Although we believe that this latter component of income is positive, it is not likely to be greater than the wage that the worker can obtain from a job. On the other hand, the current consumption is not the total consumption in the period. The current consumption is equal to the total consumption minus the minimal consumption which is necessary to survive. Therefore, this minimal consumption and the income necessary to finance it are not in the model because they cancel each other when we set the dynamic constraint. The implication of this assumption is that we can maximize the utility function subject to the condition that current consumption has to be greater or equal to zero.

As a result, the problem can be setup as:

$$\max_{c(t)} \int_0^T e^{-\rho t}\, U[c(t)]dt \tag{7.1}$$

subject to:

$$\begin{aligned} &a(t) = r * a(t) + I(t) - c(t) \quad 0 \le t \le t_M \\ &a(0) = a_0 \quad a(t_{M-}) = a_1 \ge MC \\ &a(t) \ge 0, \quad 0 \le t \le t_M \end{aligned}$$

and:

$$\begin{aligned} &a(t) = r * a(t) + I(t) - c(t) \quad t_M \le t \le T \\ &a(t_M) = a_1 - MC \quad a(T) \ge 0 \\ &a(t) \ge 0, \quad t_M \le t \le T \end{aligned}$$

Where $I(t)$ is given by:

$$I(t) = \begin{cases} I_0, & 0 \le t < t_M \\ I_d & t_M \le t \le T \end{cases} \tag{7.2}$$

and I_0 is the worker's income in the current region and I_d is the worker's income in the destination region; we assume $I_d > I_0$. In addition, t_M is the time the move is made, $a(t_{M-}) = a_1$ represents the total assets or savings accumulated just before the moving time. Also $a(t_M)$ represents the assets or savings accumulated that remain after the moving costs have been paid.

The first set of constraints describes the saving process in order to reach the necessary amount of money to pay the moving costs. The second set of constraints describes the worker's situation in the new job in the region *j*. Given the assumption that workers have perfect information, they can evaluate a *priori* the profitability of migration. Therefore, they can find the optimal path to do so, given the constraint set that they face.

Following Chiang (1991) we can solve the maximization problem, setting the following Hamiltonian:

$$L = e^{-\rho t}U[c(t)] + \lambda(t)[r * a(t) + I(t) - c(t)] + \gamma(t) * a(t); \quad 0 \leq t \leq t_M \tag{7.3}$$

The Maximum Principle implies that $c(t)$ and $a(t)$ will be optimal (e.g. $c^*(t)$, $a^*(t)$) if there exist non-negative functions $\lambda(t)$, $\gamma(t)$ such that:

$$\frac{\partial L}{\partial c(t)} = e^{-\rho t}U'[c^*(t)] - \lambda(t) = 0 \tag{7.4}$$

$$\frac{\partial L}{\partial \lambda(t)} = a^*(t) \geq 0; \quad \gamma(t) \geq 0; \quad \frac{\partial L}{\partial \gamma(t)} * \gamma(t) = a^*(t) * \gamma(t) = 0$$

$$\dot{\lambda}(t) = \frac{\partial L}{\partial a(t)} \Rightarrow \dot{\lambda}(t) = -r\lambda(t) - \gamma(t);$$

and

$$\dot{a}(t) = r * a(t) + I(t) - c(t)$$

for all $0 \leq t \leq t_M$, and $U[c(t)]$ is strictly concave (concave is enough, but we use strictly concavity to make future developments easier) with respect to $c(t)$ (see Kamien and Schwartz ,1991, and Leonard and Long, 1992).

To deduce the $c^*(t)$ path, we can evaluate equation (7.4) on any interval $[\tau_1, t]$, on which $a^*(t) > 0$. From the same equation, we have $\lambda(t) = -r\lambda(t)$ if $\gamma(t) = 0$, which may also be obtained from equation.

Therefore, integrating $\lambda(t)/\lambda(t) = -r$ on the interval $[\tau_1, t]$, we obtain the path for $\lambda(t)$, which is shown as equation (7.5):

$$\int_{\tau_1}^{t}\left[\frac{\dot{\lambda}(s)}{\lambda(s)}\right]ds = \int_{\tau_1}^{t} -rds \tag{7.5}$$

$$\Rightarrow \lambda(t) = \lambda(\tau_1)e^{-r(t-\tau_1)}$$

From equation (7.4), we know that $\lambda(\tau_1) = e^{-\rho\tau_1}U'[c(\tau_1)]$; therefore, substituting equation (7.5) into equation (7.4), and replacing $\lambda(\tau_1)$, we have:

$$U'[c(t)] = U'[c(0)] * e^{(\rho-r)^{*}t}, \forall \quad 0 \leq t \leq t_M \tag{7.6}$$

which is the path for $c^*(t)$ in implicit form.

We are assuming that the unemployed worker has an intertemporal discount rate greater than the interest rate, and that their main motivation for saving is the sole objective of reaching the necessary amount to pay the moving costs. Therefore, after moving, we can assume that $a(t_M) = 0$, based on the belief that savings' sole function was to finance the move and thus, after moving, the amount of saving is equal to zero because the unemployed worker tries to move as soon as possible.

The first assumption about the decreasing marginal utility function is a traditional one in microeconomic theory, and we can find support for it in several studies that are cited in applied textbooks. On the other hand, the assumption that the intertemporal discount rate of consumption is greater than the interest rate (r) does not have as much support from the empirical microeconomic literature. For instance, Hurd (1989) compiles two kinds of evidence. First, "Surveys and psychological experiments, in which people are asked to choose between a present and a future reward, typically show very high rates of time preferences" (Fuchs, 1982). Secondly, there is evidence that poor people save less even when they face high rates of interest. Also, this assumption is especially plausible for unemployed workers, who have low incomes, so that current consumption is more valued than future consumption.

Then, if the moving costs are the barrier to migration, she or he will move once the necessary amount has been accumulated. If $a(t)$ is equal to zero we find that:

$$c^*(t) = I_d, \forall \quad 0 \leq t \leq t_M \tag{7.7}$$

To draw the path, we will look for the behaviour of the consumption in the neighbourhood of t_M. Following Lemma 5.3 of Artle and Varaiya (1978), we can show that the consumption during the period immediately before t_M, which is called (t_{M-}), is lower than the consumption at t_M.

We know that $c^*(t_{M-})$ cannot be greater than $c^*(t_M)$ because if this occurs there is no migration. In this case, migration does not occur because $c^*(t) = I_d$ for all $t \geq t_M$. Thus, if $c^*(t_{M-})$ is greater than $c^*(t_M)$, the worker could enjoy this situation until period T (the retirement period). Therefore, $c^*(t_{M-})$ has to be lower or equal to $c^*(t_M)$. Indeed, we show that $c^*(t_{M-})$ is strictly lower than $c^*(t_M)$. We prove this by contradiction; suppose that $c^*(t_{M-}) = c^*(t_M)$.

First, we compare $c^*(t)$ with and alternative profile $c^\varepsilon(t)$ in which (t_{M-}) is extended from (t_{M-}) to [(t_{M-}) $+\varepsilon$] and where [(t_{M-}) $+\varepsilon$] is greater than t_M. It implies that in this new profile the worker stays in the origin region an additional period of length ε compared with the optimal profile. In the interval $[0, (t_{M-})+\varepsilon]$, $c^{\varepsilon}(t)$ is given by:

$$c^{\varepsilon}(t) = \begin{cases} c^{*}(t), & \text{for } 0 \leq t \leq t_{M-} \\ c^{*}(t_{M-}), & \text{for } t_{M-} \leq t \leq t_{M-} + \varepsilon \end{cases} \tag{7.8}$$

This profile is feasible because $a^{*}(t_{M-})=MC$, the income is the same that the one in period (t_{M-}) and in addition she or he has the capital gains $(a(t_{M-})*r*\varepsilon)$.

On the other hand, $\varepsilon\,[a(t_{M-})*r + Ir_0 - c^{*}(t_{M-})] = \Delta^{+} > 0$. If this is not true, the worker could move to another region before (t_{M-}) or he or she could never move to another region. Therefore, $a^{\varepsilon}(t_{M-} + \varepsilon) - MC = \Delta^{+}$.

Hence, we can write the second part of the new profile as:

$$c^{\varepsilon}(t) = \begin{cases} c^{*}(t)+\Delta^{+}, & \textit{for } t_{M-} + \varepsilon \leq t < t_{M-} + 2\varepsilon \\ c^{*}(t), & \textit{for } t \geq t_{M-} + 2\varepsilon \end{cases} \tag{7.9}$$

Finally, we compare the difference in utilities from both profile and we have:

$$\int_{0}^{T} e^{-\rho t}\{U[c^{\varepsilon}(t)] - U[c^{*}(t)]\}dt = \tag{7.10}$$

$$\varepsilon e^{-\rho(t_{M-}+2\varepsilon)}\{U[c^{*}(t_{M-} + 2\varepsilon)] + \Delta^{+} - U[c^{*}(t_{M-} + 2\varepsilon)]\}$$

Therefore, $c^{*}(t)$ cannot be optimal when $c^{*}(t_{M-})$ is equal to $c^{*}(t_{M})$. Graphically the consumption path should have a shape as the one shown in Figure 7.2 and the saving or assets path is defined by the equation of motion, and has the form shown in Figure 7.3.

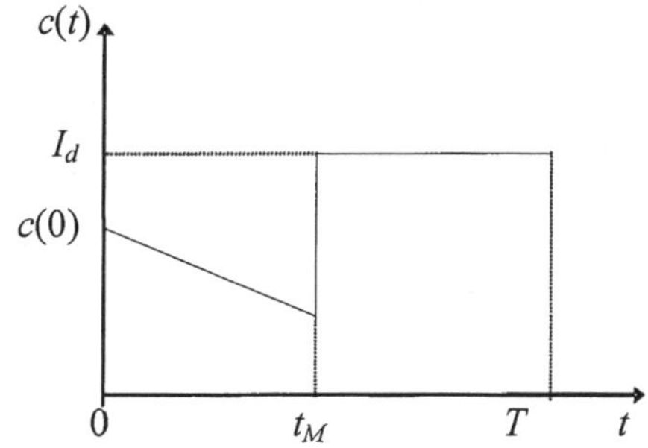

Figure 7.2 Consumption path for a migrant unemployed worker

Next, we study the case where an unemployed worker does not migrate, even though he or she has found a job in another region. This worker faces the same constraint as the one we described earlier, but his or her previous saving, initial assets, or current income are too low to make the migration possible at the time when it is still profitable.

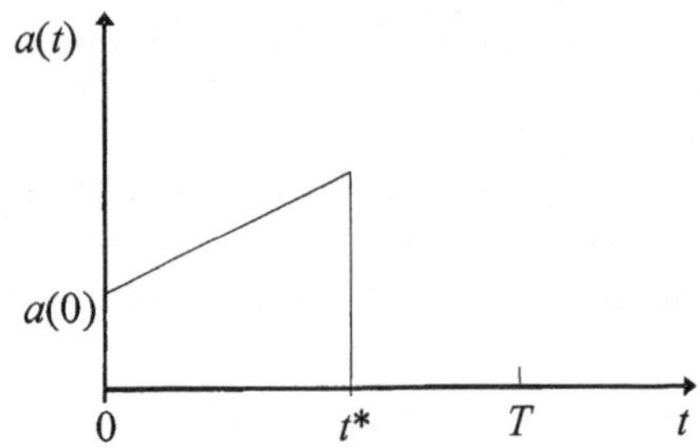

Figure 7.3 Asset path for a migrant unemployed worker

Following the previous results, and assuming that the worker has some initial assets, we can derive from equation (7.6) that:

$$U'[c(t)] = U'[c(0)] * e^{(\rho - r)*t}, \forall \quad 0 \leq t \leq t^* \tag{7.11}$$

where t^* is the period in which $a^*(t^*) = 0$. Therefore,

$$c(t) = I_0, \quad \forall \quad t^* \leq t \leq T \tag{7.12}$$

The non-migrant unemployed worker's path is different from the one we described earlier. The reason is that this worker will consume her or his assets because she or he cannot migrate and thus there is no motivation for saving. Graphically, we can see the shapes of the consumption and assets or saving path in Figures 7.4.

From the initial assumptions, we know that the worker takes the migration decision when she or he has found a job in a region that is different from the one where he or she is currently living. At that moment, an unemployed worker, who cannot pay the moving cost, starts the saving process. From the model, we can see that the time at which the move will be made (t_M) depends critically on the initial assets (a_0), and the income level in the current region (I_0). Therefore, we can deduce that the people who are most negatively affected by unemployed status are the people with lower assets level and lower income (e.g. the poorer people). While several authors have found that the transport cost plays a negative role in the migration decision (Beaudreau, 1990; Evans and Pooler, 1987; Topel, 1986), we can suggest that this problem becomes worse when the migrant workers have lower asset levels as well as lower incomes.

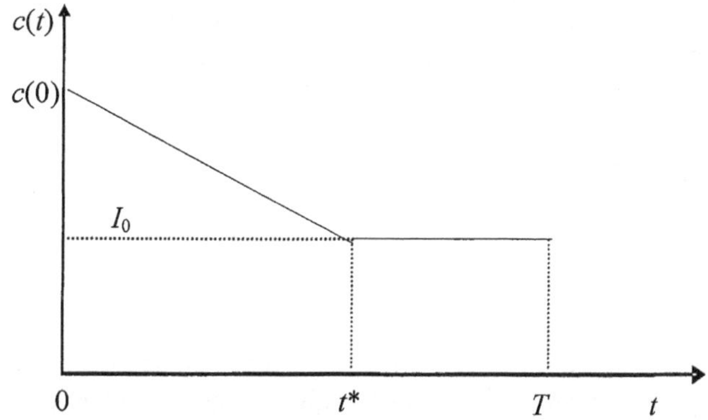

Asset path for a non migrant unemployed worker

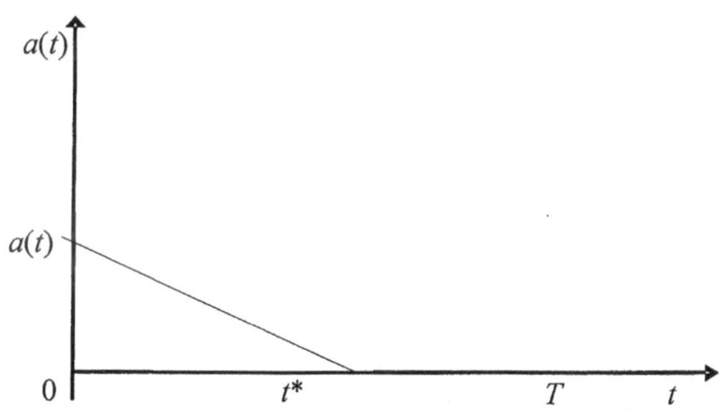

Figure 7.4 Consumption path for a migrant unemployed worker

Another common finding is that migration as an equilibrating process is less effective in depressed regions and during recession periods (Drewes, 1986; Gabriel et al., 1993; Jackman and Savouri 1992; Pissarides and McMaster, 1990). Depressed regions are characterized by high unemployment rates. Therefore, we expect that some people who are willing to migrate cannot do so because they do not have enough money to pay the transport/moving costs. Consequently, they have to wait until their savings increase enough to pay for these moving costs. Evidently, the more depressed the region, the larger the number of unemployed as well as the number willing to migrate. Therefore, there will be more people saving in order to reach the necessary amount of money to pay the moving costs. This might have a negative effect on the region, in that significant current consumption may be forgone to accumulate assets, thereby reducing income circulating in the region. Further, the accumulated assets may be used to fund capital development projects in other regions of the country, potentially further undermining the competitive advantage of the region with the unemployed workers. This situation will be inefficient compared with the alternative of having a perfect capital market, where workers could borrow independently of their employment status and moving as soon as they have found a job in other region.

THE EMPLOYED WORKER DECISION

We assumed that the worker will migrate only if he or she has a job in another region, which pays a wage high enough to make it profitable to migrate. Then, the employed migrant worker's problem can be set as:

$$\max_{c(t)} \int_0^T e^{-\rho t} U[c(t)]dt \tag{7.13}$$

subject to:

$$\begin{aligned} a(t) &= r * a(t) + I_0 - c(t) \quad 0 \leq t < t_M \\ a(0) &= a_0 \end{aligned}$$

and:

$$\begin{aligned} a(t) &= r^* a(t) + I_d - c(t) \quad t_M \leq t < T \\ a(t_M) &= a(t_{M-}) - MC, \quad a(T) \geq 0 \end{aligned}$$

The difference from the unemployed migrant worker's case is that the employed worker can obtain a loan in order to finance the moving cost as well as the existence of larger current consumption, that is reflected by the fact that $a(t)$ can be greater than, equal to, or lower than zero. The unique constraint is that when he or she will retire (at time T), all loans have to have been repaid.

From our general result, we have that the optimal path for consumption and saving or assets ($c^*(t)$, $a^*(t)$) will be:

$$\begin{aligned} U'[c^*(t)] &= U'[c^*(0)] * e^{(\rho - r)^* t}, \forall \quad 0 \leq t < t_M \\ \dot{a}^*(t) &= r * a^*(t) + I_0 - c^*(t) \quad 0 \leq t < t_M \\ a(0) &= a_0 \end{aligned} \tag{7.14}$$

and

$$U'[c^*(t)] = U'[c^*(0)] * e^{(\rho - r)^*(t - t_M)}, \forall \quad t_M \leq t < T$$

$$\dot{a}^*(t) = r * a^*(t) + I_d - c^*(t) \quad t_M \leq t < T$$
$$a(T) \geq 0$$

In this context, if moving is profitable, then the worker will do so as soon as possible. Then we can expect that the t_M will be equal to zero. Accordingly, we can say that in this case the market is highly efficient in reassigning workers. It could explain the findings that migrants have moved according to market forces, but these movements have not been strong enough to equalize regional differentials.

In addition, we can see that the employed worker has a continuous path for $c^*(t)$ and a discontinuous one for $a^*(t)$. Given the fact that $a(t)$ can be negative, then the employed worker uses the capital market to smooth his or her consumption path. The access to the credit market allows workers the possibility of separating the decision of how much to consume from the decision of how much to save. Therefore, the workers will consume more in the present and save more in the future given that $\rho > r$ was assumed.

Graphically, we can present the optimal consumption and savings or assets path shape in Figures 7.5.

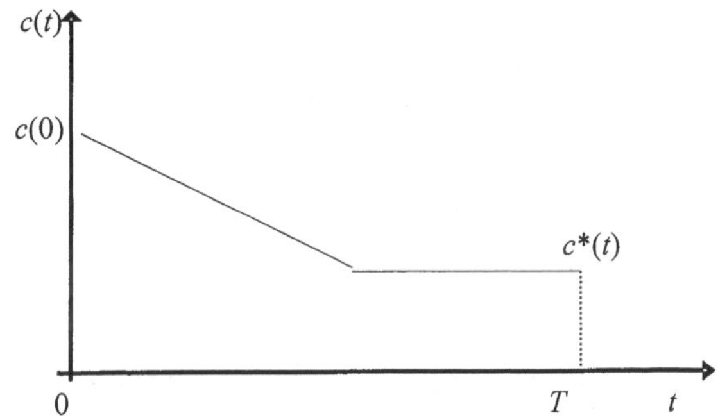

Asset path for a migrant employed worker

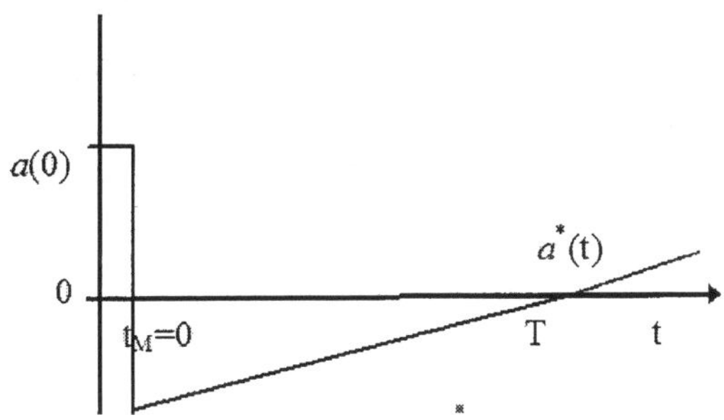

Figure 7.5 Consumption path for a migrant employed worker

POST-MIGRATION UNCERTAINTY

Until now, we have developed a model under certainty conditions, which can explain several findings in previous research. In this section, we introduce post-migration uncertainty; the worker does not know whether he or she will obtain a job in the new region. We can assume that this probability depends positively on the number of vacancies and negatively on the unemployment rate in the destination region. Furthermore, it is assumed that the worker knows the probability of obtaining a job in the new region. Therefore, in the spirit of the Harris and Todaro model (1970), the worker will maximize the expected value of his or her utility from consumption.

The uncertainty arises from the fact that the worker does not know exactly whether he or she will obtain employment or the unemployment income in the new region. Instead, he or she knows the expected value and the variance of his or her future income. Further, it would be reasonable to assume that the worker knows the situation in the current region, and whether this situation will change during the period that the worker will stay in the current region. Therefore the worker's utility maximization, for the period in which he or she stays in the current region, is given by:

$$\max_{c(t)} E_0 \left\{ \int_0^{t_M} e^{-\rho t} U[c(t)]dt + e^{-\rho t_M} V(t_M) \right\} \tag{7.15}$$

subject to:

$$\begin{aligned} \dot{a}(t) &= r * a(t) + I_0 - c(t) \quad 0 \le t \le t_M \\ a(0) &= a_0 \\ a(t) &\ge 0, \quad 0 \le t \le t_M \end{aligned}$$

where $V(t_M)$ is the expected value of a similar maximization for the post-migration period.

The second part of this maximization corresponds to the post migration period, from t_M to T. We introduce the uncertainty in the constraint. In addition, we assume that the worker obtains a job equal to a salary of I_d^e or continues in a state of unemployment with an income equal to I_d^u. The probability of finding a job in the destination region is equal to λ. With this additional information, we can write the maximization for $V(t_M)$ as follows:

$$\max_{c(t)} E_{t_M} \left\{ \int_{t_M}^{T} e^{-\rho t} U[c(t)]dt \right\} \tag{7.16}$$

subject to:

$$\begin{aligned} da(t) &= [r * a(t) + I_d^u - c(t)]dt + (I_d^e - I_d^u)dw(t) \quad t_M \le t \le T \\ a(t_M) &= a_1 - MC \end{aligned}$$

Where I_d^u and I_d^e are the incomes in the destination region for an unemployed and an employed worker respectively. In addition, we assume that *dw*(*t*) follows a Poisson Process. Therefore, *dw*(*t*) has the following behaviour:

$$dw(t) = \begin{cases} 1, \text{ with probability } \varphi dt \\ 0, \text{ with probability } (1-\varphi)dt \end{cases} \tag{7.17}$$

In the stochastic dynamic constraint, we postulate that the worker's assets, *da*(*t*), will increase as a function of the difference between the employment and unemployment income (if the worker secures a job), or otherwise it will remain constant. In this period we assumed, under certainty, that the worker will have zero assets, because the assets were accumulated with the only objective of paying the moving costs. Nevertheless, when uncertainty is introduced, the worker has other motives to accumulate assets, for example for smoothing consumption, or to save for facing unexpected changes.

This is a problem of stochastic control. In the first part of this chapter, we solve the control problem using a forward procedure. As Dreyfus and Law (1977) indicate, under suitable conditions the solutions reached by a forward procedure are equivalent to the solutions reached by a backward procedure (optimality principle). However, when a stochastic process is introduced, then only the backward procedure yields the optimal solution.

To solve this problem, we make two small changes in the setup of the problem in order to reach a closed form solution. First, we assume that $[I_d^u - c(t)]$ can be expressed as a function of *c*(*t*) given that we assumed that I_d^u is a constant. Consequently, we replace $[I_d^u - c(t)]$ by $[-k(t)^* c(t)]$, where *k*(*t*) can be greater than, smaller than or equal to zero. We can expect that in the unemployed worker case, it will be greater or equal to zero, which means that the unemployed worker consumes more than or a quantity equal to her or his unemployment income.

Secondly, we assume that we can express the difference between the employment and unemployment income in the destination region as a percentage of the assets. Therefore, we can rewrite $I_d^e - I_d^u = [z(t) - 1]a(t)$, where *z*(*t*) is greater than one. This assumption could make it appear that the differences between incomes is endogenous because *a*(*t*) is endogenous. However, *z*(*t*) is exogenously determined as is the income difference.

With the above assumption, we can state the Hamiltonian–Jacobi–Bellman equation from stochastic control theory, following Dreyfus (1965) and Malliaris and Brock (1982), for our problem:

$$0 = \max_{c \geq 0}\{e^{-\rho t}U(c) + \varphi[J(z * a,t,T) - J(a,t,T)] + J_a * (r * a - k * c) + J_t\} \tag{7.18}$$

where $J(\cdot)$ is the value of the function $V(t_M)$ with t_M replaced by t where the process starts in state a at time t and an optimal c has been chosen. J_a is the partial derivative of $J(\cdot)$ with respect to the state variable (a), and J_t is the partial derivative of $J(\cdot)$ with respect to t.

It is important to note that $J(\cdot)$ is an explicit function of t, and the reason is because T (the planning horizon) is exogenously determined by the age of the worker who is evaluating the possibility of migration. T is the time when the worker expects to retire. Therefore, the value of $J(\cdot)$ will be different depending on how near in time the worker is from T. The importance arises from the fact that if $J(\cdot)$ does not depend explicitly on t, then J_t will be equal to zero, and it will be easy to find a closed form solution for the maximization problem of equation (7.18). On the other hand, if $J(\cdot)$ depends explicitly on t, as in our case, then J_t will be different from zero and we will have a partial differential equation which can lead to a difficult or even impossible way of finding a closed form solution.

Before maximizing equation (7.18), we define $W(a,t,T) = e^{\rho t}J(a,t,T)$, and we insert this definition in equation (7.19):

$$0 = \max_{c \geq 0}\{U(c) + \varphi[W(z * a,t,T) - W(a,t,T)] + W_a * (r * a - k * c) + W_t - \rho W(a,t,T)\} \quad (7.19)$$

If we assume that $U(c)$ is strictly concave in c, then applying Theorem I from Merton (1971), we know that there exists an optimal c, which we call c^*, satisfying the constraints of our original formulation.

The first order condition from this maximization will be:

$$U_c = k * W_a \quad (7.20)$$

where U_c is the partial derivative of $U(c)$ with respect to c and W_a is the partial derivative of $W(\cdot)$ with respect to a.

To see the effect of the uncertainty on the worker's migration decision, we study the case in which the utility function reflects a constant relative risk aversion. Therefore:

$$U(c) = \frac{c^\gamma}{\gamma} \quad (7.21)$$

Taking the first order condition, and the optimal c (referred to as c^*), we can rewrite the Hamiltonian–Jacobi–Bellman equation, and posit a solution for $W(a,t,T)$ as:

$$W(a,t,T) = g(t,T)a^\varepsilon \quad (7.22)$$

Taking this possible solution, we replace each term in the revised Hamiltonian–Jacobi–Bellman equation, and then we obtain that $\varepsilon = \gamma$ for the method of equalizing exponents (see Malliaris and Brock (1982) for a detailed description).

Therefore, knowing that $\varepsilon = \gamma$, we can transform the maximized Hamiltonian–Jacobi–Bellman equation from a partial differential equation to a differential equation, which will be:

$$0 = \left(\frac{1-\gamma}{\gamma}\right)(k\gamma)^{\frac{\gamma}{\gamma-1}} * g(t,T)^{\frac{\gamma}{\gamma-1}} + [\varphi(z^{\gamma}-1) + r * \gamma - \rho]g(t,T) + g \tag{7.23}$$

Even though this differential equation looks very difficult to solve, it can be solved by letting $g(t,T)$ be equal to $y(t,T)^{1-\gamma}$. Therefore, plugging this definition in equation (7.23), yields an ordinary differential equation of the first order.

In the last period, we can say that $g(T,T)$ will be equal to zero, because if we change $a(T)$, $W(a,T,T)$ should remain constant because there is no longer uncertainty in this period (it is the last period), then there is no further utility to be derived from larger accumulations of assets. With this boundary condition, we can obtain the following solution for $g(t,T)$:

$$g(t,T) = \left\{\frac{b_0}{b_1}\right\}^{1-\gamma} \left\{e^{\frac{b_1(T-t)}{1-\gamma}} - 1\right\}^{1-\gamma} \tag{7.24}$$

where:

$$b_0 = \frac{1-\gamma}{\gamma}(k\gamma)^{\frac{\gamma}{\gamma-1}} \tag{7.25}$$

and:

$$b_1 = \varphi(z^{\gamma}-1) + r * \gamma - \rho \tag{7.26}$$

Knowing $g(t,T)$, we can obtain a solution for $W(a,t,T)$ and for W_a and inserting this result in the first order condition and using equation (7.21) we can get the optimal path for $c^*(t)$, which will be:

$$c^*(t) = (kW_a)^{\frac{1}{\gamma-1}} = [kg(t,T)a^{\gamma}]^{\frac{1}{\gamma-1}} \tag{7.27}$$

In addition, the sign of the derivative of W_a with respect to b_1 will depend on the value of b_1. To determine this, we have to make an assumption, similar but not equal to the one which we made with respect of r and ρ. We know that b_1 is equal to $\varphi(z^{\gamma} - 1) + r * \gamma - \rho$. Before, we assumed that $\rho > r$; now when we introduce uncertainty and the assumption that the worker is risk averse, we will assume that $\rho > \varphi(z^{\gamma} - 1) + r * \gamma$. We can see the effect of the uncertainty and the risk aversion of the worker on our initial assumption that $\rho > r$. First, we assume for a moment that the worker is indifferent to risk, therefore γ is equal to one. In this case, the intertemporal discount rate of the consumption has to be greater than the interest rate plus the expected value of the increasing wage rate for the worker to obtain a job in the new region. Secondly, when we add the assumption that the worker is risk averse, then our assumption is less robust, because the interest rate and the value of the increasing wage rate for obtaining a job in the new region are adjusted for risk. In addition, we can say that the more risk averse the worker, the less restrictive is the assumption. Finally, with this assumption, we note that the derivative of W_a with respect to b_1 is less than one.

With the above results and assumptions, we can see the effect on the worker's consumption path and migration decision of the postmigration uncertainty and risk aversion. First, we compare the post-migration uncertainty situation with the certainty situation, and we find that under post-migration uncertainty, the worker will be less likely to migrate compared with the certainty situation. This is true because the expected value of the income in the new region always will be lower than the income of an employment worker at the destination region. Secondly, we can see that the larger the probability of finding a job at a destination region, the more likely it is that the worker will migrate, essentially because:

$$\frac{\partial c(t)}{\partial \varphi} = \frac{\partial c(t)}{\partial W_a} \frac{\partial W_a}{\partial b_1} \frac{\partial b_1}{\partial \varphi} > 0 \tag{7.28}$$

We can also say that the larger the differential between the employment income and unemployment income at the destination region, the more likely it is that the worker migrates because:

$$\frac{\partial c(t)}{\partial z} = \frac{\partial c(t)}{\partial W_a} \frac{\partial W_a}{\partial b_1} \frac{\partial b_1}{\partial z} > 0 \tag{7.29}$$

Both conclusions are based in the fact that the expected value of the consumption, and therefore of the $V(t_M)$, will be larger as φ and z increase.

CONCLUSION

We present a migration decision model for workers under different employment status. The model has two special characteristics. First, it is a dynamic model, which is more appropriate for the analysis of an intrinsically dynamic problem such as migration. Secondly, the model considers an imperfect capital market, which is introduced as the unemployed worker's lack of access to borrowing from financial institutions. It can easily be extended to the case of poor people, when they have a job, but their salary is so small that they cannot save to accumulate wealth that they could use to pay the moving costs. The role of prior migrants helping others to move or family support to finance migration are two elements that can reduce the impacts described in this chapter, but they are issues to be investigated in future research.

This model is especially relevant to the study of migration as a mechanism for equilibrating regional disparities in unemployment rates in less developed countries, where a high percentage of the population lives in poor conditions, earns a very low wage, and does not have access to the capital market. The model considers several alternatives for workers. Unemployed and employed workers could make contracted or speculative migration, they can have different levels of wealth, independent of their employment status, but the employment status determines their access to the capital market; this is the major segmentation factor.

In this model, the migration decision of the worker not only negatively affects the development of the origin region (because he or she relocates their capacity and assets to the destination region) but also in the origin region during the period prior to migration, when part of the worker's current consumption may be foregone to accumulate assets, thereby reducing income circulating in this region.

In addition, the optimizing behaviour of the workers is modelled under perfect information, which is a close approximation of the contracted migration case, and under uncertainty, which mirrors speculative migration. Speculative migration in this model is not a privilege of the unemployed workers. The motivation to migrate is the increase of the welfare coming from a monetary benefit that is earned by migration to another region.

A summary of the conclusions derived from this theoretical model may be presented:

- Transport costs play a negative role in the migration decision and their effects are larger the poorer are the people who are contemplating migration;

- Given the fact that the people most affected by moving costs are the unemployed and poor, the migration mechanism is less effective in depressed regions;
- The migration mechanism is an efficient mechanism for reallocating people when workers have access to borrowing from financial institutions;
- Lack of information about the situation in other regions plays a negative role in the migration decision;
- The higher the probability of obtaining a job at the destination region, the greater the effect on the migration decision;
- The larger the differential between the employment income and the unemployment income at the destination region, the larger the probability of migration.

The model reproduces several conclusions that are already in the literature about migration. However, it has two additional contributions. First, there is no model in the literature that provides explanation for such a variety of alternative types of workers. Secondly, while there is consensus about the fact that the imperfect capital market makes migration an inefficient mechanism to allocate resources, there is not a formal model that interprets this phenomenon. If it is estimated that around 20 per cent of US residents face a liquidity constraint (Jappelli, 1990), then the percentage is likely to be much higher in developing economies with a consequently larger impact on migration decision-making.

REFERENCES

Aroca, P. and W. Maloney. 2005. "Migration, Trade, and Foreign Direct Investment in Mexico", *The World Bank Economic Review*, 19(3) 449–472.

Artle, R. and P. Varaiya. 1978. "Life Cycle Consumption and Homeownership", *Journal of Economic Theory*, 18, 38–58.

Becker, Charles M. and Edwin S. Mills. 1986. *Studies in Indian Urban Development*. New York: Oxford University Press.

Beaudreau, B.C. 1990. "Transport Cost and Regional Wage Differentials: Evidence from Canadian Microdata", *Canadian Journal of Regional Science*, 13, 89–98.

Bhattacharya, P.C. 1993. "Rural–Urban Migration in Economic Development", *Journal of Economic Surveys*, 7, 243–281.

Chiang, A. 1991. *Elements of Dynamic Optimisation*. New York: McGraw Hill.

Drewes, T. 1986. "Regional Unemployment Disparity and Market Adjustment Failure", *Canadian Journal of Regional Science*, 9, 323–340.

Dreyfus, S.E. 1965. *Dynamic Programming and the Calculus of Variations*. New York: Academic Press.

Dreyfus, S.E. and A.M. Law. 1977. *The Art and Theory of Dynamic Programming*. New York: Academic Press.

Evans, N.J. and J. Pooler 1987. "Distance Deterrence Effects in Constrained Spatial Interaction Models of Interprovincial Migration", *Canadian Journal of Regional Science*, 10, 259–279.

Fields, G.S. 1975. "Rural-Urban Migration, Urban Unemployment and Underemployment, and Job-Search Activity in LDCs", *Journal of Development Economics*, 2, 165–187.

Fuchs, V.R. 1982. Time Preference and Health: An Exploratory Study", in *Economic Aspects of Health*, ed. By V. Fuchs. Chicago, NBER and the University of Chicago Press.

Gabriel, S.A., J. Shack-Marquez and W.L. Wascher. 1993. "Does Migration Arbitrage Regional Labour Market Differentials?", *Regional Science & Urban Economics*, 23, 211–233.

Harris, J.R. and M.P. Todaro. 1970. "Migration, Unemployment and Development: A Two-Sector Analysis", *American Economic Review*, 60, 126–142.

Hurd, M.D. 1989. "Mortality Risk and Bequests", Econometrica, 57, 779–813.

Jackman, R. and S. Savouri. 1992. "Regional Migration in Britain: An Analysis of Gross Flows Using NHS Central Register Data", *Economic Journal*, 102, 1433–1450.

Jappelli, T. 1990. "Who is Credit Constrained in the US Economy?", *Quarterly Journal of Economics*, 55, 219–234.

Kamien, M.I. and N.L. Schwartz. 1991. *Dynamic Optimisation the Calculus of Variations and Optimal Control in Economics and Management*. New York: North Holland.

Lee, E.S. 1966. "A Theory of Migration", *Demography*, 3, 47–57.

Leonard, D. and N.V. Long. 1992. *Optimal Control Theory and Static Optimisation in Economics*. New York: Cambridge University Press.

Malliaris, A.G. and W.A. Brock. 1982. *Stochastic Methods in Economics and Finance*. New York: North Holland.

Merton, R.C. 1971. "Optimum Consumption and Portfolio Rules in a Continuous-Time Model", *Journal of Economic Theory*, 3, 373–413.

Molho, I. 1986. "Theories of Migration: A Review", *Scottish Journal of Political Economy*, 33, 396–419.

Pissarides, C.A. and I. McMaster. 1990. "Regional Migration, Wages and Unemployment: Empirical Evidence and Implications for Policy", *Oxford Economic Paper*, 42, 812–831.

Pissarides, C.A. and J. Wadsworth. 1989. "Unemployment and the Inter-Regional Mobility of Labour", *The Economic Journal*, 99, 739–755.

Topel, R.H. 1986. "Local Labour Markets", *Journal of Political Economy*, 94, 111–143.

Vanderkamp, J. 1970. "The Effects of Outmigration on Regional Employment," *Canadian Journal of Economics*, 3, 541–549.

8. Circular statistics, migration models, and the directionality of job search processes

Alessandra Faggian, Jonathan Corcoran and Philip McCann

INTRODUCTION

This chapter examines an issue which has never before been investigated empirically, namely the spatial shape and skewness of labour migration patterns. Empirical indicators of migration behaviour found in the literature are either zero-dimensional, i.e. migration probabilities, or one-dimensional, such as linear distance measures (see for example Faggian et al. 2007a,b). However, labour mobility is explicitly a two-dimensional phenomenon which is best represented as a set of vectors connecting pairs of origins and destinations. The use of vectors is typically restricted to the analysis of macro-level migration flows (e.g. Tobler 1981), where the number of observations is limited. Applying these techniques to larger micro-level datasets is, however, problematic for both visualisation (Corcoran et al. 2009) and modelling. This chapter introduces, for the very first time, the concept of circular variance (and more generally circular statistics) into migration modelling and demonstrates how this allows the simplification of complex two-dimensional phenomena into one-dimensional scalar indices suitable for econometric modelling. An important aspect of circular statistics is their ability to summarise the richness of information contained in two-dimensional vectors into an operational index, which can then be integrated into more standard econometric frameworks. This, in turn, provides further insights as to the nature of many of the zero and one-dimensional properties of migration identified in the literature. Moreover, examining the two-dimensional properties of migration also allows us to understand certain types of spatial sorting behaviour which are consistent with modern theories of agglomeration, but which previously have not been explained with existing empirical tools.

The empirical analysis is based on a unique dataset detailing employment-mobility behaviour of university graduates in Great Britain. In particular, we are interested in analysing whether different groups of migrants have different two-dimensional migration properties (as summarised by the index of circular variance) even when they exhibit similar zero or one-dimensional patterns. Finally, we investigate the determinants of these migratory dynamics through the inclusion of the circular variance index into an econometric model.

The chapter is organised as follows. The next section discusses the nature of labour migration, the role played by human capital and search processes, the relationship between these issues and urban spatial structure of the economy, and some of the aspects of these interrelationships which existing empirical tools are unable to throw light on. The third section discusses the concept of the circular variance and explains the human capital mobility data we employ. The fourth section explains the econometric methodology and provides some empirical results. The fifth section discusses the results in the light of the issues set out in the third section and the sixth and final section provides some brief conclusions.

HUMAN CAPITAL, LABOUR MIGRATION AND SPATIAL EMPLOYMENT SEARCH AREAS

The relationship between the distance moved by a migrant for employment reasons and the migrant's personal characteristics is well understood. In general both human capital migration theory (Sjaastad 1962) and job-search migration theory (Lippman and McCall 1976a,b, 1979; Pissarides 1976) predict not only that the probability of migration is positively related to the level of human capital, but also that the expected migration distance increases with higher levels of human capital. Yet, in order to reach this conclusion regarding the relationship between the migration distance and the level of human capital, these models make two assumptions. Firstly, they assume that employment opportunities are randomly distributed over space and secondly, that the average geographical distances between such opportunities is inversely related to the wage associated with each job. As such, the fact that higher-skilled and -wage jobs are expected to be more sparsely, but randomly, distributed across space than lower-skilled and -wage jobs, means that the expected search area can be considered to be approximately circular, with the expected "search radius" increasing for higher-skill and -wage jobs. This positive relationship between the level of human capital and the expected migration distance therefore arises from the fact that higher-skilled workers seek higher-wage jobs than lower-skilled workers, under conditions where there are also differences in the densities of the spatial distribution of different job opportunities.

However, these assumptions appear to conflict with one another somewhat given what we know from modern economic geography and urban economics. Employment opportunities are clearly not randomly distributed in space. While many low-wage and low-skill occupations can be approximated as being more or less ubiquitously and randomly distributed across space, high-wage and high-skilled jobs are definitely neither ubiquitously nor randomly distributed across space. As the modern understanding of agglomeration economies emphasises, there is a tendency for high-wage and high-skill occupations to be concentrated in specific cities (Glaeser and Mare 2001), rather than simply cities in general. As the skill and wage level of the occupation increases, the number of potential employment locations falls. In many countries, the number of locations offering the highest skill and wage occupations becomes extremely small, and for most countries, this often reduces to just one or two dominant urban centres. This is true not only in small countries such as Denmark, New Zealand and Ireland, but also in many large countries such as the UK (CoL, 2006), France and Japan (Nakamura, 2008). When the number of urban centres offering the highest skill and wage occupations falls, the use of "search radius" becomes inappropriate, as the spatial employment distribution is clearly not random. Instead, employment searches for high-skilled and high-wage jobs are biased towards a particular location. Assuming that highly skilled workers receive the best job offers first (Bartel 1979), and also assuming that the high-skill workers originate from all over the country, then we ought to observe explicitly a spatial sorting in a directional sense amongst migrants, according to their skill levels. High-skilled workers should be not only more mobile (Faggian et al. 2007a), but also display a more unidirectional migratory pattern towards the dominant centre(s). In contrast, lower-skilled workers should have more multidirectional mobility patterns. This is because they are lower down the queue of job offers (Bartel 1979) and hence less likely to find jobs in the high-skill centre. These arguments suggest that the employment-migration search area for highly skilled workers would tend to be non-circular, asymmetric and elongated towards the dominant urban centre, whereas for low-skilled workers it would tend to be more evenly distributed across space. This issue has never before been analysed empirically for labour migration behaviour and this underpins the main impetus of this chapter.

In one sense, the degree of this asymmetry can be understood as being somewhat analogous to the gravity "pull" effect of the dominant city within a gravity model of migration. Yet, the usefulness of gravity models for analysing these spatial employment search issues is limited for two reasons. Firstly, while gravity models indicate the strength of inter-urban or inter-regional employment-migration interactions between particular points across space, they have no two-dimensional spatial properties and, as such, cannot tell us anything about the directionality of such inter-urban or inter-regional interactions. Secondly, gravity models have no links with the human capital and search theories of migration, or the spatial sorting arguments. A more nuanced understanding of the spatial features of the employment search area is, therefore, called for. An ideal empirical characterisation of the migration search area needs to be consistent with the various insights of the human capital model, the search model and the spatial sorting argument.

This chapter aims to demonstrate that the use of circular statistics (and more specifically the circular variance measure) represents a major step in this direction as it allows the integration of information on both the distance and the frequency of migration moves, as conventionally contained in gravity models, with additional information on the direction of such movements. The combination of distance, frequency and the direction of migration movements captured by a single circular statistic index provides for both a much richer and a more nuanced characterisation of the geography of spatial employment search areas. Moreover, it also allows for a much sophisticated econometric modelling of the determinants of human capital migration than conventional gravity models.

CIRCULAR VARIANCE AND THE GRADUATE MIGRATION DATA

Circular statistics have been described as a "curious byway of statistics ... somewhere between the analysis of linear and the analysis of spherical data" (Fisher 1993, p.1). Their application originally dates back to the mid-eighteenth century (Bernoulli 1734). Ever since then, circular statistics have generally developed as a set of techniques mainly used in the physical, ecological and biological sciences to study phenomena better represented by two-dimensional vectors. In the context of this chapter, one of the key features of circular statistics is that it allows the use of indices that reduce the dimensionality of certain phenomena. It is relatively easy, for instance, to summarise certain two-dimensional spatial and directional characteristics into a one-dimensional point value. Despite the fact that these techniques have been primarily, if not solely, applied to physical and natural phenomena, their potential usefulness in social research is undoubtedly clear. Migratory patterns, travel and commuting behaviour and transport movements (just to name a few) are all phenomena which could provide a fertile ground for the application of these techniques. Yet, the use of circular statistics in social sciences so far has been extremely limited. Notable exceptions are Brunsdon and Corcoran (2006) and Corcoran et al. (2009) which represent a first application of these techniques to social sciences phenomena.

In this chapter we progress beyond these initial attempts by integrating circular statistics into a more traditional econometric modelling framework in order to provide a systematic study of the issue of migration and spatial job-search. The availability of geo-coded data where both the coordinates of origin and destination are known is crucial to the applications of these techniques. The Higher Education Statistics Agency (HESA) Student Records and First Destination Survey are perfect examples of such datasets. When combined, they provide detailed postcode information on both the university attended and the first employment location after graduation[1] for a large proportion of British university graduates. For the year 2000, which is the basis of this study, it includes 47,485 valid observations on British university graduates. Personal characteristics such as gender, ethnicity, age and professional family background of the students are also recorded together with their final grade of the degree awarded, the subject studied and the distribution of subjects taught across the whole university.

The particular circular statistic used here, i.e. circular variance, is an index of circular spread, and captures the spatial dispersion of graduate employment opportunities relative to a common migration origin point, namely the university attended by the student. The index of circular variance ($\bar{R}$) is defined as:

$$\bar{R} = 1 - \frac{1}{n}(A^2 + B^2)^{\frac{1}{2}} \tag{8.1}$$

where:

$$A = \sum_{i=1}^{n} \sin(\omega\theta_i) \tag{8.2}$$

and:

$$B = \sum_{i=1}^{n} \cos(\omega\theta_i) \tag{8.3}$$

where $\{\theta_i,\ldots,\theta_n\}$ are the n observed cyclic data values or, in other words, the computed angle that a graduate migrates from their university to their first employment, and ω is a scaling constant to place the values in the cyclic range defined as $(0,2\pi)$.[2]

The interpretation of the circular variance is that lower values of $\overline{R}$ indicate a circular distribution with less spread, while higher $\overline{R}$ values are associated with greater levels of spread. One limitation of $\overline{R}$ is that when values approach zero, thereby indicating a very low geographical dispersion, there is no guarantee that the circular variance is a true measure of the distribution, because two clusters of directions that are 180 degrees apart will also yield a very small value of $\overline{R}$. To remedy this situation a circular plot can be generated for each of the cases where very low values of $\overline{R}$ are computed in order to investigate whether the circular variance is a true measure of the distribution.

In order to compute the circular variance measure it was first necessary to pre-process the data using bespoke tools built within the ArcView GIS environment (ArcView 2002). Here, the postcode data for both the university and first–employment locations for each graduate record was imported into the geographical information system. Using the *x,y* co-ordinates of the centroid of the relevant postcode area a polyline was generated for every record which connected a single university to a single first employment location. This process resulted in the creation of 47,485 polylines for the movements from university to first employment. For each university location, an angle of travel was then computed for each graduate's migration to first employment, and using the angle of travel the measure of circular variance, as defined in equations (8.1) to (8.3), was calculated for each university.

Figure 8.1 The location of ten universities selected for a preliminary circular statistic analysis

Several diagrammatic examples can be used to illustrate the application of circular variance statistics. Figure 8.1 shows the location of ten higher education institutions (HEIs) selected from a preliminary circular variance statistical analysis of their graduate employment-migration patterns.

Figure 8.2 maps for each university the total number of such movements and the direction of each movement outside of a local area, whereby the local area is defined according to three different areal unit scales, namely a 15km radius and a 50km radius of the university, as well as the travel to work area (TTWA) containing the university. The reason for comparing different definition of "migrants" is to identify the extent to which the migratory patterns are themselves sensitive to the spatial areas employed which is commonly known in geography as "modifiable area unit problem" (or MAUP).

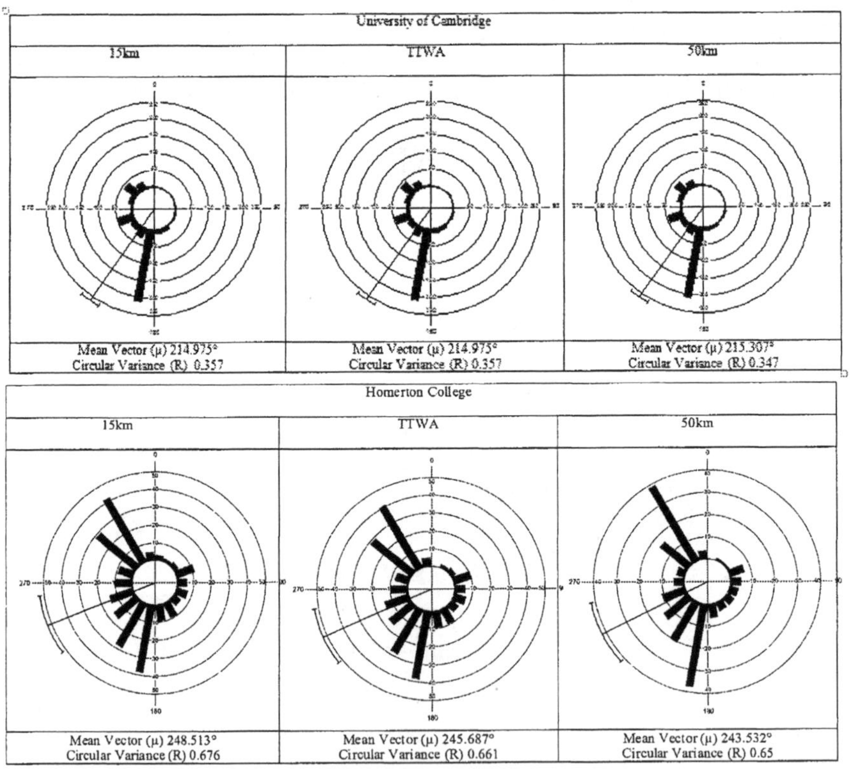

Figure 8.2 The university to first-employment migration patterns for the graduates of ten selected universities

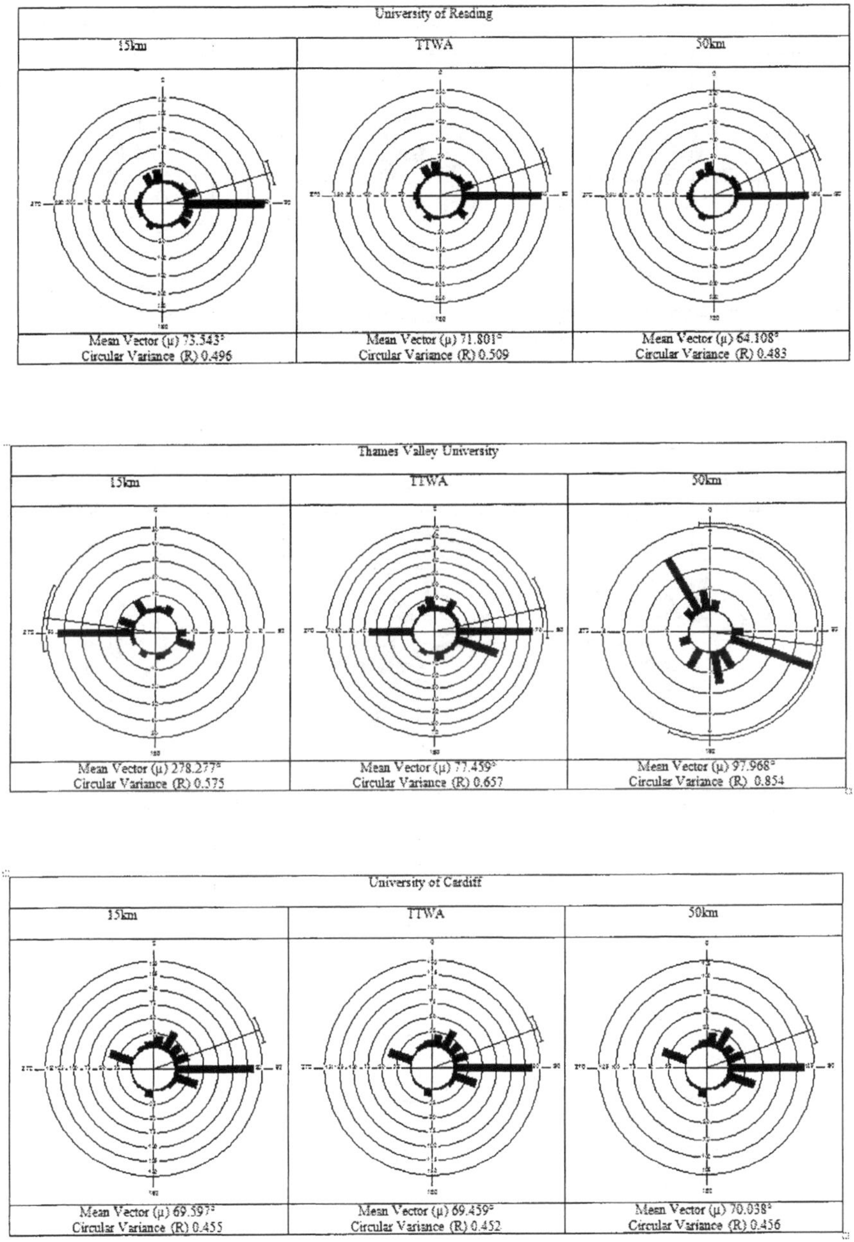

Figure 8.2 (continued)

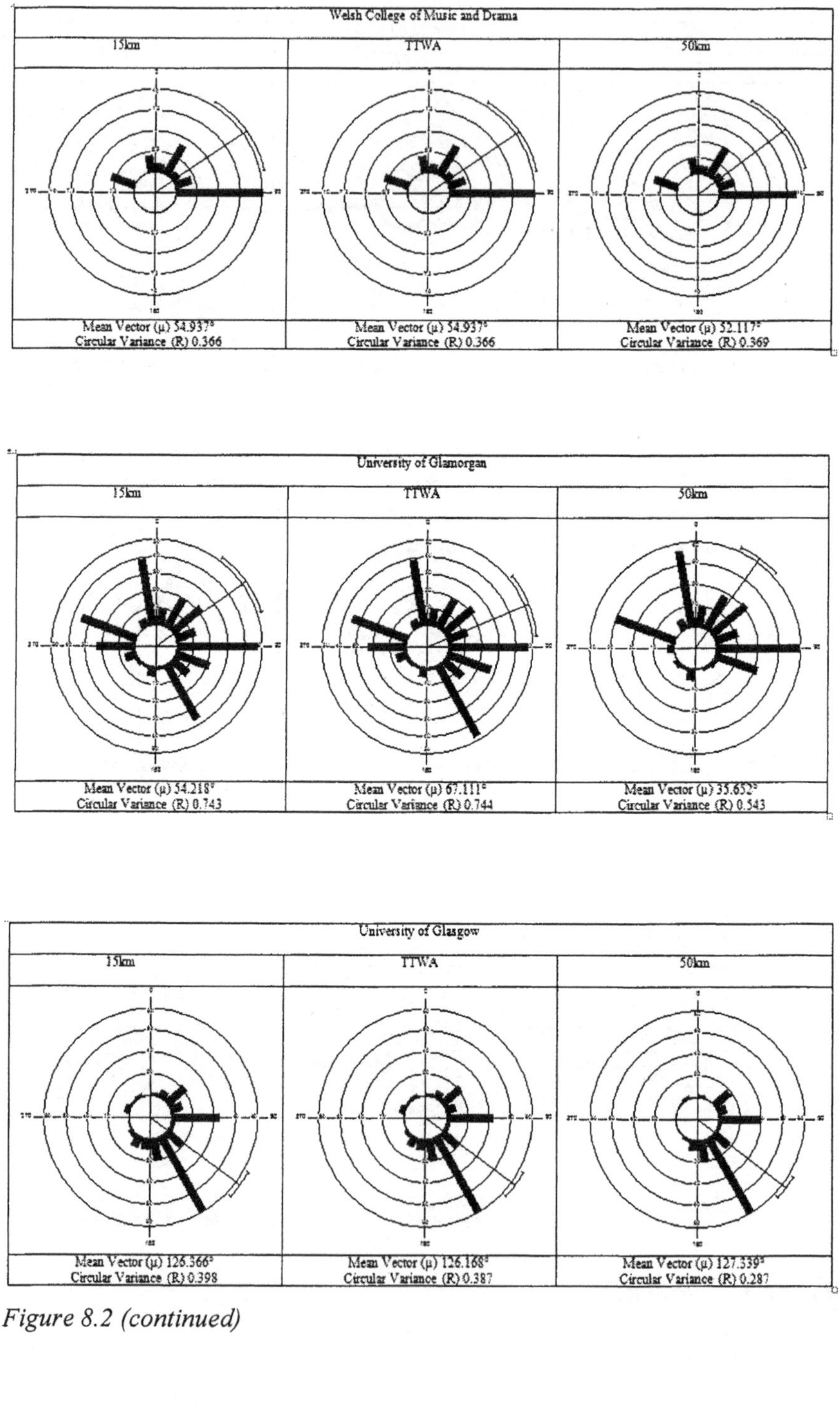

Figure 8.2 (continued)

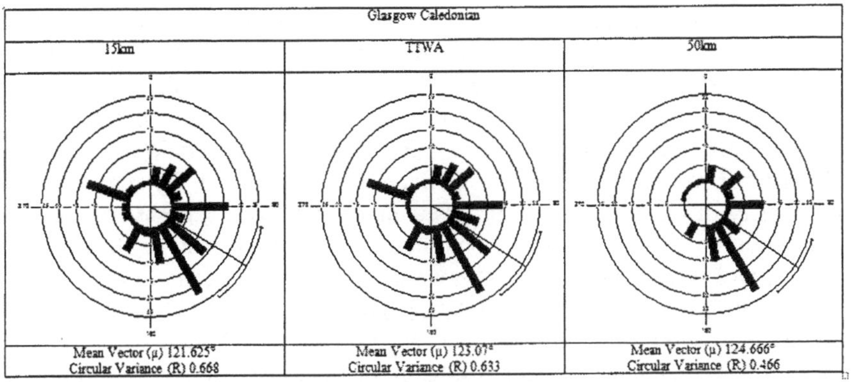

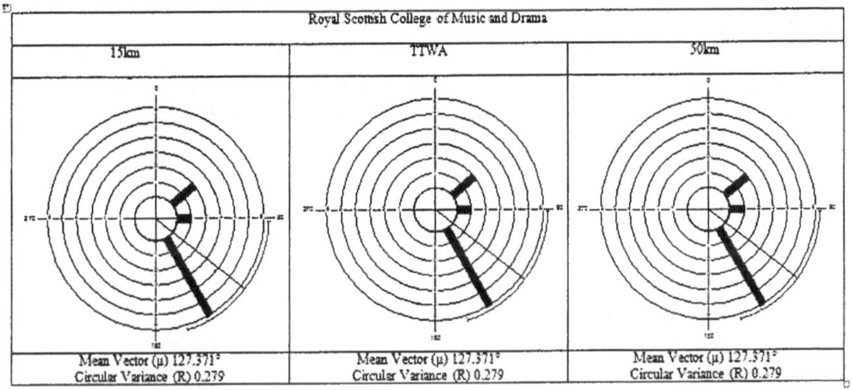

Figure 8.2 (continued)

As Figure 8.2 shows, many of the employment-migration moves of graduates from the University of Cambridge are directed towards the west, but by far the largest single group are movements in the direction of London. At the same time, Homerton College, a teacher training college which is also part of the University of Cambridge, displays a very different spatial pattern of graduate migration, with many more of its graduates moving north west, south west and east. As a result, the circular variance of Homerton College is much higher than that of the University of Cambridge and this holds irrespectively of the definition of "local area". If we compare the University of Reading with its neighbour Thames Valley University, we notice how the University of Reading has a small number of graduates entering employment north and east of the university with a very large proportion of graduates heading in the direction of London. In contrast, although at the 15km radius and the TTWA scales both universities exhibit similar graduate mobility patterns, when we consider a 50km radius of the university, Thames Valley

University has a much more geographically dispersed pattern of graduate employment. The result is that the circular variance of the University of Reading is slightly lower than that of Thames Valley University when we use the 15km and TTWA definition of local are but very marked when the 50km is employed. In the case of three HEIs located in South Wales, namely Cardiff University, the Welsh College of Music and Drama and the University of Glamorgan, we observe similar migratory patterns for the first two (with movements to employment in the direction of west, north and north east, as well as a large number of movements in the direction of London), but a very different pattern for the latter. Graduates from the University of Glamorgan exhibit a much broader spatial distribution of graduate employment outcomes and this results in a higher circular variance value. In the case of three Scottish HEIs located in Glasgow, namely the University of Glasgow, Glasgow Caledonia University and the Royal Scottish College of Music and Drama, we see that the graduate employment pattern of the latter institution is totally dominated by movements in the direction of London, as to a lesser extent is the one of the University of Glasgow, whereas Glasgow Caledonia University has a much more spatially dispersed pattern of graduate employment mobility. The result is that the Royal Scottish College of Music and Drama has the lowest circular variance, followed by the University of Glasgow and then Glasgow Caledonia University.

At this stage we can make some tentative suggestions as to why the graduate employment-migration patterns differ. The very spatially skewed graduate employment mobility patterns for the two specialised colleges of music and drama are not surprising, given that London totally dominates these industries. In contrast, the public sector orientation of education, in which teaching jobs are dispersed approximately according to the population distribution, probably accounts for the diverse spatial pattern of Homerton College's graduates relative to that of the University of Cambridge as a whole. Meanwhile, the strong local orientation of the University of Glamorgan, Thames Valley University and Glasgow Caledonian University probably accounts for the diverse spatial structure of their graduate employment patterns. In contrast, the University of Cambridge, the University of Reading, the University of Glasgow and Cardiff University all exhibit similar graduate employment mobility patterns, with some students entering their first employment in the immediate vicinity of the university, but with very large numbers of graduates migrating to jobs in the specific direction of London. However, these explanations are, as yet, somewhat speculative, and this type of analysis is rather ad hoc and heuristic. It is therefore necessary to develop our empirical analysis on the basis of detailed data and the insights of migration models and local labour market analysis.

MODEL AND EMPIRICAL RESULTS

The econometric model estimated here aims to establish a more robust explanation for the observed spatial employment-migration patterns of a university's graduates. It has already been demonstrated elsewhere that the propensity of a university's graduate cohort to enter first employment in the vicinity of or far away from their university depends on the social, ethnic (Faggian et al. 2006) and educational characteristics (Faggian et al. 2007a) of both the university and its student body (Faggian and McCann 2009a). The independent explanatory variables we have chosen for our econometric model here reflect therefore these characteristics.

We estimate the following econometric model:

$$\bar{R} = \alpha + \beta_1 DIST_i + \beta_2 RUSSELL_i + \beta_3 ASIAN_i + \beta_4 BLACK_i + \beta_5 FMRATIO_i + \beta_6 PROFMAN_i + \beta_7 MATURE_i + \beta_8 GOODDEG_i + \beta_9 LONDON_i + \beta_{10} SPEC_i + \beta_{11} COAST_i + \beta_{11} EDUC_i + \beta_{12} ENG_i + \beta_{13} MATHS_i \quad (8.4)$$

in which the dependent variable, the circular variance $\bar{R}$, is estimated as a function of a range of independent explanatory variables which are listed in Table 8.1.

As in Figure 8.2, the circular variance statistic $\bar{R}$ is calculated using three different specifications of local areas, i.e. a 15km radius from the university's full postcode, the travel to work area (TTWA) within which the university is located, and a 50km radius from the university. For each definition of the local area of the university, we exclude those graduates who moved into employment within the area, because for our purposes they represent non-migrants. We then calculate the circular variance statistic on the basis of the students who moved outside of the local area in order to enter employment. The basic estimated model defined by equation (8.4) therefore takes three slightly different forms, Model 1, Model 2 and Model 3. These three models reflect the fact our estimations take place over three different spatial definitions, in that Model 1 is estimated within a 15km radius of the university, Model 2 within the university TTWA and Model 3 within a 50km radius of the university. This allows us to identify the extent to which the estimated results are themselves sensitive to the spatial areas employed. The three models are estimated both with standard OLS and with White adjusted standard errors to control for heteroscedasticity. The results of the six estimated models are given in Table 8.2. All six models are robust to different specifications and the robust version correcting for outliers is very similar to the normal OLS. They pass all the standard tests in that the errors are normally distributed, the Ramsey Reset test confirms that there are no omitted variables problems, there is no significant residual heteroscedasticity, the explanatory variables do not suffer from multicollinearity (as measured by the VIF) and a linear specification is justified.

Table 8.1 List of variables

Variable	Definition
DIST	Distance travelled (in km) from the university attended to the first–employment location
RUSSELL	Dummy variable for universities being a member of the Russell Group of elite research universities (=1)
ASIAN	Percentage of the university student body which is of Asian ethnicity
BLACK	Percentage of the university student body which is of Afro-Caribbean ethnicity
FMRATIO	The ratio of females to males in the university student body
PROFMAN	Percentage of the university student body whose domestic social background is of a professional or managerial status
MATURE	Percentage of the university student body which is defined as being "mature" i.e. over the age of 22 when first attending university.
GOODDEG	Percentage of the university student body achieving a class 1 or class 2.1 final degree classification.
LONDON	Percentage of the universities graduates entering first–employment in Greater London
SPEC	Herfindahl Index of the concentration of range of subjects taught at the university
COAST	Dummy for a university being in a local area which is on the coast, as this artificially reduces the value of the circular variance.
EDUC	Percentage of the university student body studying education
ENG	Percentage of the university student body studying engineering
MATHS	Percentage of the university student body studying maths

The majority of the results presented in Table 8.2 conform to our expectations. Students who moved farther from their initial domicile to enter university (*DIST*) are then more likely to be more "focused" in their subsequent migratory move to enter employment (lower circular variance) and so are students graduating from an HEI belonging to the Russell Group of elite research universities (*RUSSELL*). Not surprisingly, correcting for the proximity of the HEI to the coast also decreases the spread of migration movements (*COAST*). The circular variance also tends to be negatively

Table 8.2 Regression results

	Dependent Variable: Circular Variance					
	Model 1: 15 km radius		Model 2: TTWA		Model 3: 50 km radius	
	OLS	Robust	OLS	Robust	OLS	Robust
DIST	-1.25e-06***	-1.39e-06***	-105e-06**	-1.22e-06***	-1.37e-06***	-1.45e-06***
RUSSELL	-.149***	-.141***	-.121**	-.118**	-.130**	-.133**
COAST	-.133***	-.130***	-.143***	-.131***	-.169***	-.163***
LONDON	-.144*	-0.105	-.205**	-.193**	-.165**	-0.132
GOODDEG	-.270*	-.307*	-0.073	-0.067	-.322**	-.319*
SPEC	-.385***	-.414***	-.263**	-.261**	-0.086	-0.09
EDUC	.362***	.358***	.326***	.328**	0.107	0.101
ENG	.327*	.373**	0.184	0.214	0.101	0.147
MATHS	1.47*	1.37	0.996	1.07	0.728	0.823
ASIAN	-0.005	-0.003	0.000	0.000	-0.005	-0.007
BLACK	.012*	0.008	0.003	0.001	.014*	0.014
FMRATIO	.068*	.073**	0.013	0.014	0.041	0.043
MATURE	-.635**	-.646**	-0.014	-0.033	-.600**	-.640**
PROFMAN	-0.452	-0.406	-0.469	-0.49	0.06	0.063
CONST	.839***	.856***	.729***	.733***	.794***	.797***
Number of obs.	*105*	*105*	*105*	*105*	*102*	*102*
R-squared	0.588	0.499	0.530	0.490	0.490	0.553

related to both the proportion of graduates entering into employment in London (*LONDON*) and the proportion of graduates gaining a class 1 or 2.1 final degree classification (*GOODDEG*). Assuming that high-skilled and -wage jobs tend to be clustered in particular urban areas, we would expect high ranking research universities with many highly qualified graduates to have a very skewed spatial employment pattern which is reflected by a low circular variance statistic. Our econometric results for *RUSSELL*, *GOODDEG* and *LONDON* demonstrate exactly that. The reason we observe this is that highly qualified individuals are in general more geographically mobile in search of jobs and wages consistent with their skills (Faggian et al. 2007a,b; Faggian and McCann 2009a), and in the UK their employment-migration behaviour is primarily directed towards London (Faggian and McCann 2009b). In contrast, lower-skilled graduates from lower ranking universities tend to travel shorter distances and in different directions in search of employment. As such, the employment-migration behaviour of higher-skilled graduates is more focused, in that they are targeting specific local labour markets, and this is true irrespective of whether their mobility is speculative or contracted (Molho 1986; Basker 2003), whereas the employment migration behaviour of the lower-skilled graduates is more general.

HEIs which are highly specialised in certain subjects and offer a limited number of specific degree programmemes (SPEC) also exhibit lower circular variance, while universities with a high proportion of students in education (*EDUC*), engineering (*ENG*) and to a lesser extent mathematics (*MATHS*), have higher circular variance values. Including the degree of specialisation of an HEI and the subjects studied is crucial as it provides a way of assessing the degree of geographical targeting which takes place in the employment search process of a university's graduates, and the success of this targeting. This argument is entirely consistent with the human capital and search model theories but the existing empirical tools are unable to capture how the degree of targeting distorts the geography of the search process. While the human capital and search model theories assume that the mobility distance increases, our results are consistent with the argument that search outcomes are also a result of spatial sorting (Bartel 1979) whereby the most highly skilled are those who are offered the highest quality jobs first, and many of these high quality jobs are in the dominant city.

Finally, there is evidence that the percentage of a university's student cohort who are of Afro-Caribbean ethnicity (*BLACK*), who are women (*FMRATIO*) or who are mature students (*MATURE*) move more broadly, entering into employment in a range of directions and in a less targeted manner. These findings are consistent with other empirical evidence which suggests that after graduation a higher percentage of women and mature students stay within the vicinity of the university for employment (Faggian et al. 2007a), and that students of Afro-Caribbean ethnicity also have a low migration propensity (Faggian et al. 2006). No role seems to be played by the socio-economic background (*PROFMAN*) or Asian ethnicity.

CONCLUSIONS

Our chapter represents the first attempt to integrate circular statistics into the study of economic geographical phenomena such as migration.

In the last few years, the role of graduate migration flows in fostering local economic development has become increasingly clear. Previous studies (see Faggian and McCann, 2006; Faggian et al. 2006, 2007a,b; Faggian and McCann 2009a,b) have shown that the primary role of the university system appears to be as a conduit for attracting potential high quality undergraduate human capital to a region and that this is far more important than direct effects of local university–industry spillovers. Yet, the phenomenon of graduate migration is still relatively unexplored and the traditional statistical techniques applied in economics do not seem to fully account for the complexity of this phenomenon. In this chapter we introduced the use of circular statistics to the study of interregional graduate migration flows in Great Britain. These techniques, applied previously in fields such as biology, meteorology and geology, allowed us to better understand the directionality and spread of student flows after graduation in an effort to link these dynamics to local economic variables and/or characteristics of higher education institutions.

Our analysis suggests that students with higher levels of human capital not only tend to be more migratory as previously demonstrated (e.g Faggian and McCann 2009c) but they also tend to be more focused in their job search process post-graduation. Other factors being equal, having a good degree classification or studying in more prestigious HEIs means graduates are more likely to target specific job markets rather than search in multiple directions. Other personal characteristics also appear to play a role. Mature students seem to be directionally more selective, while certain ethnic minorities and female students seem to travel shorter distances (see Faggian and McCann 2007a,b) but in a broader range of directions.

This chapter has demonstrated that the combination of non-traditional circular statistics tools with more traditional econometric modelling enhances our understanding of migratory dynamics. More fundamentally, it also lays the foundation for future extensions and applications to other contexts and to different economic geographical phenomena.

APPENDIX 8A
THE DEFINITION OF THE LOCAL SPATIAL AREA

First of all we tested how sensitive retention rates within the local area are to different definitions of a local area and therefore also of a migrant who moves outside of the local area. We defined local area as our usual 15km radius and then compared it to a 50km radius and, finally, to TTWA areas.

A comparison between the retention rates based on an area of 15km radius and the ones based on TTWAs shows that a circular area of 15km is actually a very good approximation for TTWAs, thereby justifying the approach adopted in other papers (Faggian et al. 2006, 2007a,b; Faggian and McCann 2006, 2009a,b). The average geographical size of a TTWAs is 772.55km^2 with an average radius length of 15.68km, as compared with an area of 706.5km^2 for a circular area with a radius of 15km. The smallest TTWA is Torquay (26.89 km^2) and the largest is Sutherland (5,542 km^2).

The migration patterns and hence the calculated circular variances are found to be very similar irrespective of how we define local areas. Values of the circular variance calculated using the 15km radius or TTWAs are very close for almost all universities. Circular variances calculated using the 50km radius are also generally similar to the other two, but there are more differences here, both because the movements are much longer and also because the number of observations classified as migrants is lower, and hence the values of the circular variance are less robust.

The only major exception here is the case where a university is located along the borders of two adjacent TTWAs, such as is the case for Thames Valley University depicted in Figure 8A.1.

In fact, for universities located at the borders of two or more TTWAs, a circular area of 15km radius seem to be a preferable definition of the local area. The reason is that in our models using a TTWA as the definition of a local area, the definition of a migrant is a graduate who enters into employment outside of the TTWA. In the case of TVU the fact that the university is located on the border of three TTWAs means that with our definition of a migrant, we will underestimate the true local retention rate of TVU. The reason is that all the graduates moving east will be included into the migrant category irrespective of the length of their movement, even though many of them will actually be entering into local employment. In fact, the retention rate calculated using TTWAs is found to be 19% lower than that which is based on a circular 15km area. We repeated this exercise for all universities in the UK in order to examine how sensitive our results are to these border problems. Overall, our results are found to be very robust.

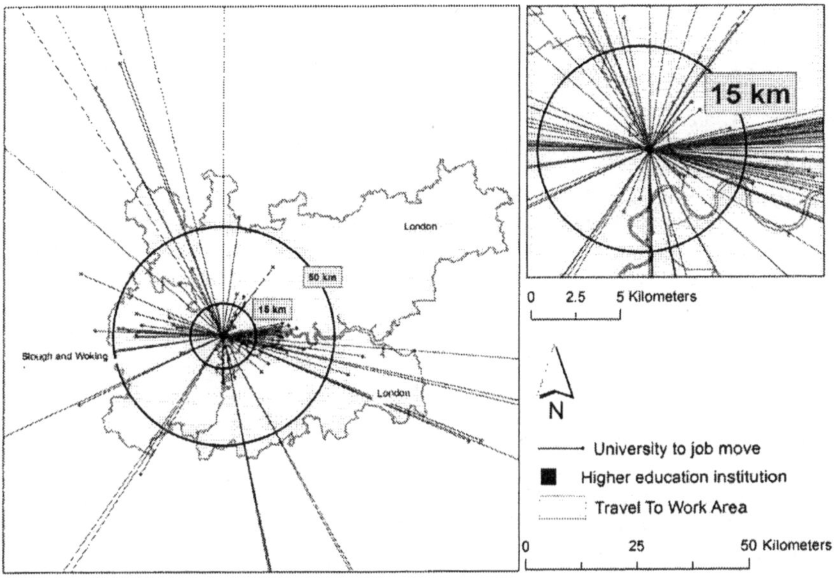

Figure 8A.1 Modifiable areal unit problem (MAUP): The case of Thames Valley University

NOTES

1. This information is recorded between six and eighteen months after graduation.
2. For the purposes of our graduate migration data in which the direction data are calculated in degrees, the scaling constant is defined as $\omega = \pi/180$.

REFERENCES

ArcView, 2002, Version 3.3, Environmental Science Research Institute, Redlands, CA.

Bartel, A., 1979, "What Role does Job Mobility Play?", *American Economic Review*, 69, 775–786.

Basker, E., 2003, "Education, Job Search and Migration", University of Missouri Department of Economics Working Paper 02-16.

Bernoulli, D., 1734, "Récherches Physiques et Astronomiques, Sur Le Problème Proposé Pour La Seconde Fois Par l'Academie Royale des Sciences des Paris", Récuiel des Pièces Qui Ont Remporté Le Prix de l'Académie Royale des Sciences, *Tome III*, 95–134.

Brunsdon, C. and Corcoran, J., 2006, "Using Circular Statistics to Analyse Time Patterns in Crime Incidence", *Computers, Environments and Urban Systems*, 30, 300–319.

CoL, 2006, *Graduate Skills and Recruitment in the City*, Corporation of London, London.
Corcoran J., Faggian A. and McCann P., 2010, "Human capital in remote and rural Australia: the role of graduate migration", *Growth and Change*, 41(2), 192-210.
Faggian, A. and P. McCann, 2006, "Human Capital Flows and Regional Knowledge Assets: A Simultaneous Equation Approach", *Oxford Economic Papers* 58(3), 475–500.
Faggian A. and McCann P., 2009a, "Human capital, graduate migration and innovation in British Regions", *Cambridge Journal of Economics*, 33, 317–333.
Faggian A. and McCann P., 2009b "Human Capital and Regional Development" in Capello R. and Nijkamp P. (eds.), *Handbook of Regional Growth and Development Theories*, Edward Elgar, Cheltenham, 131–151.
Faggian A. and McCann P., 2009c, "Universities, Agglomerations and Graduate Human Capital Mobility", *Journal of Economic and Social Geography (TESG)*, 100(2), 210–223.
Faggian, A., McCann, P. and Sheppard, S., 2006, "An Analysis of Ethnic Differences in UK Graduate Migration Behaviour", *Annals of Regional Science*, 40(2), 461–471.
Faggian, A., McCann, P. and Sheppard, S., 2007a, "Some Evidence that Women are More Mobile than Men: Gender Differences in UK Graduate Migration Behaviour", *Journal of Regional Science*, 47(3), 517–539.
Faggian, A., McCann, P. and Sheppard, S., 2007b, "Human Capital, Higher Education and Graduate Migration: An Analysis of Scottish and Welsh Students", *Urban Studies*, 44(3), 2511–2528.
Fisher, N.I., 1993, *Statistical Analysis of Circular Data*, Cambridge University Press, New York.
Glaeser, E.L. and Mare, D.C., 2001, "Cities and Skills", *Journal of Labor Economics*, 19, 316–342.
Hornsby, K. and Egenhofer, M., 2002, "Modelling Moving Objects Over Multiple Granularities", *Annals of Mathematics and Artificial Intelligence*, 36(1–2), 177–194.
Kwan, M.P., 2000, "Interactive Geovisualisation of Activity-Travel Patterns in Three-Dimensional Geographical Information Systems: A Methodological Exploration with a Large Data Set", *Transportation Research Part C: Emerging Technologies*, 8(1–6), 185–203.
Lippman, S.A., and McCall, J.J., 1976a, "The Economics of Job Search: A Survey, (Part I)", *Economic Inquiry*, 14, 155–189.
Lippman, S.A., and McCall, J.J., 1976b, "The Economics of Job Search: A Survey, (Part II)," *Economic Inquiry*, 14, 347–368.
Lippman S.A. and McCall J.J., 1979, Studies in the Economics of Search, North-Holland, Amsterdam.
Molho, I., 1986, "Theories of Migration: A Review", *Scottish Journal of Political Economy*, 33(4), 396–419.
Nakamura, R., 2008, "Agglomeration Effects on Regional Economic Disparities: A Comparison between the UK and Japan", *Urban Studies*, 45(9), 1947–1971.
Pissarides, C.A., 1976, "Job Search and Participation", *Economica*, 43(169), 33–49.
Sjaastad, L.A., 1962, "Costs and Returns of Human Migration", *Journal of Political Economy*, 70(5), 80–93.
Tobler, W., 1981, "A Model of Geographic Movement", *Geographical Analysis*, 13(1), 1–20.

9. Commuting and migration: can they be complementary?

Raphael Bar-El and Miki Malul

INTRODUCTION

The Spatial Mismatch Hypothesis

The "spatial mismatch hypothesis" (Kain, 1992) states that the dispersal process of jobs from the centre to the suburbs creates a distorted equilibrium, with high levels of employment and income in the suburbs and unemployment and low income in the centre. This is attributed mainly to the existence of a "housing segregation" effect, which prevents the poor population from migrating to richer areas. Bar-El (2006) attempted to identify a mechanism of inter-regional labour markets that would explain the existence of these apparent regional imbalances as a function of the influence of variables such as land costs, commuting behaviour and economic structures. He presented a survey of the theoretical background behind the "spatial mismatch hypothesis", and developed a general theoretical model based on two regions (the centre and the suburbs) that would contribute to the explanation of labour market behaviour. We first bring here a summary of Bar-El's model, and then illustrate it with empirical data from the centre and suburban regions in Israel for the period 1970–2005.

The spatial mismatch hypothesis basically states that the process of job dispersal from the centre to the suburbs induces a "people follow jobs" effect, leading to migration of the labour force to the suburbs. However, people who can afford to migrate are mostly richer than the poorer population that remains in the centre (Selod and Zenou, 2006). This effect is further reinforced by a housing segregation effect, preventing the poor population from migrating to richer areas. Commuting is also limited for those living in the centre because of the cost it implies, and because employers prefer local workers. Therefore, in spite of an increasing labour demand in the suburbs, we find unemployment and low levels of income in the centre (Ihlanfeldt and Sjoquist, 1998). This implies the existence of a deadlock situation, where an external obstacle (segregation) impedes the action of the free market process that could lead to a clean equilibrium with equal levels of employment and income in both regions. The consequence is a need for government intervention with compensatory measures such as social assistance and transfer payments (Coulson et al., 2001).

Empirical evidence is not always clear (see Holzer, 1991; Taylor and Ong, 1995). However, we find some evidence in the literature that migration to a given region is not necessarily driven solely by economic activity. Assuming the existence of a labour demand-driven regional growth as shown by Mathur and Song (2000), it can be seen that increasing labour demand in neighbouring regions (suburbs) can actually attract population to the centre (Khan et al., 2001). An important aspect of this finding is the potential for the achievement of equilibrium through commuting as an alternative to migration. However, the main conclusion is that commuting costs discriminate against the poor population in the centre (Zenou, 2002). Most of the labour force is unskilled and commuting costs are higher in proportion to their salary, leading to an increase in the gap between the centre and the suburbs.

In the next section we present an alternative theoretical model to the classical spatial mismatch model, which would be expected to lead to completely different outcomes. This is followed by an empirical testing in two regions in Israel. Finally we derive conclusions for public policy.

A THEORETICAL ALTERNATIVE MODEL

The Israeli experience in two regional levels as reported in the literature leads us to a set of rules or principles that may explain the empirical situation of a regional equilibrium where various locations coexist at different levels of employment and income, with no "free market" movement of population or of economic activity from one region to another. We present here some of those principles, described in a more formal manner by Bar-El (2006), and in the next section provide some empirical findings from the Israeli case that tend to support them.

The Model

The model lays out a framework for a theoretical general equilibrium model that may explain the existence of a spatial equilibrium with inequalities in employment and income in various regions:

Labour demand

Labour demand is defined as the number of jobs offered within each region. For the sake of simplification, we assume that all jobs offered are actually taken, meaning that labour demand is actually the number of employed persons within a region (however, they are not all necessarily residents of the region).

The labour demand (LD_r) in each region (r) depends upon four variables:

AE_r – A measure of agglomeration economies in each region. Agglomeration economies are defined as the benefits of agglomeration of economic activity, and are assumed to increase rapidly in the first stages of economic agglomeration and at a decreasing rate in the later stages, as a response to diseconomies resulting from increasing levels of crowdedness (see also Malul and Bar-El, 2009).

LC_r – Land costs in each region, which depend upon the land density, including both the agglomeration of economic activities and the agglomeration of population in each region (assuming increasing marginal cost as presented in Malul and Bar El, 2009).

W_r – A measure of the level of wages in each region.

P_r – The population in each region.

Thus the labour demand equation is:

$$LD_r = f(AE_r, LC_r, W_r, P_r) \tag{9.1}$$

Labour demand in each region responds positively to an increase in agglomeration economies and to an increase in population, and negatively to the cost of the production factors of land and labour (wage level).

Labour supply

Labour supply (LS_r) represents the number of individuals that participate in the labour force in each region.

The factors that determine the labour supply are:

POP_r – The population in region r. The size of the population can change either by natural increase or by migration to/from the region. Migration is influenced by the relative cost of land (housing) in both regions, the gap between the wages, and the cost of commuting as an alternative to migration and a constant value for migration cost.

P_r – The participation rate in the labour force in each region.

NC_r – The gap between the number of in-commuters and out-commuters in the region. The scope of commuters from one region to another depends upon the differences in the wages between the regions and the commuting cost between the regions. Commuting becomes more attractive when the commuting cost is lower and when the wage differences between the two regions is bigger.

Thus the labour supply equation could be written as follows:

$$LS_r = f(P_r, POP_r, NC_r) \tag{9.2}$$

The main patterns of behaviour could stem from the above equations, as follows:

1. This model actually supports the existence of a process of job dispersal from the centre to the suburbs. This is instigated by the decreasing return to agglomeration economies in the centre, the increasing return to agglomeration economies in the suburbs and the changing proportion of land costs. According to the current model, such a process may explain why the spatial mismatch does not necessarily lead to labour market instability, as the process of job dispersal stabilizes at a certain level.
2. The spatial mismatch hypothesis is generally explained by the non-adaptability of the labour supply. Migration is hindered by so-called housing segregation and by commuting of the labour force from poor centre areas to richer suburbs, and is reduced because of the relatively high cost of commuting for unskilled workers. Reverse commuting from the suburbs to the centre is generally assumed to be non-existent. This model tries to provide a more detailed explanation of migration and commuting and their influence on the equilibrium of population distribution and labour markets. The traditional assumption that serves as a basis for the spatial mismatch hypothesis, claiming that people follow jobs, is complemented in this model by the assumption that people follow low housing costs as well. Furthermore, the generally accepted assumption that migration and commuting are necessarily alternatives is rejected here: they may be also complementary.

We can therefore expect the existence of three patterns based on combinations of job-driven and housing-driven migration:

(a) Regular migration for jobs, as in the spatial mismatch hypothesis.
(b) Migration of the population from one region to another, mainly driven by the increasing relative cost of housing in the first region, even if the expectation for a job in the second region is quite low. This is the case when the coefficient of the influence of housing is

stronger than the coefficient of the influence of the labour demand. This generally occurs with workers with high job mobility (workers who are confident in finding jobs at any place) or with new entrants to the labour force who are not yet strongly linked to a job.

(c) Migration caused by changes in housing price proportions as above, but strengthened by low commuting costs that permit keeping a job at the region of origin. Here we have a case of complementarity (instead of a trade-off) between migration and commuting: the low cost of commuting enables migration and benefits from lower housing costs.

EMPIRICAL EVIDENCE FROM TWO REGIONS IN ISRAEL

In this section we present some empirical evidence for the case of Israel, in order to illustrate at least some aspects of market behaviour, as shown above.

Israel has a population of about seven million inhabitants within an area of about 20,000km^2, distributed across six statistical districts (the West Bank and Gaza are not included in Israeli statistics, with the exception of East Jerusalem, which is included in the Jerusalem District). For the purposes of this chapter, we consider two districts: the metropolitan centre (C) and the suburbs (S).

The metropolitan centre (C) is defined as the Tel-Aviv District; it includes the main urban centre Tel-Aviv, which, together with a few smaller cities, forms the continuous urban area of Greater Tel-Aviv (Bar-El and Parr, 2003). This has a population of about 1.1m, concentrated in an area of 170km^2. Thus, 16% of the population is concentrated in less than 1% of the area of the nation, representing a high density of 6,700 persons/km^2.

The suburbs (S) region is defined as the so-called Centre District, and includes a wide area surrounding the Tel-Aviv District. Its population is comparable to (but somewhat greater than) that of the Tel-Aviv District, although it is distributed over an area almost eight times larger. Most of the population is urban, and is distributed among centres located within commuting distance of Tel-Aviv, the largest being Rishon Lezion, which has a population of around 175,000. Compared to the Tel-Aviv District, population density is relatively low (about 1,000 persons/km^2), although it is still greater than that of all other districts with the exception of Jerusalem.

Labour Demand: Dispersal of Jobs

Empirical findings actually support the existence of a process of job dispersal from the centre to the suburbs (as in the spatial mismatch hypothesis), apparently instigated by three main factors: the decreasing return to agglomeration economies in the centre, the increasing return to agglomeration economies in the suburbs and the changing proportion of land costs. However, according to our approach, the spatial mismatch does not necessarily lead to labour market instability. Given the quality of the influence of agglomeration economies (first increasing return to agglomeration, later decreasing return), the process of job dispersal stabilizes after a certain level is reached. The changes that are induced by the proportion of land costs between the regions also play a stabilizing role.

Table 9.1 and Figure 9.1 show the changes in labour demand in the two regions during a period of almost three decades, divided according to the business cycles in Israel: 1977–1990 and 1997–2002 as periods of relative recession, and 1991–1996 and 2003–2005 as periods of fast growth.

Table 9.1 Average annual growth rates of labour demand by period and region (in percent)

Region/years	1977–1990	1991–1996	1997–2002	2003–2005	Total
C	1.5	4.3	0.3	2.4	1.9
S	2.0	6.9	4.9	3.3	3.7
I (Israel)	2.0	5.1	2.1	3.0	2.8

Source: *Statistical Abstract of Israel* (1977–2005), Table 12.16.

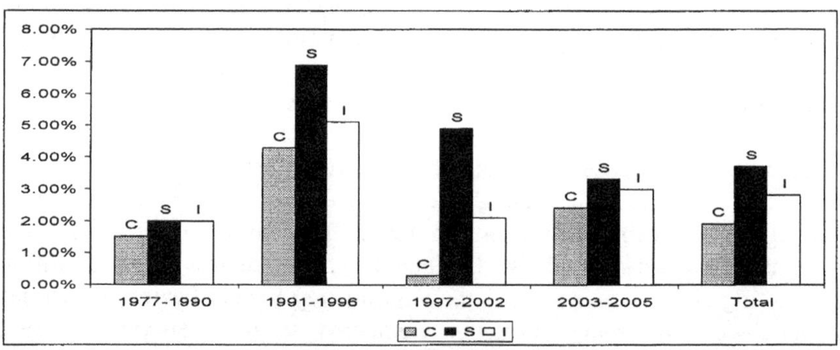

Figure 9.1 Average annual growth rates of labour demand by period and region

An indication of the existence of such processes can be seen in the case of the two regions in Israel. The labour demand in the suburbs (the Centre District) grew rapidly (3.7% a year in average during the whole period 1977–2005) in comparison to that in the centre (the Tel-Aviv District), which was quite stable during this period and hardly followed the population growth.

The growth period 1991–1996, which was also characterized by a heavy flow of migration to Israel from the former USSR, reflected a rapid growth in the whole country. However, during the following period of recession, from 1997 to 2002, the increase in labour demand was more concentrated in the suburbs.

Labour Supply: Migration and Commuting

Tables 9.2, 9.3, 9.4 and Figure 9.2 show a picture of the changes in population and labour supply along the period analysed.

The data in the tables above indicate that:

1. Labour supply increases very slowly in C (the district of Tel-Aviv), excluding the period 1991–1996 that was characterized by a strong flow of immigration, mostly from the former Soviet Union (although it was weaker in C), indicating a steady process of out-migration, mostly to S.
2. Migration to S during the 13 years from 1977 to 1990 is concomitant with an increasing gap between labour supply and labour demand in that region. During this period, unemployment levels remained as low as in C, indicating that migrants from C to S continued mostly in their employment at C, as will be seen next in the commuting figures.
3. This process creates agglomeration economies in S, which is explained by the fact that since the 1990s the increase of labour demand has been higher than the increase in labour supply in S.

Table 9.2 Average annual rates of population growth, by period, and region (in percent)

Region/ years	1977–1990	1991–1996	1997–2002	2003–2005	Total
C	0.9	0.7	0.3	0.8	0.7
S	2.9	3.6	3.2	2.3	3.1
I	2.2	3.0	2.4	1.8	2.3

Source: *Statistical Abstract of Israel* (1977–2005), Table 2.10

4. This process of adaptation seems to have stabilized during the last few years, with a stable population in C and a balance between the growth of labour demand and labour supply in S.

Table 9.3 Average annual migration balance per thousand population, by period and region

Region/years	1977–1990	1991–1996	1997–2002	2003–2005	Total
C	-3	-15	-13	-1	-7
S	7	7	14	9	9

Source: *Statistical Abstract of Israel* (1977–2005), Table 2.5

Table 9.4 Average annual growth of labour supply, by period and region (in percent)

Region/years	1977–1990	1991–1996	1997–2002	2003–2005	Total
C	1.2	2.9	0.9	1.6	1.5
S	3.2	4.8	4.9	2.8	3.8
I	2.5	4.5	2.8	2.5	3.0

Source: *Statistical Abstract of Israel* (1977–2005), Table 12.4

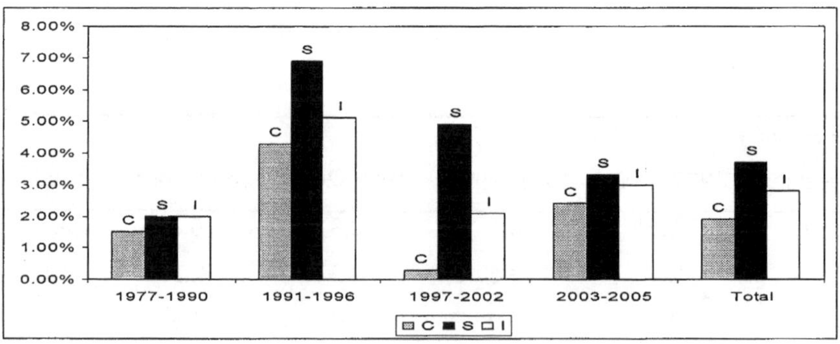

Figure 9.2 Average annual growth of labour supply, by period and region

The following tables show the changes in commuting patterns and in unemployment rates.

Table 9.5 Average share of out-commuters from the region out of labour supply, by period and region (in percent)

Region/years	1977–1990	1991–1996	1997–2002	2003–2005	Total
C	13	13	16	17	14
S	30	30	30	28	30

Source: *Statistical Abstract of Israel* (1977–2005), Table 12.15

Table 9.6 Average share of in-commuters to the region out of labour demand, by period and region (in percent)

Region/years	1977–1990	1991–1996	1997–2002	2003–2005	Total
C	21	27	34	41	27
S	17	19	21	21	19

Source: *Statistical Abstract of Israel* (1977–2005), Table 12.15

Table 9.7 Average unemployment rates, by period and region (in percent)

Region/years	1977–1990	1991–1996	1997–2002	2003–2005	Total
C	5.2	7.7	7.9	8.6	27
S	5.2	8.4	8.0	9.5	6.9

Source: *Statistical Abstract of Israel* (1977–2005), Table 12.24

The data in the above tables support some of the principles of our approach:

1. The heavy migration out of the centre as shown above does not lead to any diminution in its economic role: the share of commuters to the centre increases constantly from an average of 21% in the 1970s and 1980s, to an average of 41% of the labour demand in recent years. This supports the thesis of a "housing cost driven" migration out of the centre to the suburbs, which is enabled by the ability to commute back to jobs in the

centre. Recall that the share of the centre in total labour demand did not grow and even decreased during this period.

2. The share of commuters from the suburbs out of the labour supply was quite stable over the years (30%), showing a good integration into the regional economy: the heavy migration to the suburbs was followed by a parallel growth in labour demand, generated by the creation of agglomeration economies and by increasing market demand. Another factor is the growing attraction of commuters from other regions to the suburbs. Still, in absolute terms, the number of commuters from the suburbs (with a growing labour supply) to the centre (with a stable labour supply) is growing constantly, as a consequence of the "migration with commuting" pattern.

CONCLUSIONS

The empirical findings seem to support our argument that spatial mismatch does not necessarily lead to labour market instability. Given the quality of the influence of agglomeration economies (first increasing returns to agglomeration, later decreasing returns), the process of job dispersal stabilizes after a certain level. The changes that are induced in the proportion of land cost between the regions also play a stabilizing role.

The generally accepted assumption that migration and commuting are necessarily alternatives is rejected here: they may be also complementary. The heavy migration out of the centre does not lead to any diminution in its economic role: the share of commuters into the centre has increased constantly, from an average of 21% in the 1970s and 1980s, to an average of 41% of the labour demand in more recent years. This supports the thesis of a housing cost driven migration out of the centre to the suburbs, which is enabled by the ability to commute back to jobs in the centre.

Finally, the public policy issue is of major importance. The spatial mismatch hypothesis claims that the problems caused by the process of job dispersal should be solved by the adoption of "compensatory" measures that help the poor. This model implies that public policy measures can be used as instruments for the solution of bottlenecks in labour market dynamics and to help achieve a more efficient and economically sound equilibrium. The measures that can be considered as viable are public investment in transportation costs to facilitate commuting solutions, public investment in infrastructure to stimulate economic development, facilitating agglomeration economies, and land-allocation policies to alleviate distortions in land prices.

REFERENCES

Bar-El R. (2006), "Inter-regional Labour Market Equilibrium: Another Pattern of Spatial Mismatch", *Annals of Regional Science* 40(2): 393–05.

Bar-El R. and Parr J. (2003), "From Metropolis to Metropolis-based Region: The Case of Tel-Aviv", *Urban Studies* 40(1): 113–25.

Coulson N.E., Laing D. and Wang P. (2001), "Spatial Mismatch in Search Equilibrium", *Journal of Labour Economics* 19(4): 949–72.

Holzer H.J. (1991), "The Spatial Mismatch Hypothesis: What Has the Evidence Shown?", *Urban–Studies* 28(1): 105–22.

Ihlanfeldt K.R. and Sjoquist D.L. (1998), "The Spatial Mismatch Hypothesis: A Review of Recent Studies and Their Implications for Welfare Reform", *Housing Policy Debate* 9(4): 849–92.

Kain J.F. (1992), "The Spatial Mismatch Hypothesis: Three Decades Later", *Housing Policy Debate* 3(2): 371–40.

Khan R., Orazem P.F. and Otto D.M. (2001), "Deriving Empirical Definitions of Spatial Labour Markets: The Roles of Competing versus Complementary Growth", *Journal of Regional Science*, 41(4): 735–56.

Malul M. and Bar-El R. (2009), "The Gap Between Free Market and Social Optimum in The Location of Economic Activity", *Urban Studies*, 46(10): 2045–2059.

Mathur V.K. and Song F.M. (2000), "A Labour Market Based Theory of Regional Economic Development", *Annals of Regional Science* 34(1): 131–45.

Selod H. and Zenou Y. (2006), "City Structure, Job Search and Labour Discrimination: Theory and Policy Implications", *The Economic Journal*, 116(514): 1057–87.

Statistical Abstract of Israel (1977–2005), Jerusalem, Israel: Central Bureau of Statistics, Government of Israel.

Taylor B.D. and Ong P.M. (1995), "Spatial Mismatch or Automobile Mismatch? An Examination of Race, Residence and Commuting in US Metropolitan Areas", *Urban Studies* 32(9): 1453–73.

Zenou Y. (2002), "How Do Firms Redline Workers?", *Journal of Urban Economics* 52(3): 391–08.

10. Globalisation, the CEECs and European policy

Daniela Constantin, Zizi Goschin and Philip McCann

INTRODUCTION

This chapter discusses the issues raised by flows of foreign direct investment (FDI) into the Central and Eastern European countries (CEECs). These inward investment flows take place as part of the processes of internationalisation and globalisation currently operating in all parts of the world. However, the fact that these countries are the beneficiaries of such flows raises issues regarding the role of EU policy towards these countries. In the next section we review the major themes of globalisation as they impact not just on Europe, but on all parts of the world. In the third section we discuss the recent experience of the CEECs, focusing in particular on the shifts of FDI eastwards within and across the European Union into the newly integrating economies. While the experience of the individual countries differs markedly, as a group we see that prior to the 2008 financial crisis there was a steady increase in the stocks of FDI in these countries as foreign investment inflows continued apace. The mobility of capital into the CEECs obviously provides a major stimulus for the growth in these countries, whose growth experience can be characterised broadly as reflecting slow convergence with the richer EU countries, a trend reflected in other parts of the world. Yet, the fact that these countries benefit from the market responses to EU integration itself poses questions regarding the justification for, the role of, and the nature of EU regional policy in these countries. This is the context in which EU cohesion policy operates. EU cohesion policy is starting to undergo a period of fundamental reforms, and in the case of the CEECs the market-driven inflows of capital also imply that structural reforms of the policy are necessary to enhance the capability of these countries to absorb such flows. These issues are discussed in the fourth section, accompanied by some very brief conclusions.

GLOBALISATION AND REGIONS UNDER PRESSURE

The modern era of globalisation emerged during the five-year period of rapid institutional and technological changes which took place between 1989 and 1995. Since the 1960s the world economy had been gradually re-opening up to international integration processes following half a century of anti-globalisation, driven by the two world wars, the inter-war Depression and the Bretton-Woods system. The rise of the Eurodollar markets in London in the 1960s was the precursor to the era of modern global finance, and advances in information and communications technologies in the 1970s and 1980s further drove these processes as increasing numbers of countries floated their currencies. In manufacturing, the rise of Japanese industry in automobiles, electronics and power equipment re-shaped the global competition for both household and industrial goods, and the development of global supply chains was further enhanced by containerisation and satellite global positing systems. At the same time, the geography of global markets shifted as China started to open up rapidly from 1989 onwards after Tiananmen Square, the transition economies converted to market systems after the fall of the Berlin Wall, South Africa remerged into the community of nations after the release of Mandela in 1990, India and Indonesia both restructured and internationalised after the second wave of industrial reforms in both countries in 1991, the European Single Market was established at the turn of 1991–92, Brazil floated its new currency in 1994 at the same time as the North American Free Trade Agreement took place. Taken together, the impact of these institutional changes was that one third of the world's labour force for the first time entered into the global marketplace. Most importantly, however, at the very time that all of these institutional changes were taking place, the invention of the internet as we know it in 1991 meant that all different types of electronic technologies could be combined onto a single platform.

The outcomes of these globalisation changes have been very profound. Many types of economic and commercial activities and operations have been out–sourced and off–shored to the CEECs recently joining the European Union as well as to the large newly emerging BRIICS countries of Brazil, Russia, India, Indonesia, China and South Africa (McCann 2009). In particular, a large number of technical operations have been off–shored, and managed over long distances using ICTs. These off–shored activities have largely been in the middle-skills categories of jobs, rather than high-skilled or very low-skilled jobs, and tend to impact adversely and disproportionately on male workers. At the same time, the distribution of earnings and wealth in all societies has become increasingly skewed in almost all societies with an increasing proportion of both income and wealth being accounted for by the higher skills groups, as the gains from global markets are increasingly reaped by high human capital elites.

In terms of economic geography and regions the outcomes of globalisation have also been very different from what was anticipated by most observers in the 1980s and 1990s. Predictions that distance and geography were becoming increasingly irrelevant as the world was flattening in economic terms have proved to be groundless (McCann 2008). Location is becoming more important than ever as a competitive issue. The most obvious manifestation of this is that the importance of densely populated cities and regions in economic activity has increased all over the world, in order to take advantage of the agglomeration economies associated with industrial clustering (World Bank 2009).

These clustering and localisation processes are increasingly associated with knowledge-driven societies and increasingly require face-to-face interaction between people in order to undertake higher value knowledge-intensive activities. Ironically, for high knowledge-intensive activities, the highest levels of interaction between people which are mediated via information and communications technologies are between the same people who meet regularly face to face. The result of this is also that the geography of knowledge indicators, such as patents, copyrights, licenses, inventions, innovations, entrepreneurship and creativity indices, are all becoming increasingly localised. In economic geography terms, the world is actually becoming what we can characterise as being more spiky, more uneven, rather than more flat, and this is true for a whole range of economic measures including value-added per capita, incomes, business formation rates and wages. Moreover, once agglomeration effects promote competiveness, the processes tend to become long-lasting, as the combined local agglomeration effects of all of the myriad networks of ancillary services, expertise and human capital are what make agglomerations "sticky" and so resilient even in adverse circumstances.

However, in many countries these agglomeration processes are not simply a matter of scale. While cities are generally growing in the USA, Japan and Korea, and knowledge is therefore increasingly contained in large agglomerations, in many of the other OECD countries things look rather different (McCann and Acs 2011). In Europe, in particular, cities in general, and dominant city sizes per se, are not increasing, but rather the development of inter-urban networks is increasing, as is the complex pattern of shifting fortunes of particular cities and regions. Rather than the urban scale, the European experience appears to be dominated by the degree of connectivity between locations, in that the connectedness of the knowledge generated and transmitted between the locations dominated by high-skilled people and occupations appears to be critical in the emerging global human and communications networks.

As such, both globalisation, in the sense of economic activity spreading to newly emerging countries, and localisation, in the sense of the growing importance of particular knowledge centres, are increasing at the same time, and this tendency towards localisation is most noticeable within individual countries. While globalisation is driving a process of slow convergence between countries, in that poorer countries are slowly catching richer countries, localisation is also meaning that regions are tending to diverge within individual countries. In addition to the face-to-face requirements associated with knowledge-intensive activities, these processes of regional divergence are also driven by two related but quite distinct demographic phenomena, namely ageing and migration.

All advanced economies are becoming older, in the sense that their population profiles are ageing due to both health-care improvements afforded by the large cohort of baby-boomers, and also the slowdown in the subsequent replacement ratios. These intergenerational effects imply that the working life expectancy of future generations will increase, while the pension returns are likely to fall. These population and demographic changes have no spatial implications per se. However, another modern demographic phenomenon has very profound regional implications. One of the key features of modern globalisation is that people are becoming increasingly geographically mobile, both internationally (OECD 2007, 2008a) and also, more noticeably, inter-regionally.

People are increasingly accustomed to moving for work, and also to moving multiple times during their working lives. However, given the nature of globalisation, in advanced economies, the most important group is the highly skilled. Here, young highly educated university graduates increasingly move to particular places, or rather to particular types of places, which are those locations dominated by knowledge-intensive activities and occupations. As this happens, the areas from which they move away naturally age in profile. Moreover, to the extent that the out-migrants are also the highly skilled and highly educated (OECD 2007, 2008a), the areas from which they originate experience an adverse combination of an increasingly ageing population, a declining population, a population whose wealth and incomes are declining (OECD 2008b) and problems of dereliction. In contrast, the locations into which highly educated groups move exhibit an increasingly highly-skilled and educated population, a growing population, and an increasingly youthful population. Yet, these same areas also tend to face increasing land prices, housing costs and congestion problems. In addition, to the extent that the middle-skill activities in these locations have been outsourced and offshored, these same locations also tend to exhibit increasingly segregated communities.

As people become more mobile, and as mobility is increasingly associated with skills, the result is that places become increasingly different. The long-term provision of public services such as health and education therefore becomes ever more complex, and different needs are manifested by increasingly different population profiles. For five decades the widespread existence of social safety nets in industrial countries has helped to cushion many of the worst effects of economic restructuring, but the modern era of globalisation implies more profound changes, as individual countries, and national and regional authorities, have less and less influence over the destinies of their own local constituencies and communities. The global economy today is very different from what it was in earlier decades. As we have seen, societies are far more interconnected via transportation technologies, supply chains and information technology than in previous eras, and people are increasingly mobile and less tied to particular localities. On the one hand, this technology connectedness contributes to a much greater degree of globally distributed knowledge and learning, and for more deeply integrated and sophisticated supply systems. At the same time, it also ensures that local and regional impacts increasingly depend on external events taking place in other parts of the world.

The long-run growth impacts of infrastructure, education and health-care investments are essential elements in helping us to achieve our future potential, but not only the level of these investments, but also the location of them is critical (EC 1998a). Ostensibly the right investment but in the wrong place or with the wrong scale is actually the wrong investment. As regions and localities become more different from each other, the nature of the issues associated with the provision of these public goods and services will become ever more complex, and for regions which are sparsely populated, declining or ageing the challenges appear to be particularly significant, as the globalisation processes appear to be systematically favouring other types of localities (Barca 2009).

THE EXPERIENCE OF CENTRAL AND EASTERN EUROPEAN COUNTRIES

The features described above are common to many areas of the so-called CEECs, the Central and Eastern European countries within the EU. Many regions within these countries face severe problems of a combination of low fertility and the out-migration to other parts of Europe (EC 2008) primarily by the youngest and the most highly educated cohorts, and this is particularly marked in the case of rural regions (Eurostat 2010). Capital city regions in some of the CEECs are becoming increasingly dominant by OECD standards (McCann and Acs 2011), leading to interregional disparities within CEECs which are very marked by EU standards. At the same time as cities such as Warsaw and Prague are relatively very buoyant, many of the outer and rural regions are declining rapidly. The impacts of the adverse effects of globalisation and the EU single market integration processes are seen to be very severe in certain localities. Moreover, there is no assumption that these processes of decline can be arrested within a relatively short period.

Such experiences suggest that these regions ought to be the major focus for EU cohesion policy, as indeed argued by both the two major influential reports on these subject, namely the 2003 Sapir Report *An Agenda for a Growing Europe* (Sapir et al. 2003) and more recently by the 2009 Barca Report *An Agenda for a Reformed Cohesion Policy* (Barca 2009). However, exactly how and why this should be the case is viewed in a fundamentally different way by the two reports, and the implications of these different viewpoints will be discussed in the latter part of this chapter. At this point it is sufficient to note that the experiences of globalisation in the CEECs within the EU have been very severe in certain regions. That of itself, however, is not a sufficient argument for policy favouring these countries. The reason why is that as EU integration continues, these countries have already benefited significantly from capital inflows. Whether the financial crisis of 2008 will change any of these historical patterns is still an open question, and this will be commented on in the final section of the chapter. However, because of the macroeconomic complexities associated with the financial crisis of 2008, we confine ourselves here to the period before that and observe the flows of inward investment into these countries as reflecting the processes of integration. Yet, the point remains that EU integration has significantly benefited the CEECs in terms of capital inflows.

Figures 10.1–10.11 describe the stocks of FDI in the CEECs as a whole, and then for each of the individual new member states of Bulgaria, Cyprus, the Czech Republic, Estonia, Hungary, Latvia, Lithuania, Poland, Romania, Slovakia and Slovenia, while Figures 10.12–10.22 describe the level of new inflows of foreign direct investment capital. All figures are calculated in Euro equivalent terms.

What becomes immediately obvious from the figures is that the stocks of FDI in these countries have continued to increase more or less continually during the decade prior to the financial crisis. To the extent to which some of these countries experienced a slight dip in inflows in 2002, this probably reflects the fall-out of the Asian financial crisis, but the long-term observed upward trajectory was once again re-established shortly afterwards. The outcome of these processes has been that foreign investment flows were moving eastwards within Europe. These flows have been well documented in the automobile industry in particular, but actually reflect a more general pattern of a drift eastwards of FDI flows within Europe. The eastwards mobility of FDI capital reflects the business and commercial decisions undertaken by many firms in order to take advantage of the significant labour and land price differentials between eastern and western Europe. The increased transportation costs and loss of accessibility with western EU markets associated with these shifts are often more than compensated for by the lower factor prices in these localities. In addition, these FDI movements will also be associated with the opening up of new markets in the former transition economies, and an increasing presence of multinational establishments in these CEECs provides for increased market access opportunities in these newly-emerging EU markets.

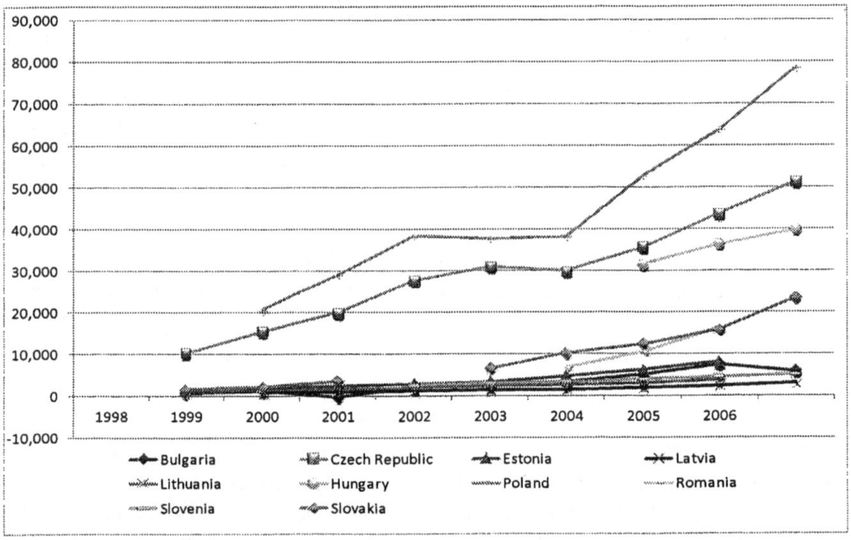

Figure 10.1 Foreign direct investment stocks in the new member states

The picture in Central and Eastern Europe therefore emerges as being very complex. On the one hand, many of the regions in these countries have experienced significant outflows of labour to other parts of Europe, and in particular to northern and western Europe. At the same time, these countries have also benefited from significant and fairly consistent capital inflows. In general, a neo-classical adjustment model would suggest that this combination of labour outflows and capital inflows would imply rising real wages, and indeed, this is exactly what we observe amongst these countries.

GDP per capita has been increasing significantly in many CEECs, the result of which is that the EU as a whole has experienced a period of slow convergence between its member states. This therefore begs the following question – if market adjustment processes are leading to rising real wages in CEECs, then why is cohesion policy needed in these countries if the single market itself can achieve the desired result? Answering this question is the focus of the next section.

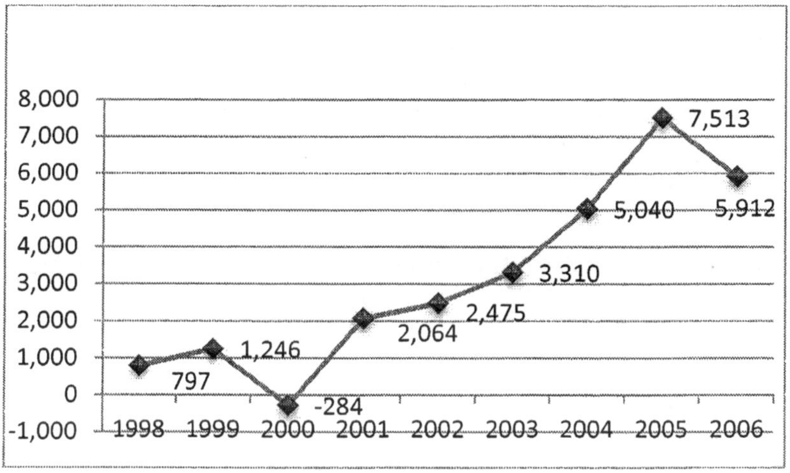

Figure 10.2 The stock of FDI in Bulgaria

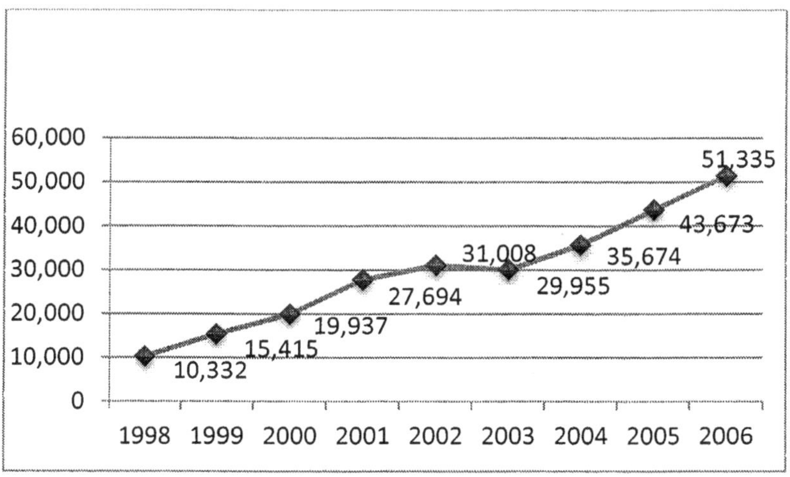

Figure 10.3 The stock of FDI in the Czech Republic

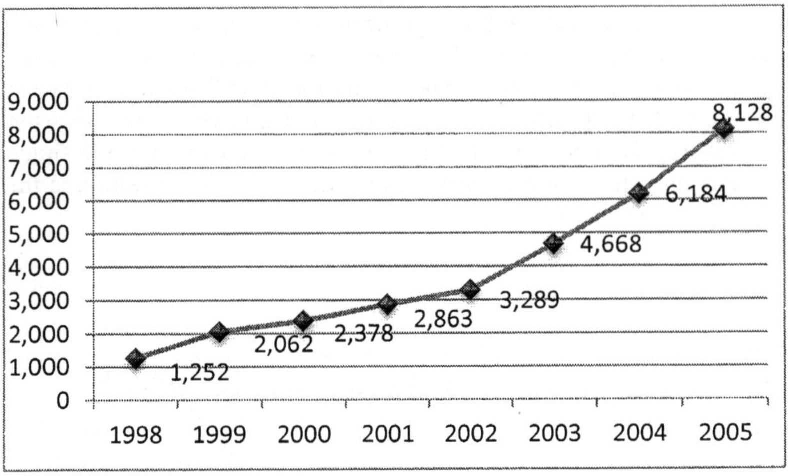

Figure 10.4 The stock of FDI in Estonia

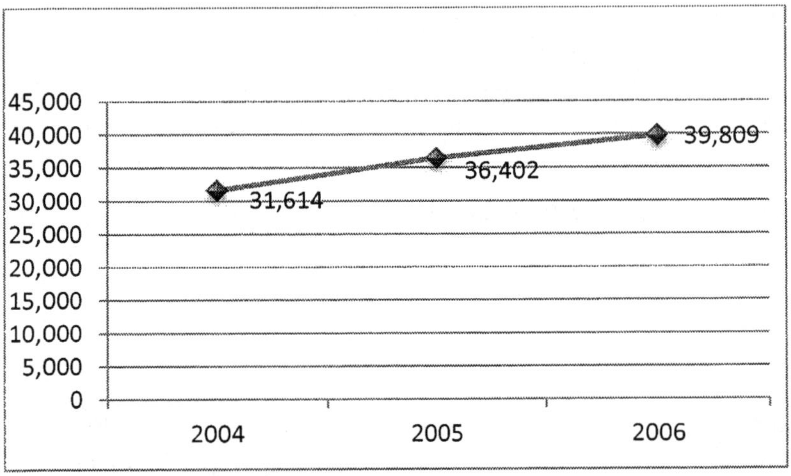

Figure 10.5 The stock of FDI in Hungary

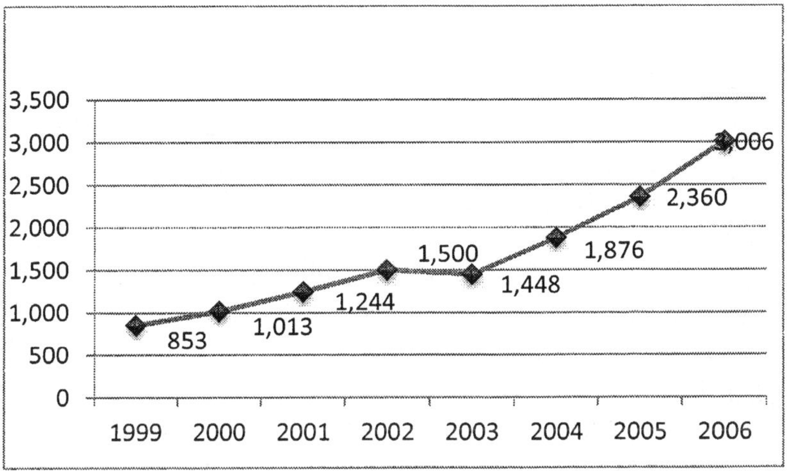

Figure 10.6 The stock of FDI in Latvia

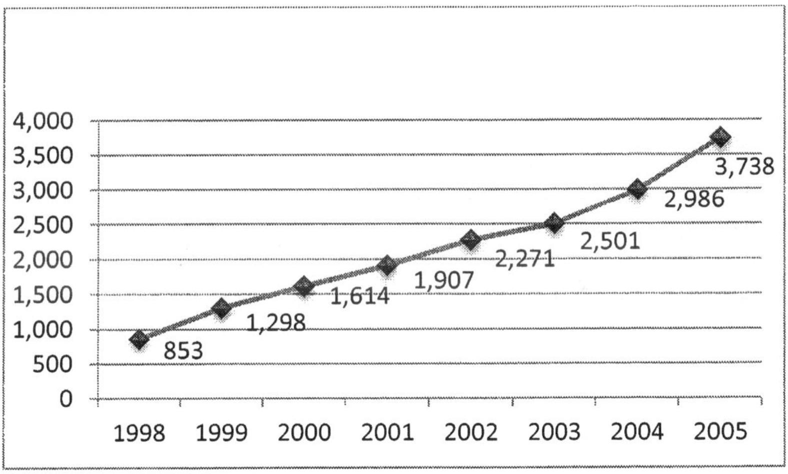

Figure 10.7 The stock of FDI in Lithuania

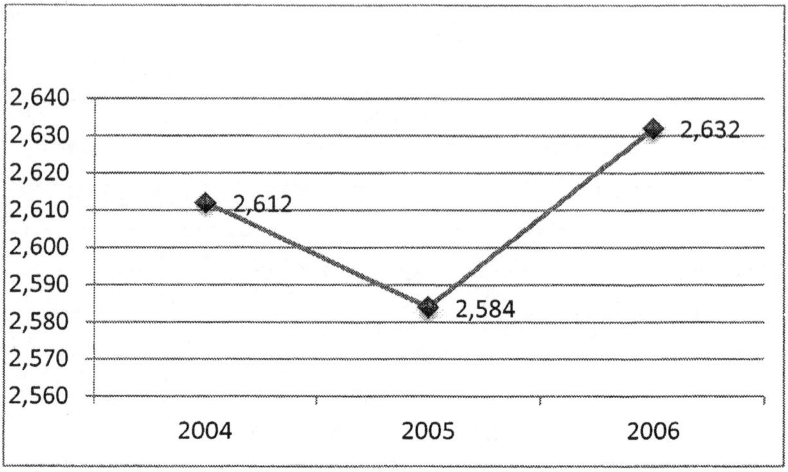

Figure 10.8 The stock of FDI in Poland

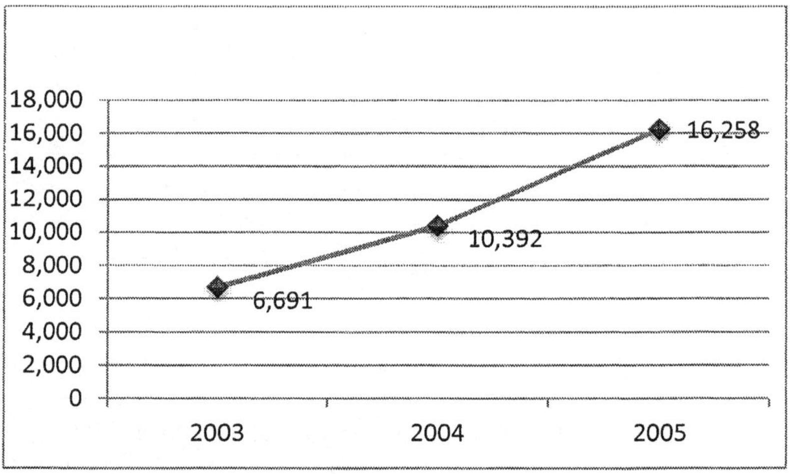

Figure 10.9 The stock of FDI in Romania

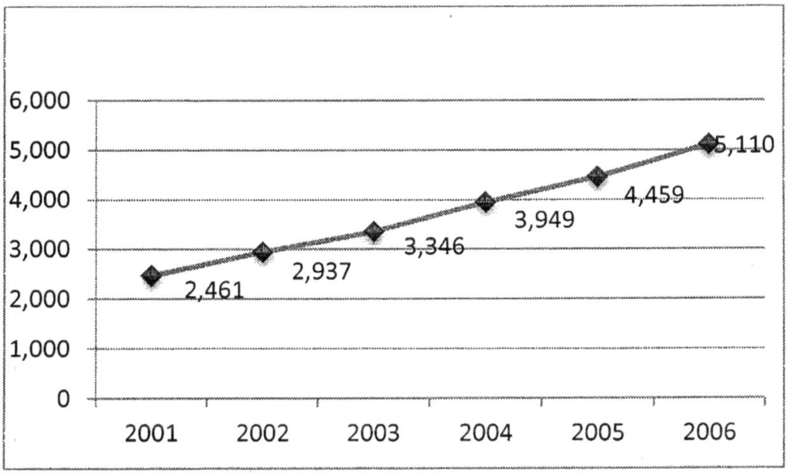

Figure 10.10 The stock of FDI in Slovenia

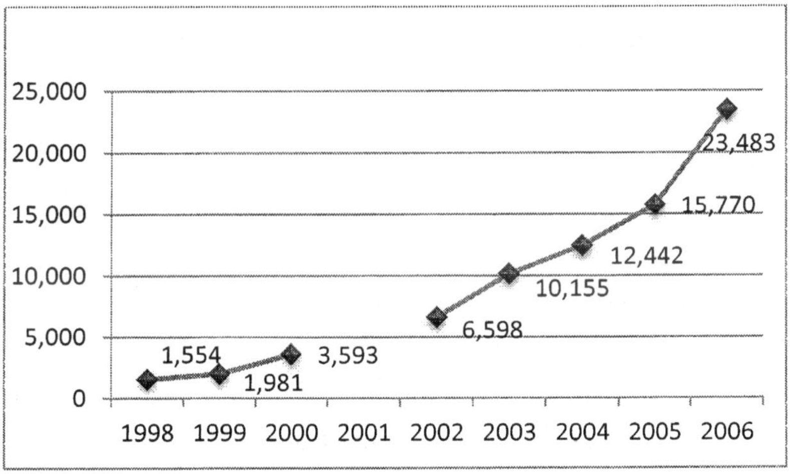

Figure 10.11 The stock of FDI in Slovakia

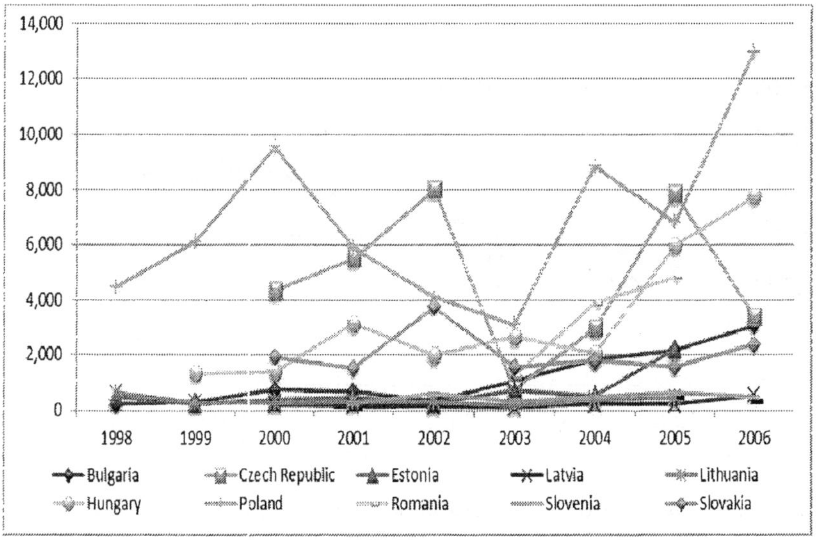

Figure 10.12 Foreign direct investment inflows into the new member states

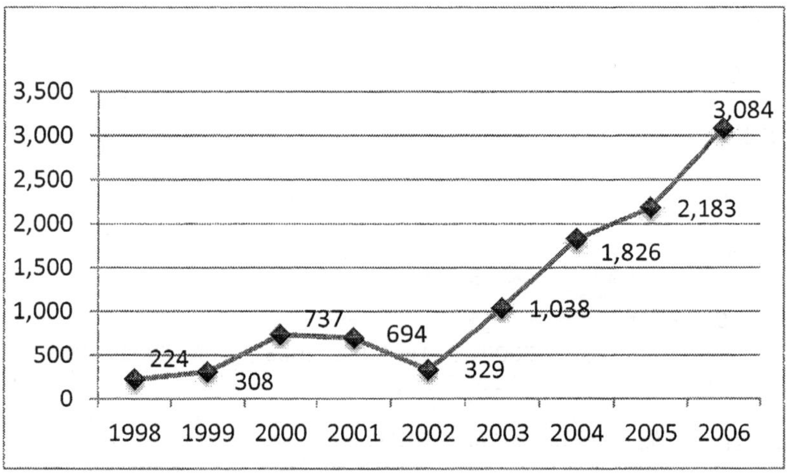

Figure 10.13 The inflows of FDI into Bulgaria

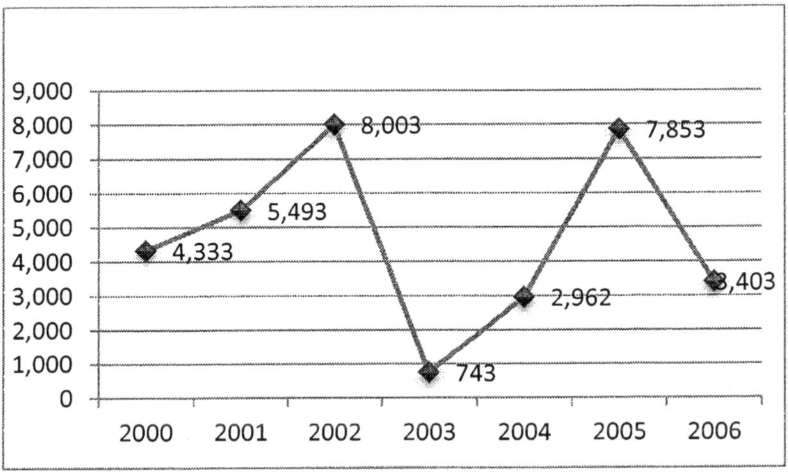

Figure 10.14 The inflows of FDI into the Czech Republic

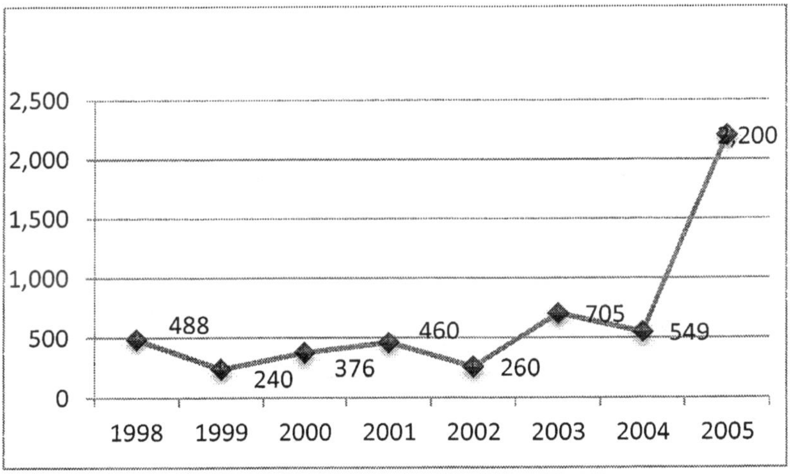

Figure 10.15 The inflows of FDI into Estonia

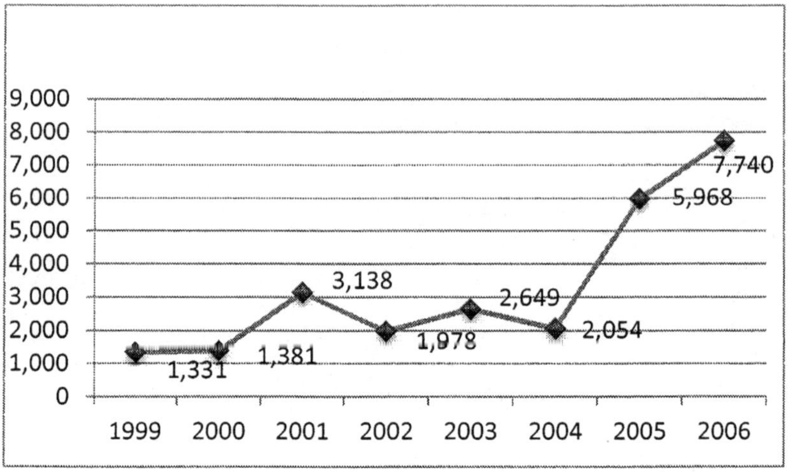

Figure 10.16 The inflows of FDI into Hungary

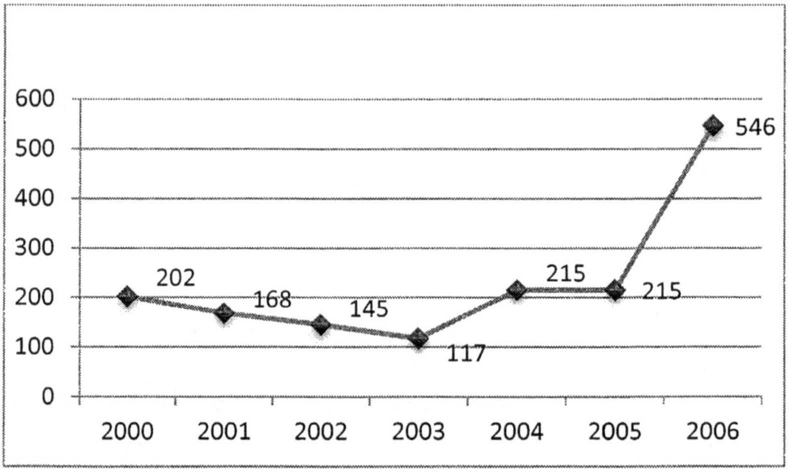

Figure 10.17 The inflows of FDI into Latvia

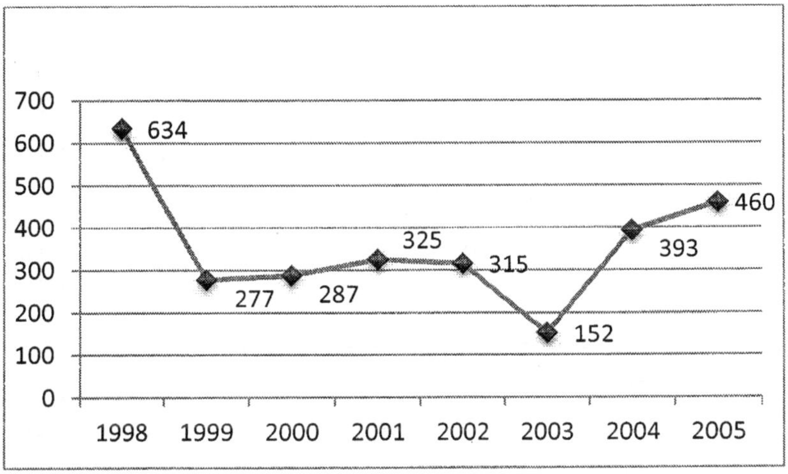

Figure 10.18 The inflows of FDI into Lithuania

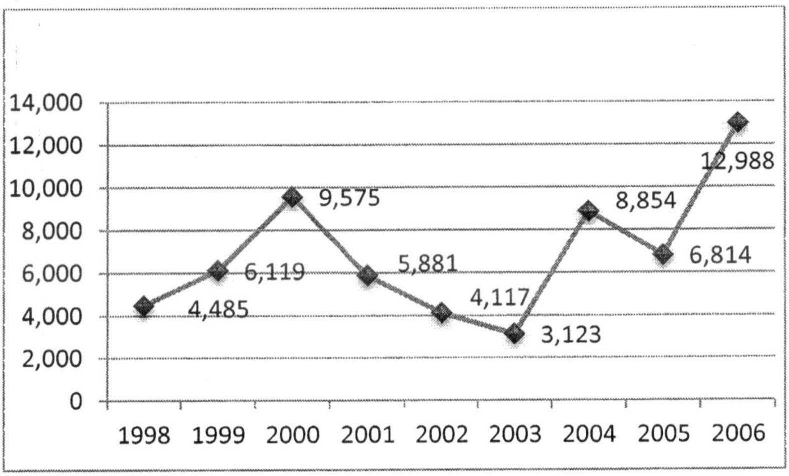

Figure 10.19 The inflows of FDI into Poland

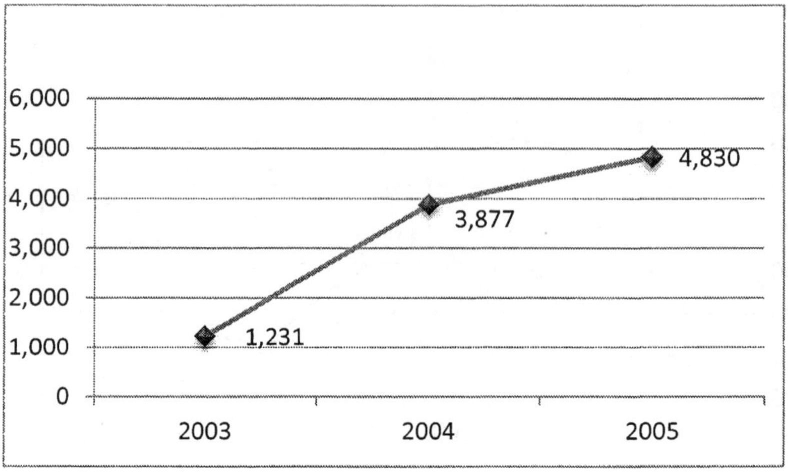

Figure 10.20 The inflows of FDI into Romania

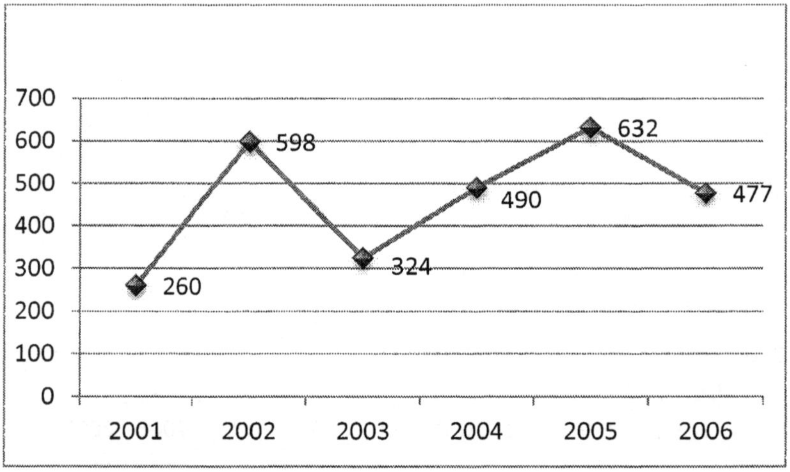

Figure 10.21 The inflows of FDI into Slovenia

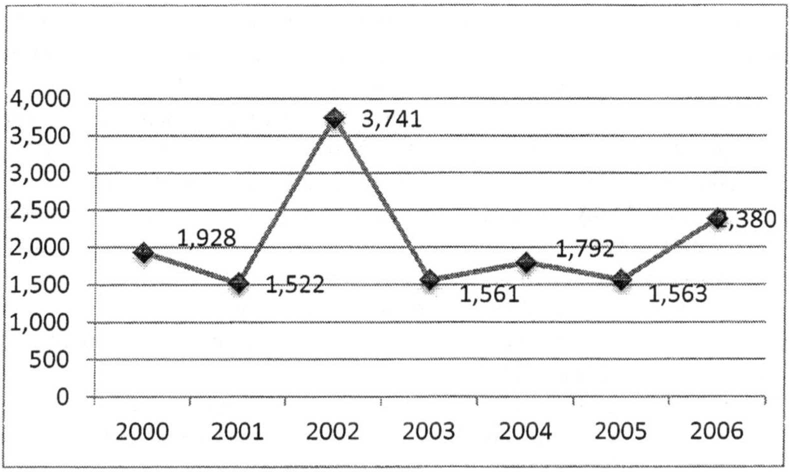

Figure 10.22 The inflows of FDI into Slovakia

EUROPEAN UNION COHESION POLICY

The starting point for answering the question as to why European regional policy, or more accurately, European Union cohesion policy, should cater to the needs of the CEECs, it is necessary to start from the perspective that market adjustment processes, and particularly in the context of the EU Single Market which itself knits together a diverse group of economies and their associated cultural and institutional frameworks, are themselves often inefficient at particular times and places. These inefficiencies, and the resulting bottlenecks and missing links associated with them, arise because of the classic market failure and externality issues (Barca 2009). However, this is even more complex in the case of the CEECs because the process of economic transition in these countries involves the countries re-orientating their economies so as to better fit with those of Western Europe. All of these policies for promoting the quality of market mechanism can also be seen as being consistent with policies which promote the quality of the economic activities underpinning the market system. However, one of the most important aspects of this reorientation process is regarding the development of appropriate institutions for facilitating, promoting and managing the market mechanism. This is where appropriately designed regional policies can play a constructive role.

As the CEECs, which were the former transition economies, continue to move towards the traditional capitalist and mixed economy models of western Europe, clues as to how these countries will develop over the medium term, and which policies should be adopted in order to encourage these processes, can be found by observing the development of thinking in European regional and cohesion policy itself.

Across the EU there is a wide range of evidence of successful regional policy initiatives in the provision of infrastructure, support for enterprises, education and training, environment, R&D and local development (EC 1996; EC 2010a,b,c). Many of these positive developments have taken place in regions which have been undergoing fundamental and difficult structural transformations, and the policies are designed to foster transformation and change in declining or problematic regions. Such policies for transformation, however, can only be successful if the processes of change that they engender lead to innovation and growth.

Since 1997, the focus of EU cohesion policies, and the thinking which underpins them, has shifted increasingly towards the role of cities and city-regions (EC 1997). In part this is because an increasing share of the population in all EU countries is living in urban areas (EC 1998b), and cities are increasingly seen as drivers of innovation and growth. But also, this is because many of the most severe economic, social and environmental issues in the EU are concentrated in cities (EC 1998c). Policies for sustainable urban development are increasingly coordinated under four themes (EC 1998b): strengthening prosperity and employment in cities; promoting equality, social inclusion and regeneration in urban areas; protecting and improving the urban environment; good local governance and local empowerment. However, the priority is that local solutions must be sought out of local situations and local knowledge (EC 1998b).

This emphasis on locality as a source of solutions is based on the principle that local contextual knowledge of the economy and consideration of the specific economic, social and environmental features of the region will produce the most appropriate policies. Such an approach underpins the thinking behind the Barca Report (Barca 2009) and the recent work of the OECD (2009). While the EU has a range of policy initiatives each differentiated in terms of logic, structure, scale and scope (EC 1998d), and focussed on a wide range of sectors including the industrial, scientific, educational and creative sectors (EC 1998e; EC 2000a), once the urban and regional programmemes are adopted the individual member states designate the authority responsible for the implementation of initiatives (EC 1998d). The emphasis on locality is also allied with an emphasis encouraging widespread regional or local partnerships in order to produce policy solutions based on the integration and coordination of job creation measures (EC 1999a). As well as these themes of local decision-making and partnerships for coordination, another theme which is central to EU regional policy thinking is that of the importance of evaluation. In particular, the theme is how the process of evaluation can be best employed for future policy programming (EC 1999b; EC 2005a).

These themes, which emphasise the importance of a local knowledge of the context, a coordinated approach to policy development which is backed up by evaluation of the impacts as a guide to long-term development, fits neatly with an approach to policy development and evaluation which requires a broad holistic consideration of the nature of the regional economy. In particular, no "off-the-shelf" approach to policy is possible.

Regional policy has played a role in the slow process of EU convergence through the 1990s (EC 1999b). Since then, EU regional policy has entered a new phase with the entry of the new accession countries to the EU, most of which have undergone a process of transition (EC 1999b). Broadly, since the end of the 1980s, the disparities between the EU member states have narrowed by 33%, and those between the different regions by 20%. However, the expansion of the EU by 12 accession states increased disparities markedly, in that while the EU population increased by 33%, GDP rose by only 5% (EC 2001a). This expansion also radically changed the regional policy context, in that the areas eligible for EU regional policy have changed and now include many regions within accession countries (EC 2001b; EC 2004a). EU regional policy is being reshaped (EC 2004b). Between 2000 and 2006 almost €213bn was allocated by the structural funds and cohesion funds (EC 2000b), whereas between 2007 and 2013, some €336.3bn will be allocated across the EU27, and this will require a sound financial base and an efficient use of resources. This is because the cohesion policy is now seen fundamentally as a development policy (Barca 2009), a policy for wealth creation, rather than a policy for redistribution (EC 2004b).

During the period of increasing FDI inflows into CEECs, as described by Figures 10.1–22, the largest share of regional policy expenditure was accounted for by the former Objective 1 regions (lagging economically) and then the former Objective 2 regions (undergoing conversion), with much smaller sums allocated for other initiatives (EC 2000b), including those aimed at the accession countries (EC 2004c). Under the current 2007–2013 arrangements, the former Objective 1 regions are now seen as convergence regions, and the former Objective 2 regions are seen as competitiveness and employment regions. For regions which are transitioning between the thresholds of these two groupings over time, there are additional sets of arrangements. The cohesion policy funds are also spent in a variety of different and specific economic environments. As well as economically lagging regions, there are also other regions with specific problematic issues, ranging from problem agricultural areas in mountainous regions (EC 2000c; 2002a), to cross-border regions (EC 2002b,c), and to peripheral coastal, maritime and island regions (EC 2001c; 2002d).

A consistent feature of EU regional policy is that of connectedness; the improved linking of the EU regions via improved transportation systems and networks. These networks are of four main types: the trans-European networks; secondary, regional and local networks; transport in distant and outermost regions; and urban transport networks (EC 2000d). Over recent years, the network approach to regional integration has also been associated with a much more explicitly spatial approach to EU regional policy thinking (EC 2000e). This is known as the European Spatial Development Perspective (ESDP), and represents a new, and explicitly spatial, approach to EU economic and territorial planning. The genesis of the ESDP process dates back at least to the first meeting of the Council of Ministers meeting in Nantes in 1989, which affirmed the need for forward-looking reflection on territorial planning in Europe. This culminated in the 1991 *Europe 2000* report and in 1994 in the *Europe 2000+* report. The ESDP structure was agreed in 1993 and approved in 1994, and following further meetings was formally adopted in 1999 in Potsdam (EC 2000e). The EU structural and cohesion funds are aided by loans from the European Investment Bank (EIB), the majority of whose funding streams are directed towards the weaker EU regions (EC 2001d).

Since the completion of the single market, the introduction of the Euro, and the advent of more explicitly spatial approaches to regional policy thinking and regional inter-connectedness, EU regional policy and the use of structural funds has increasingly shifted in focus to five major themes: regional competitiveness; sustainable regional development; innovation and the information economy; territorial cohesion; and the exchange of good practices and regional networking (EC 2003a). In addition, although an increasing EU focus on human capital and the networking of higher education institutions is not specifically a regional issue, it is likely to have profound implications for regions (EC 2004d), and there are two reason for this. Firstly, increasing higher education attainment is likely to make people become more geographically mobile, thereby reducing the local aspects of skills and labour. Secondly, such networks, if properly implemented, may also contribute to local regional knowledge spillovers.

Many of these themes mirror the themes which underlie the current thinking on regional innovation systems and regional innovation policies. However, over recent years, as already mentioned, these themes have also been increasingly set in the context of urban areas. Just under 80% of EU citizens live in urban areas and just under half live in cities of over 50,000 (EC 2001e). For this reason, most of the regional measures under the 2000–2006 Structural Funds Programmeme included an urban development strand. In adopting the ESDP in May 1999, the heads of state and government were recognising the need to harness the innate strengths of the urban areas to spur growth in the surrounding areas, redressing the balance of economic development (EC 2001d).

Moreover, in keeping with the themes of competitiveness and sustainability outlined above, EU regional and structural policies are increasingly focusing on three major urban themes: competitiveness, social exclusion, and physical and environmental regeneration (EC 2003b). The themes of the Urban Community Initiative are that different policies are integrated in a flexible manner, programmes (UCI) are run at a local level, there is close involvement from local communities and partnerships with local associations are prioritised, and there is a built-in "learning cycle". Within UCI there is also the URBACT programme which helps foster learning and good practice exchanges between some 200 EU cities (EC 2003b), and the implementation and performance of these urban policies are monitored by the EU Parliament as part of the continuing evaluation process applied to regional policies (EC 2001f).

These developments in EU regional and urban policy initiatives have also been accompanied by a rise in networks designed to promote the interest of various regions' cities. For example, EURADA, the European Association of Regional Development Agencies, was established in December 1991 (EC 2003c). Its current membership includes some 150 development agencies from 25 EU and CEECs. The network facilitates information exchanges and policy analysis. Similarly, the "Cities for Cohesion" network functions as a platform for exchanges and joint actions between Europe's local authorities (EC 2003d). In June 2002, 16 urban areas in Europe supported the "Cities for Cohesion" manifesto. In this document the signatories stress the ten key principles for ensuring that the urban dimension is better catered for in economic and social cohesion policy. In particular, the one key theme is the need for better integration of policy through the structural funds (EC 2003d). Finally, the Eurocities network was founded in 1986, and brings together 120 cities in 31 European countries (EC 2003e). This network argues for cities to be given a more central role in the future of the structural funds.

In the immediate post-Lisbon Treaty period, during the early part of this decade the three major guidelines which determined the priorities for the next generation of EU regional policies post-2007 (EC 2005b). These are: guideline I – making Europe and its regions more attractive places to invest and work; guideline II – improving knowledge and innovation for growth; and guideline III – generating more and better jobs. Each of these three guidelines had sub-themes attached to it which are the means by which such guideline themes are to be achieved. In the case of the first guideline which focuses on making Europe a more attractive place to invest, the major emphasis of policy is on improving the attractiveness, accessibility and environment of regions and cities by improving transportation infrastructure and the environment (EC 2005c). For the second guideline which focuses on knowledge and innovation, the policy emphasis is on improving R&D, entrepreneurship, ICT and access to finance. The third guideline theme which focuses on improving the number and quality of jobs, prioritises human capital development and labour market flexibility (EC 2005b,c). Continuing the trends over recent years, these priorities are increasingly being focussed at the level of cities, with increased knowledge exchanges taking place within the URBACT programmeme, so as to maximise best practice knowledge across the European networks (EC 2006a) and to maximise transparency (EC 2006b).

These themes were subsequently re-focused in 2010 into the *Europe 2020* strategy, which has as its main elements three central cross-cutting themes, namely smart growth, sustainable growth and inclusive growth. All future EU policies will be required to adopt the *Europe 2020 Agenda* and to develop policies which are not only consistent with these themes, but which also drive the transformational processes needed to achieve these goals. These principles were made clear in *Investing in Europe's Future: Fifth Report on Economic, Social and Territorial Cohesion* (EC 2010d).

For CEECs the Europe 2020 strategy poses major challenges. These countries still face very major difficulties regarding environmental issues, because of the legacy of the former planned socialist economic systems. At the same time, the market responses to liberalisation have been dramatic in many of these countries, with a high level of factor mobility to the capital cities and large outflows from many other regions, both rural and urban regions. As such, the FDI inflows have tended to be to the dominant centres, and these flows are also mirrored by inflows to these same regions by high human capital individuals, most notably university graduates.

This creates major problems in terms of dealing with the other non-core regions, because large-scale factor outflows are often associated with the degradation and deterioration of the origin regions, particularly in terms of the built environment infrastructure. As such, this makes the rejuvenation of these regions a much more difficult proposition, and the highly selective nature of interregional migration flows implies that the social inclusion objectives of Europe 2020 will be harder to achieve in these localities.

One way of driving the Europe 2020 agenda in these countries is to promote innovation, and in the Europe 2020 thematic terms, smart growth. The EU approach to this is that of smart specialisation, whereby regions are encouraged to identify potential areas of competitive advantage and to aim to add value in those sectors, or technological fields which complement the already-existing areas of competitive advantage. Adopting these policies, however, requires sufficient local capabilities to identify such niches and also the absorptive capacity to embody new ideas, related both to private market developments and also to institutional and governance reforms. These are all central elements of the conclusions of the Fifth Cohesion Report, and relate primarily to the need to foster realistic and long-lasting multi-level governance reforms as are required in many of the CEECs in order to successfully implement the place-based model of EU Cohesion Policy advocated by the 2009 Barca Report *An Agenda for a Reformed Cohesion Policy*.

For a reformed cohesion policy to be successful post-2013 governance and institutional reform are central to the priorities highlighted by the 2009 Barca Report. However, responding to such reforms will pose major challenges for the CEEC countries. The successful adaptation to these realities will also improve the absorptive capacity of these countries in terms of market inflows of factors, and most notably FDI inflows of capital. At the same time, such governance reforms will also allow the CEECs to better respond to the internal interregional and structural adjustment issues that they face. There are major challenges ahead for the CEEC countries to adapt to the new realities of cohesion policy post-2013, but the governance and institutional reforms required by a reformed cohesion policy will actually produce much greater and more widespread benefits to these countries than simply that of policy concordance. Rather these policy-led reforms will foster reform of the whole economic system of these countries and improve their absorptive capacity to respond both to EU policies as well as international private sector FDI. These are exactly the public good and externality effects cohesion policy is designed to foster.

The final issue to note briefly is the question of the impact of the 2008 global financial crisis on the EU and the CEECs in particular. The data here refer to the period before the crisis, and the long-run outcomes of the crisis are still evolving. However, the countries which are perceived to be more risky investments have suffered significantly in the crisis and the urgency of the need for appropriate institutional reform has increased within the EU. This urgency in many ways has accelerated the case for EU cohesion policy reform rather than stalled or altered it.

REFERENCES

Barca, F., 2009, *An Agenda for a Reformed Cohesion Policy: A Place-Based Approach to Meeting European Union Challenges and Expectations*, Independent Report Prepared at the Request of Danuta Hubner, Commissioner for Regional Policy, Brussels

EC, 1996, *Regional Success Stories Profiles of 36 Projects in Europe*, European Commission, Luxembourg

EC, 1997, *Europe's Cities: Community Measures in Urban Areas*, European Commission, Luxembourg

EC, 1998a, *Driving Regional Development: EU Structural Funds Programmemes in the UK*, European Commission, Luxembourg

EC, 1998b, *Sustainable Urban Development in the European Union: A Framework for Action*, 28.10.98 – COM/98/605, European Commission, Luxembourg

EC, 1998c, *Urban Pilot Projects Phase II, 1997–99*, European Commission, Luxembourg

EC, 1998d, *Guide to Community Initiatives 1994–99 Volume 2*, European Commission, Luxembourg

EC, 1998e, *Investing in Culture: An Asset for All Regions*, European Commission, Luxembourg

EC, 1999a, *Guide to Territorial Employment Pacts 2000–2006*, SEC(99) 1933, European Commission, Luxembourg

EC, 1999b, *Better Management Through Evaluation: Mid-Term Review of Structural Funds Programmes Objectives 1 and 6 (1994–99)*, European Commission, Luxembourg

EC, 1999b, *Sixth Periodic Report on the Regions: Summary of Main Findings*, Fact Sheet 04.02.99 EN, European Commission, Luxembourg

EC, 2000a, *Community Rules on State Aid: Vade-Mecum*, European Commission, Luxembourg

EC, 2000b, "Europe is Being Built on the Ground", *Inforegio Panorama* 1, European Commission, Luxembourg

EC, 2000c, *Structural Policies and European Territory: The Mountains*, European Commission, Luxembourg

EC, 2000d, *Europe on the Right Track: Transport Projects Supported by the European Regional Development Fund and the Cohesion Fund*, European Commission, Luxembourg

EC, 2000e, *TERRA: An Experimental Laboratory in Spatial Planning*, European Commission, Luxembourg

EC, 2001a, "Cohesion Policy: What Future for an Enlarged Europe", *Inforegio Panorama* 3, European Commission, Luxembourg

EC, 2001b, *Working for the Regions*, European Commission, Luxembourg

EC, 2001c, *Structural Policies and European Territory: Islands and Coastal Regions*, European Commission, Luxembourg

EC, 2001d, "Where Next for Cohesion Policy? The Debate Takes Shape", *Inforegio Panorama* 4, European Commission, Luxembourg

EC, 2001e, "Driving Development in Europe's Cities", *Inforegio Panorama* 5, European Commission, Luxembourg

EC, 2001f, "Spain: The Peninsular Gets Connected", *Inforegio Panorama* 2, European Commission, Luxembourg

EC, 2002a, "Solidarity and Cohesion", *Inforegio Panorama* 8, European Commission, Luxembourg

EC, 2002b, "Cooperation Without Borders", *Inforegio Panorama* 6, European Commission, Luxembourg

EC, 2002c, "Structural Policies and European Territory: Cooperation without Frontiers", European Commission, Luxembourg

EC, 2002d, "Bringing the Regions Closer Together", *Inforegio Panorama* 7, European Commission, Luxembourg

EC, 2003a, *Competitiveness, Sustainable Development and Cohesion in Europe: From Lisbon to Gothenburg*, European Commission, Luxembourg

EC, 2003b, *Partnership with the Cities: The Urban Community Initiative*, European Commission, Luxembourg

EC, 2003c, "Second Progress Report on Cohesion", *Inforegio Panorama* 9, European Commission, Luxembourg
EC, 2003d, "Towards Enlargement", *Inforegio Panorama* 10, European Commission, Luxembourg
EC, 2003e, "Exchanging Good Practice", *Inforegio Panorama* 12, European Commission, Luxembourg
EC, 2004a, *Working for the Regions*, European Commission, Luxembourg
EC, 2004b, "Spotlight on Enlargement: A New Partnership for Cohesion", *Inforegio Panorama* 13, European Commission, Luxembourg
EC, 2004c, "The Cohesion Fund: A Boost for Solidarity", *Inforegio Panorama* 14, European Commission, Luxembourg
EC, 2004d, *The Magazine: Education and Culture in Europe*, European Commission, Luxembourg
EC, 2005a, "Best Practice in Regional Development", *Inforegio Panorama* 16, European Commission, Luxembourg
EC, 2005b, "Cooperation at the Heart of Cohesion: Interreg in Action", *Inforegio Panorama* 17, European Commission, Luxembourg
EC, 2005c, "Cohesion Policy in Support of Growth and Jobs: Community Strategic Guidelines 2007–2013", European Commission, Luxembourg
EC, 2006a, "Cities for Growth, Jobs and Cohesion", *Inforegio Panorama* 19, European Commission, Luxembourg
EC, 2006b, "A New Transparent European Cohesion Policy", *Inforegio Panorama* 21, European Commission, Luxembourg
EC, 2008, *Regions 2020: Demographic Challenges for European Regions*, Directorate General for Regional Policy, European Commission, Brussels
EC, 2010a, *Ex-Post Evaluation of Cohesion Policy Programmes 2000–06 Co-Financed by the ERDF (Objective 1&2): Synthesis Report*, European Commission, Regional Policy Evaluation Unit, Brussels
EC, 2010b, *Ex-Post Evaluation of Cohesion Policy Programmes 2000–06: The Urban Community Initiative. Evaluation Report*, European Commission, Regional Policy Evaluation Unit, Brussels
EC, 2010c, *Ex-Post Evaluation of INTERREG III 2000–06: Evaluation Report*, European Commission, Regional Policy Evaluation Unit, Brussels
EC, 2010d, *Investing in Europe's Future: Fifth Report on Economic and Social and Territorial Cohesion*, European Commission, Brussels
Eurostat, 2010, "Ageing in the European Union: Where Exactly?", *Eurostat Statistics in Focus*, 26/2010
McCann, P., 2008, "Globalisation and Economic Geography: The World is Curved, Not Flat", *Cambridge Journal of Regions, Economy and Society*, 1(3), 351–370
McCann, P., 2009, "Globalisation, Multinationals and the BRIICS Countries", in Lattimore, R. and Safadi, R. (eds), *Globalisation and Emerging Economies*, OECD, Paris
McCann, P. and Acs, Z.J., 2011, "Globalisation: Cities, Countries and Multinationals", *Regional Studies*, Forthcoming
OECD, 2007, *Gaining From Migration: Towards a New Mobility System*, Organisation for Economic Cooperation and Development, Paris
OECD, 2008a, *The Global Competition for Talent: Mobility of the Highly Skilled*, Organisation for Economic Cooperation and Development, Paris
OECD, 2008b, *Growing Unequal: Income Distribution in OECD Countries*, Organisation for Economic Cooperation and Development, Paris

OECD, 2009, *How Regions Grow: Trends and Analysis*, Organisation for Economic Cooperation and Development, Paris

Sapir, A., Aghion, P., Bertola, G., Hellwig, M., Pisani-Ferry, J., Rosati, D., Vinals, J. and Wallace, H., 2003, *An Agenda for a Growing Europe: Report of a High-Level Study Group Established on the Initiative of the President of the European Commission*, Brussels

World Bank, 2009, *World Development Report 2009: Reshaping Economic Geography*, Washington DC

11. Spatial filtering methods for tracing space–time developments in an open regional system: experiments with German unemployment data

Roberto Patuelli, Daniel A. Griffith, Michael Tiefelsdorf and Peter Nijkamp

INTRODUCTION

Over the decades, economic location theory and modern economic geography have highlighted the relevance of space–time factors in (socio-) economic development. Spatial matters may be regarded as being of critical importance when investigating socio-economic (and other) phenomena (see, for example, Bockstael 1996; Weinhold 2002), including their implications for policymaking (Lacombe 2004). To account for the presence of spatial structures that influence (positively or negatively) observable economic entities, such as unemployment or trade, calls for a rigorous and systematic assessment of their impact and extent. Spatial autocorrelation (SAC) represents the correlation, computed among the values of a single georeferenced variable, that is attributable to the geographic proximity of the objects to which the values are attached. Introduction of the SAC concept is a departure from the classical assumption of independence of observations constituting a single variable. SAC also complements the concept of temporal autocorrelation, which has been extensively studied and dealt with in time-series econometrics. SAC measures are used to quantify the nature and degree of the spatial correlation contained in data for a given variable, or to test the assumption of independence or randomness. From a statistical analysis viewpoint, spatial correlation patterns may become problematic, since they make standard statistics, such as correlation coefficients or ordinary least squares (OLS) estimates, potentially inappropriate.

This chapter aims to provide an assessment of how important spatial effects are in explaining unemployment levels in Germany and, particularly, to show that these (or, more precisely, a subset of these) patterns are consistent over time. The definition of stable and recognizable spatial patterns enables one to observe systematic differences in regional unemployment. Such findings may have clear implications for policy evaluation and strategic planning. This chapter presents statistical analyses carried out by means of a semi-parametric 'spatial filtering' technique, described in Griffith (2003), which is based on the decomposition of spatial weights matrices. In our analysis, these matrices are defined for 439 German districts (i.e., *Kreise*), according to both topological and distance-based criteria – such as shared boundaries or centroid distance – and economic flows. In this regard, journey-to-work flows are used as a proxy for economic linkages.

Earlier work on pattern identification in German labour markets was carried out by, amongst others, Kosfeld and Dreger (2006), who investigated Verdoon and Okun's laws for German regional labour markets in the period 1992–2000. Their approach, however, involves computing spatial filters for each year within the framework of a spatial seemingly unrelated regression (SUR) model. Our approach differs from theirs in that, in addition to limiting ourselves to autoregression, we focus on the search for a set of spatial filters that are significant and consistent over time and, therefore, can be employed for the entire time period considered (that is, 1996–2002). Also, we employ data at a more refined level of disaggregation (439 districts versus 180 regions), which enables a more detailed analysis of the underlying spatial patterns.

SPATIAL FILTERING TECHNIQUES

Recent years have witnessed an increasing popularity for the use of spatial-econometric tools in regional research employing georeferenced data (see, e.g., Anselin 1988, 2007; Griffith 1988; Anselin et al. 2004). Among standard spatial econometric methods, spatial autoregression techniques have become powerful methods, in particular through the use of spatial weight matrices that mirror the intensity of spatial linkages (dependences) between spatially-referenced data. The insight has matured that – due to efficiency and normality problems – OLS may not be carried out with spatially dependent data. Furthermore, maximum likelihood (ML) estimators of spatial regression models are based on restrictive assumptions. These recognitions have led to the search for appropriate statistical estimation techniques. An alternative approach to spatial autoregression is the use of spatial filtering techniques, described inter alia in Griffith (1981), Haining (1991), Getis and Griffith (2002), and Tiefelsdorf and Griffith (2007). The advantage of these filtering procedures is that the variables studied (which, initially, are spatially correlated) can be decomposed into spatial and non-spatial components, which then can be used in an OLS or ML modelling framework. Filtering out spatially autocorrelated patterns also enables one to reduce the stochastic noise in the residuals of conventional statistical methods such as OLS. This conversion procedure requires the computation of 'spatial filters.' The approach developed by Griffith (1996, 2004) is adopted in our study. This approach is preferred in our case study to the one by Getis (1990, 1995), which requires variables with a natural origin. This limitation would not allow us to apply the same method to the analysis of other labour market variables, such as employment growth rates.

The spatial filtering technique introduced by Griffith is based on the computational formula of Moran's I (MI) statistic.[1] This methodology utilizes eigenvector decomposition techniques, which extract orthogonal and uncorrelated numerical components from an $n \times n$ matrix (see, for details, Tiefelsdorf and Boots 1995).[2] We extract the eigenvectors of the modified spatial weights matrix $\left(\mathbf{I} - \mathbf{1}\mathbf{1}^T/N\right)\mathrm{W}\left(\mathbf{I} - \mathbf{1}\mathbf{1}^T/N\right)$ where $\mathbf{I}$ is an identity matrix of dimension $n \times n$, and $\mathbf{1}$ is an $n \times 1$ vector containing ones. Such eigenvectors maximize the sequential residual MI values, so that the numerical values of the first computed eigenvector, E1, generate the largest MI value among all n eigenvectors. All subsequently extracted eigenvectors again maximize the MI values while being orthogonal and uncorrelated with the preceding ones, eventually giving the complete set of all possible (mutually) orthogonal and uncorrelated map patterns posited by the spatial weights matrix used (Getis and Griffith 2002).

When employed as regressors, these eigenvectors may function as proxies for missing explanatory variables. A smaller set of m ($< n$) 'candidate' eigenvectors can be selected for parsimony reasons, on the basis of their MI values. For example, an MI threshold value of 0.25 can be specified for selection screening purposes. The final set of eigenvectors describing the data at hand can be selected (out of the m candidates) by means of stepwise regression. In this regard, the orthogonality of the eigenvectors makes selection easier, because no partial correlations exist between the eigenvectors. The linear combination of the eigenvectors belonging to the final selection can be defined as the 'spatial filter' for the variable examined.

The eigenvector components may be regarded as independent map patterns, and represent the latent SAC of a georeferenced variable, according to a given spatial weights matrix. They also can be interpreted as redundant information due to spatial interdependencies, in the framework of standard regression methods. Also relevant to the use of the eigenvector decomposition process is the choice of the matrix to be used, particularly regarding: (a) the definition of proximity; (b) the variable chosen (if necessary) to indicate proximity; and (c) the coding scheme employed in the calculation of the weights matrix. While points (a) and (b) are discussed subsequently in this chapter, the latter point is briefly addressed now.

The previously discussed spatial filters are computed on the basis of a modified spatial weights matrix. It is straightforward that the choice of the matrix to be used is critical in defining the set of spatial filters. Many coding techniques for spatial weights matrices can be found in the literature (Tiefelsdorf et al. 1999; Getis and Aldstadt 2004). The main feature that discriminates between the different schemes is the way in which each scheme treats the spatial links between georeferenced objects (like regions).

Generally speaking, we can define a *family* of coding schemes based on the following expression (Tiefelsdorf and Griffith 2007, with practical details in Chun et al. 2005):

$$\mathbf{V}_{[q]} = \frac{n}{\sum_{i=1}^{n} d_i^{q+1}} \cdot \mathbf{D}^q \cdot \mathbf{B}, \tag{11.1}$$

where $\mathbf{B}$ is a binary spatial weights matrix, and $\mathbf{D}^q$ is a diagonal matrix that contains d_i^q components $(d_1^q, \ldots, d_n^q)$, belonging to vector $\mathbf{d} = \mathbf{B} \cdot \mathbf{1}$, and representing the degree of 'linkage' of spatial object i. Different coding schemes are obtained by varying the q parameter. In particular, the following schemes can be obtained:

$q = 0$: *C-coding* (globally standardized); this scheme is commonly used in spatial statistics, and tends to emphasize spatial objects with a greater linkage degree. The C-coded matrix is symmetrical;
$q = -0.5$: *S-coding* (variance stabilized); this scheme tends to even the variation levels of weights assigned to spatial objects; and,
$q = -1$: *W-coding* (row-sum standardized); this scheme is mostly used in autoregressive response (AR) and simultaneous spatial autoregressive (SAR) model specifications and, contrary to the C-coding scheme, tends to emphasize the weight of objects with small spatial linkages.

Different individual spatial patterns may result from the calculation of the eigenvectors of the different coded matrices. For instance, a W-coded matrix can be expected to show more 'extreme' values along the edges of a study area, while, conversely, a C-coded matrix is expected to present stronger patterns in the inner study area. Figure 11.1 presents an illustrative example, for the case of German unemployment, of the first two eigenvectors generated from the adjacency matrix coded in the different coding schemes.

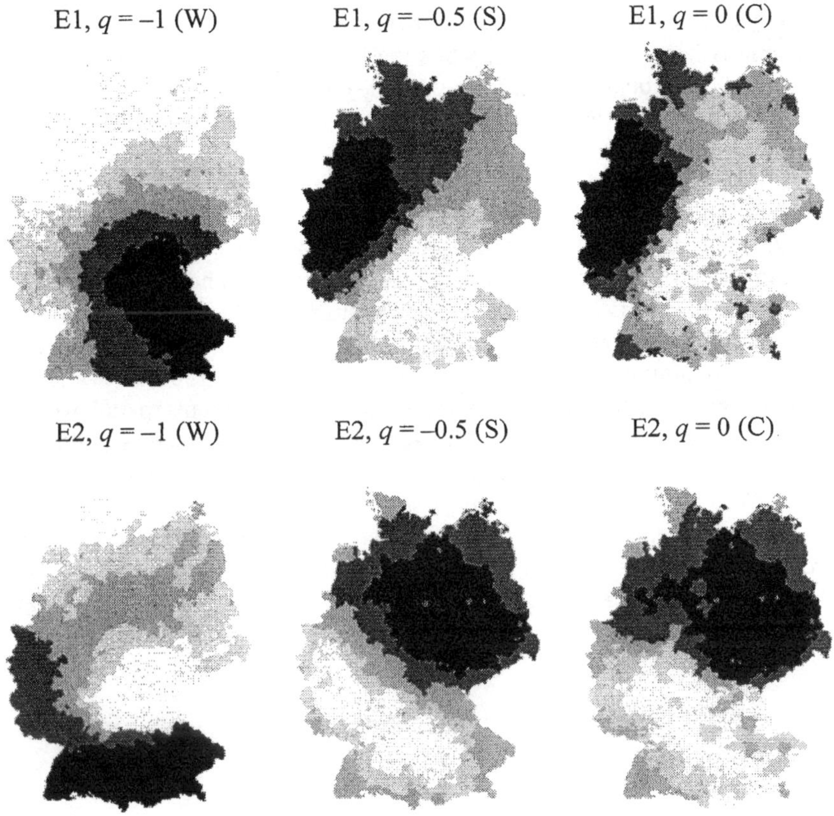

Figure 11.1 Eigenvector variation for different coding schemes, the case of German unemployment

The choice of coding scheme, and therefore of the spatial weights matrix, although yielding a different set of eigenvectors from which the spatial filters are selected, still produces a linear combination account for SAC in a spatial econometric or spatial statistics context; the differentiating feature tends to be parsimony. In the empirical application presented in this chapter, both W-coding and C-coding are employed (see the third section). Results of a correlation analysis of the spatial weights matrices used also are presented, in order to compare the various approaches.

Next, the preceding spatial filtering techniques are illustrated empirically using German unemployment data. The dataset consists of cross-sectional data about the number of employed and unemployed individuals, collected by the (German) Federal Employment Services (*Bundesanstalt für Arbeit*, BA), for 439 German districts (*Kreise*). The time period for which the data are available is from 1996 to 2002, while the level of aggregation of the dataset is NUTS-3. The unemployment rates employed in our analysis are computed as a ratio between the number of unemployed individuals and the active workers population.

A further spatial relationship matrix, namely German commuting flows, is employed in our analysis. The data comprise, for each couple (*i*, *j*) of NUTS-3 origins and destinations, the number of employees that live in district *i* and work in district *j*. Therefore, we can treat these flows as home-to-work trips. The data used in this chapter refer to the year 2002, and are employed in the computation of an 'economic flows' spatial weights matrix (see the third section). Commuting data for one year only are employed in our case study, because varying commuting data would generate different spatial weights matrices and, consequently, different sets of eigenvectors (although techniques exist to handle a set of such matrices; e.g., Rogerson and Plane 1984). Furthermore, one can assume some spatio-temporal persistence with respect to the local commuting patterns. The daily commuting flows between two districts are transformed to satisfy the statistical symmetry requirement of spatial link matrices. This transformation represents the daily to-work and back-to-home flows.

COMPUTATION AND CHOICE OF SPATIAL FILTERS FOR GERMAN UNEMPLOYMENT DATA

Spatial Weights Matrices: The Different Approaches Used

As previously mentioned, the spatial filtering methods employed in this case study are based on the decomposition of a spatial weights matrix. Therefore, in addition to matrix computation methods, carefully considering the concept of proximity employed, and its consequences, is important. In our case study, we present a set of different definitions of the spatial weights matrix:

- *economic flows*: based on patterns of symmetrized commuting flows;
- *shared boundaries*: based on geographical contiguity, which by definition is symmetric; and,
- *distance*: based on symmetric distances separating district centroids.

The three definitions highlighted here enable one to observe the influence of different operational definitions of proximity on the final results. First, commuting flows are employed as a proxy of the economic intertwining among districts. Second, shared boundaries utilize the topology of administrative boundaries in defining proximity. Third, distance-based matrices calculated using district centroids define proximity in terms of geographical distance decay relationships.

A total of five spatial weights matrices are employed. The matrices are computed as follows:

a) A journey-to-work flows matrix is computed according to the $q = -1$ scaling scheme (W-coding); this matrix is based upon the location-to-location commuting data described above.
b) Two matrices based upon shared boundaries, constructed by defining contiguity according to the so-called 'rook' rule, and then computed according to the C- and W-coding schemes; results from the application of a 'queen' contiguity rule are not considered here, since the two specifications of adjacency differ only by 25 neighbour links.
c) Two distance-based matrices derived from a spatial interaction model (SIM);[3] the variables used for the estimation of the model are district employment data (discussed above), and the distance between the centroids of each district:
   - First, the distance decay exponent of –2.7 is taken from the estimated SIM, and then converted to the W-coding scheme;
   - Second, this distance decay exponent is increased to –6.3 in order to obtain the same number of candidate eigenvectors as are obtained with the shared boundaries W-coding scheme.

The following unconstrained gravity model is the SIM used to describe flows and estimate distance decay parameters:

$$F_{ij} = \kappa W_i^{\alpha} J_j^{\beta} e^{-\gamma d_{ij}} + \varepsilon_{ij}, \tag{11.2}$$

where: F_{ij} is the quantity of flows between areal units i and j; W_i is the number of workers residing in origin areal unit i; J_j is the number of jobs located in destination areal unit j; κ, α, β and γ are parameters; and ε_{ij} is a random error associated with flows between origin i and destination j. The distance decay parameter, $\hat{\gamma}$, estimated with the log-linear form of equation (11.2) – but with a multiplicative rather than an additive error structure – was used to define the W-coding scheme, resulting in the matrix components being defined as:

$$w_{ij} = e^{-\hat{\gamma} d_{ij}} \Big/ \sum_{j=1}^{n} e^{-\hat{\gamma} d_{ij}} \tag{11.3}$$

Next, $\hat{\gamma}$ was incrementally increased until the resulting $(\mathbf{W}^T + \mathbf{W})/2$ matrix yielded the same number of candidate eigenvectors as are obtained with $(\mathbf{W}^T + \mathbf{W})/2$ constructed as the row-standardized version of the topological-based binary 0/1 adjacency matrix **C**. Of note is that the eigenvectors for all W-coding schemes are extracted from $(\mathbf{W}^T + \mathbf{W})/2$ in order to convert the matrix from an asymmetric to a symmetric one.

Computation and Selection of the Spatial Filters over Time

The first step in the construction of a spatial filter to be applied to the variable of study is the computation of the eigenvectors of a spatial weights matrix, followed by the choice of a set of candidate eigenvectors from which selection is made. Eigenvectors are selected for inclusion on the basis of their MI values and their correlations with the georeferenced data about regional unemployment. A minimum MI/max(MI) value of 0.25 has been used in our case study to identify the candidate set to be evaluated for inclusion. The results of this process, carried out for the matrices presented in the preceding section, are presented in Table 11.1.

Table 11.1 Candidate eigenvectors selected and maximum MI values

Spatial weights matrix	No. of candidate eigenvectors	Max(MI)
Journey-to-work flows matrix	78	2.92
Rook matrix (S-coding)	130	1.07
Rook matrix (C-coding)	98	1.24
Distance-based matrix ($\beta = -2.7$)	36	0.97
Distance-based matrix ($\beta = -6.3$)	97	1.02

Once the sets of 'candidate' eigenvectors have been selected, the statistical significance of each, as an explanatory variable for German regional unemployment, has to be established. This process was carried out by means of a stepwise logistic regression analysis. The stopping condition employed is a 10% level of significance for both inclusion and retention. In addition to the stepwise regression analysis, a further manual backward elimination of regressors was carried out through the sequential estimation of a logistic regression model, in order to reduce over-correction for SAC. Marginal eigenvectors were excluded as long as their χ^2 values remained non-significant at the 5% level.

The same process was repeated for all years of available data – from 1996 to 2002 – and for each spatial weights matrix. Consequently, seven sets of 'significant' eigenvectors (one set for each year) have been selected for each of the employed spatial relationship matrices. These are the 'spatial filters' uncovered for each year and matrix.

Next, for each matrix we pinpointed a subset of eigenvectors that is common to the years 1996 to 2002. Results of the preceding analyses are summarized in Table 11.2, while details about the eigenvectors selected in each context and year are shown in the Appendix (Table 11A.1). In Table 11A.1, in all cases, the sum-of-squared prediction error (SSPE) divided by the mean squared error (MSE) is roughly 1 (that is, $\sqrt{SSPE/MSE} \approx 1$; in other words, a jack-knife type of cross-validation assessment of the selected eigenvectors yields prediction error that is almost identical to the OLS error minimization results, validating the constructed spatial filters.

The results summarized in Table 11.2 show that we found sets of eigenvectors (whose linear combinations produce spatial filters) that are significant, as explanatory variables of regional unemployment, over the entire time period considered. Of note here is that all contexts (i.e., economic flows, shared boundaries and distance) enable us to define sets of common spatial filters.

In terms of statistical relevance, the amount of variance explained by the spatial filtering regressors is fairly consistent over the years (reasonably, unemployment patterns do not change much from year to year), and at comparable levels, for all of the *geographic* contexts (that is, shared boundaries and distance). The adjusted pseudo-R^2 values found for these analyses are around 0.60–0.80, with the S-coded rook weights matrix approach being the most significant. The results obtained for the commuting flows matrix approach are not as encouraging. The amount of variance explained by the model, in this case, is only in the 0.29–0.35 range.

A plot of the real and estimated unemployment values is shown in Figure 11.2. The plots refer to the rook adjacency matrix S-coding scheme, and to the years 1996 and 2002, and show a fairly good fit, though a tendency toward underestimation can be observed, particularly for the year 2002, which exhibits more 'extreme' unemployment percentages.

Table 11.2 Amount of variance explained by the selected eigenvectors and the number of common eigenvectors, 1996–2002

		Ajusted pseudo - R^2						
Spatial weights matrix	No. of common eigenvectors	1996	1997	1998	1999	2000	2001	2002
Journey-to-work flows matrix	14	0.300	0.291	0.330	0.314	0.338	0.345	0.328
Rook matrix (S-coding)	17	0.648	0.682	0.729	0.745	0.794	0.802	0.791
Rook matrix (C-coding)	15	0.593	0.642	0.685	0.707	0.748	0.768	0.755
Distance-based matrix ($\beta = -2.7$)	6	0.621	0.597	0.652	0.693	0.729	0.745	0.738
Distance-based matrix ($\beta = -6.3$)	11	0.623	0.607	0.650	0.682	0.725	0.744	0.733

The results summarized in Table 11.2 show that we found sets of eigenvectors (whose linear combinations produce spatial filters) that are significant, as explanatory variables of regional unemployment, over the entire time period considered. Of note here is that all contexts (i.e., economic flows, shared boundaries and distance) enable us to define sets of common spatial filters.

In terms of statistical relevance, the amount of variance explained by the spatial filtering regressors is fairly consistent over the years (reasonably, unemployment patterns do not change much from year to year), and at comparable levels, for all of the geographic contexts (that is, shared boundaries and distance). The adjusted pseudo-R^2 values found for these analyses are around 0.60–0.80, with the S-coded rook weights matrix approach being the most significant. The results obtained for the commuting flows matrix approach are not as encouraging. The amount of variance explained by the model, in this case, is only in the 0.29–0.35 range.

A plot of the real and estimated unemployment values is shown in Figure 11.2. The plots refer to the rook adjacency matrix S-coding scheme and to the years 1996 and 2002, and show a fairly good fit although a tendency toward underestimation can be observed, particularly for the year 2002 which exhibits more 'extreme' unemployment percentages.

(a)

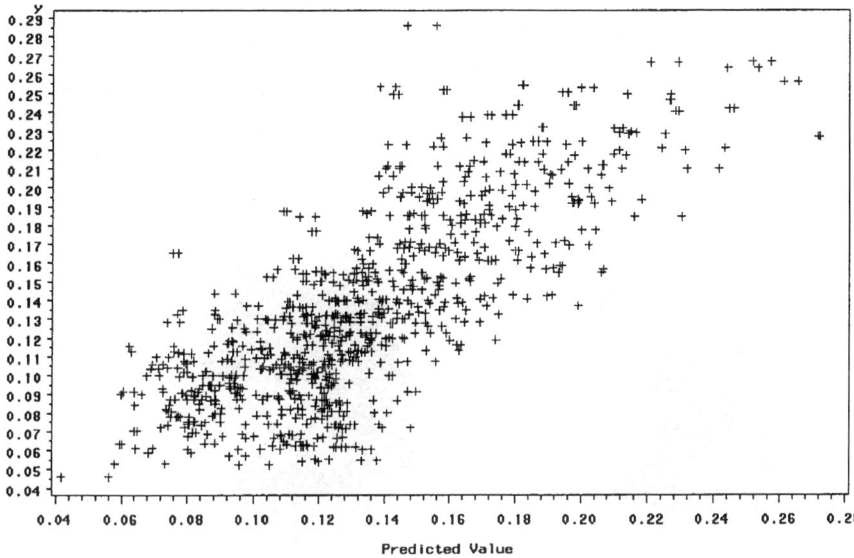

(b)

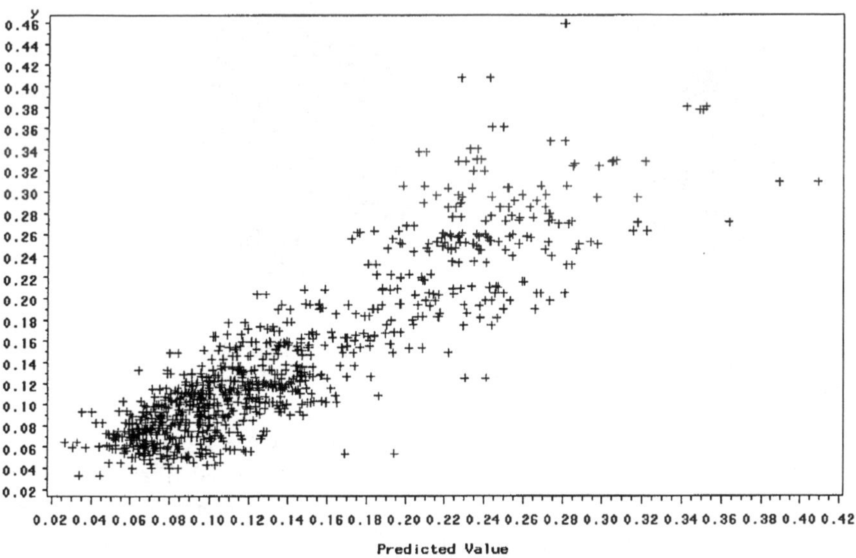

Figure 11.2 Predicted and observed unemployment values (rook adjacency matrix (S-coding scheme), years 1996 (a) and 2002 (b): black crosses denote model-predicted values, and grey crosses denote cross-validation-predicted values)

As mentioned in the section on Spatial Filtering Techniques, the constructed spatial filters can be interpreted not only as potential explanatory variables substituting for missing ones, but also as map patterns. A graphical visualization of the spatial filters uncovered by our analysis provides an example of the map features embedded in the eigenvectors' values. Figure 11.3 shows the four eigenvectors with the largest MI values computed for the rook adjacency matrix S-coding scheme, and that are common to all the years examined.

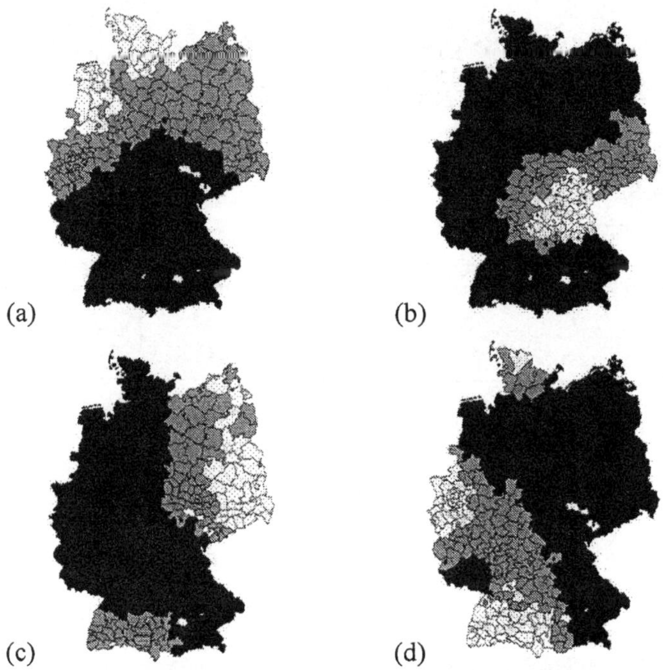

Figure 11.3 Eigenvectors computed for the rook matrix (S-coding): (a) E2; (b) E3; (c) E5; and (d) E6 (see Appendix, Table 11A.1)

As noted previously, the first two eigenvectors for adjacency matrices usually show East–West and North–South patterns. Spatial filter (a) (E2) in Figure 11.3 seems, in fact, to be characterised by a North–South pattern. When we observe the subsequent spatial filter components (b, c and d), the geographic patterns mapped relate to characteristics of smaller geographical scale, showing patterns that can be categorized first as 'regional' and then as 'local'. Although they may contain some common map patterns (for example, North–South and East–West patterns), spatial filters computed with different spatial weights matrices will vary to some degree. Meanwhile, an assessment of the statistical significance of the spatial filters (shown in Table 11.2) enables us to assess the utility of the different proximity approaches employed.

Synthesis: Results for Different Proximity Approaches

The preceding section reveals that all of the definitions employed in this chapter in order to operationalize proximity have been found to generate sets of eigenvectors (whose linear combinations are spatial filters) that are significantly correlated with the dependent variable, regional unemployment, and for all the years examined. Consequently, our focus is on similarities and differences in the statistical performance of the different definitions used.

In order to understand the descriptive performance associated with different spatial weights matrices, we need to compare the matrices themselves. Therefore, a correlation analysis of the matrices employed in our chapter has been carried out. Results of this analysis appear in Table 11.3 (for details about the computation of correlation between matrices, see Oden 1984, and Tiefelsdorf 2000).

Table 11.3 Correlations of spatial weights matrices

	Journey-to-work flows matrix	Rook matrix (S-coding)	Rook matrix (C-coding)	Distance-based matrix ($\beta = -2.7$)	Distance-based matrix ($\beta = -6.3$)
Journey-to-work flows matrix	1.0000	0.5641	0.5102	0.4919	0.5949
Rook matrix (S-coding)	0.5641	1.0000	0.9152	0.6892	0.7923
Rook matrix (C-coding)	0.5102	0.9152	1.0000	0.6533	0.6879
Distance-based matrix ($\beta = -6.3$)	0.5949	0.7923	0.6879	0.8775	1.0000

Several features of Table 11.3 are noteworthy. The most conspicuous result pertains to the low correlations between the journey-to-work flows matrix and the remaining matrices (that is, shared boundaries and distance-based). The correlation values found are plausible and, to a certain degree, to be expected. The flows matrix differs from the other matrices in that it is not based on closeness, but is a proxy for economic links between the districts. These links are, in fact, not fully limited by geographic contiguity, embracing hierarchical components of the geographic landscape, as well. With regard to the remaining matrices, they all seem to have fairly high correlations, which would be consistent with the similarities seen in the statistical performance of their computed eigenvectors (see Table 11.2).

Finally, we also note that:

- matrices based on more similar definitions tend to be more strongly correlated with each other than with those based on less similar definitions;
- the correlation between the two rook adjacency-based matrices is higher than the one between the two distance-based matrices, in spite of the different coding schemes employed; and,
- both distance-based matrices, which have been constructed with the W-coding scheme, seem to be more strongly correlated with the S-coded than with the C-coded rook matrix.

These findings call for a more in-depth analysis of the issues related to the choice of a coding scheme, particularly in view of the type of data patterns that a spatial analyst wants to emphasize. The discussion of such problems goes beyond the scope of this chapter; an interesting treatment can be found in Tiefelsdorf et al. (1999).

CONCLUSIONS

Our study aimed to map out German regional unemployment patterns by means of spatial filtering techniques, so as to uncover spatial structures underlying the georeferenced unemployment data. Several definitions of proximity have been employed, in order to operationalize spatial linkages according to geographic and non-geographic criteria. Each of these definitions has yielded a set of time-stable spatial filters, though at different levels of statistical significance.

Initial sets of eigenvectors have been selected on the basis of the SAC they accounted for (that is, by decreasing MI values), only later to be reduced in size by means of stepwise regressions followed by manual backward elimination. The final subsets of eigenvectors used to construct spatial filters render fairly satisfactory statistical descriptions. In the shared boundaries- and distance-based approaches, the spatial filters explain 60 to 80 per cent of the total variance displayed by unemployment when utilized as the sole regressors in a logistic regression model. But the 'economic flows' approach, based on a journey-to-work flows matrix, fails to produce the same encouraging results. This finding might be due to the artificial nature of the data used (logical connections between districts) and to the lack of a more proper measure of regional economic linkages.

A correlation analysis of the spatial weights matrices employed in our analysis (see Synthesis: Results for Different Proximity Approaches) shows that matrices computed on the basis of the same proximity measure tend to be highly correlated, regardless of the coding scheme applied in their standardisation. Also, the journey-to-work matrix seems to be much less correlated with the topological-based matrices. This result is consistent with the varying statistical performance of the spatial filters computed.

Future research will start from this preliminary estimation in order to carry out more detailed experiments about the dynamics of unemployment patterns. However, the joint employment of spatial filters and other explanatory variables involves further attention to spatial filtering, since eigenvectors that are significant both to the explained and (an) explanatory variable(s) imply filtering also of the latter. This research challenge may be addressed in the framework of new dynamic modelling experiments.

APPENDIX:
DETAILED EIGENVECTOR SELECTIONS, BY YEAR

Table 11A.1 Common and year-specific eigenvectors selected, years 1996–2002

Year	No. of eigenvecs	Year-specific eigenvectors			Common eigenvectors			Scale	Adj. pseudo R^2	$\sqrt{SSPE/S}$
		Global	Regional	Local	Global	Regional	Local			
Eigenvectors extracted from the journey-to-work flows matrix (78 candidate eigenvectors)										
1996	20		E10	E35, E52, E62, E63, E69	E1	E4, E5, E7	E13, E18, E19, E25, E38, E44, E48, E50, E54, E77	26.6747	0.3004	1.0438
1997	18			E31, E62, E63, E78				31.3204	0.2911	1.0407
1998	22		E10	E31, E62, E63, E68, E70, E71, E78				31.0519	0.3305	1.055
1999	20		E10	E31, E62, E63, E71, E78				32.1972	0.3142	1.0442
2000	21		E10	E31, E62, E63, E69, E71, E78				35.2523	0.3379	1.0487
2001	21		E10	E31, E62, E63, E69, E70, E78				37.9095	0.3453	1.0507
2002	19		E10	E31, E68, E71, E78				37.9052	0.3285	1.0411

			Eigenvectors extracted from the rook matrix (S-coding) (130 candidate eigenvectors)							
1996	23	E1	E24, E25, E60	E113, E124	E2, E3, E5, E6, E7, E8, E9, E10, E11	E15, E16, E22, E39, E41, E52, E71	E130	20.8775	0.64766	1.0226
1997	24		E17, E25, E28, E70, E82, E97	E113				23.3176	0.68207	1.0395
1998	20		E14, E25, E28, E36, E60, E70, E82	E113, E129				22.0736	0.72933	1.0476
1999	26		E14, E23, E36, E38, E70, E82	E113, E115, E129				22.0632	0.74533	1.0438
2000	31		E14, E25, E28, E33, E36, E38, E40, E50, E70, E82, E85	E113, E115, E129				22.0026	0.79449	1.0702
2001	28		E14, E18, E23, E32, E36, E38, E40, E82	E110, E115, E129				23.8654	0.80218	1.0489
2002	25		E14, E23, E36, E38, E40, E82	E115, E129				24.8114	0.79092	1.0387

Eigenvectors extracted from the rook matrix (C-coding) (98 candidate eigenvectors)										
1996	24		E9, E16, E21, E25, E41, E52, E53, E64	E89	E2, E3, E4, E5	E6, E7, E8, E11, E18, E24, E28, E30, E39, E60	E74	21.9823	0.5929	1.0232
1997	23	E1	E15, E19, E21, E34, E38, E64	E93				24.3798	0.6425	1.0412
1998	27		E13, E15, E16, E19, E21, E34, E38, E42, E52, E66	E68, E93				23.5231	0.6846	1.0438
1999	27		E9, E13, E15, E16, E19, E21, E34, E38, E42, E52, E66	E93				23.2457	0.7068	1.0364
2000	30		E9, E13, E15, E16, E19, E21, E25, E34, E38, E42, E51, E52, E66	E93, E97				23.8338	0.7483	1.0507
2001	30		E9, E12, E13, E15, E16, E19, E34, E42, E52, E56, E65, E66	E68, E93, E97				25.1826	0.7683	1.0489
2002	29	E1	E9, E12, E13, E15, E16, E19, E20, E25, E38, E42, E52, E65, E66					26.0816	0.7549	1.0459

Eigenvectors extracted from the distance-based matrix ($\beta = -2.7$) (36 candidate eigenvectors)								
1996	18	E7, E11, E12, E14	E26, E29, E30, E32, E34	E1, E2, E3	E5, E6, E16, E23	21.5904	0.6215	1.0018
1997	13	E7, E12, E14, E17	E31, E32			25.6508	0.5968	1.0063
1998	14	E11, E12, E17, E21	E26, E31, E32			24.8788	0.6519	1.0079
1999	14	E11, E14, E17, E21	E26, E30, E31			24.406	0.693	1.004
2000	15	E11, E14, E17, E21	E26, E30, E31, E32			25.5416	0.7296	1.0092
2001	14	E11, E17, E20, E21	E26, E30, E31			26.6811	0.7448	1.0087
2002	13	E11, E17, E20, E21	E26, E30			26.871	0.7382	1.0101

Eigenvectors extracted from the distance-based matrix ($\beta = -6.3$) (97 candidate eigenvectors)										
1996	24		E13, E17, E26, E27, E29, E31, E34, E35, E39, E52	E90, E96	E1, E2, E3, E5, E6, E8	E15, E32, E55, E64	E91	20.7515	0.6233	1.046
1997	20	E7	E13, E17, E20, E23, E34, E35	E79, E96				24.6605	0.6067	1.0334
1998	21		E13, E17, E20, E23, E24, E26, E31, E34	E79, E96				24.2862	0.6501	1.0424
1999	19		E13, E20, E23, E26, E39, E63	E79, E96				24.0949	0.6818	1.0248
2000	24		E13, E17, E20, E23, E24, E26, E29, E39, E40, E63	E71, E79, E96				24.6917	0.7247	1.0371
2001	23		E13, E17, E20, E23, E24, E25, E26, E39, E40, E63	E79, E96				26.0291	0.7442	1.0392
2002	21		E17, E20, E23, E25, E26, E27, E39, E40, E63	E79				26.9843	0.7331	1.1589

ACKNOWLEDGMENTS

The authors wish to thank Uwe Blien (IAB, Nuremberg, Germany) for providing the labour market dataset, as well as Günter Haag (STASA, Stuttgart, Germany) for the commuting flows dataset. The authors are grateful to participants of the 52nd Annual North American Meetings of the Regional Science Association International (Las Vegas, NV) and the International Workshop on Spatial Econometrics and Statistics (Rome), and to an anonymous referee, for useful comments on earlier versions of the chapter.

NOTES

1. Moran's I coefficient is defined as:
$$I = \frac{n \sum_i \sum_j w_{i,j} (x_i - \bar{x})(x_j - \bar{x})}{(\sum_i \sum_j w_{i,j}) \sum_i (x_i - \bar{x})^2},$$
Where: n is the number of cases; x_i is the value of variable X at location i; and $w_{i,j}$ is the cell (i, j) of the geographic weights matrix $\mathbf{W}$. Positive values $[I > -1/(n-1)]$ imply that geographical proximity tends to produce similar values of the variable examined.
2. Griffith's spatial filtering techniques may be compared to principal components analysis (CPA), since both methodologies generate orthogonal and uncorrelated new 'variables' that can be employed in regression analyses. However, the components derived in PCA have an economic interpretation because eigenvectors are used to construct linear combinations of attribute variables, whereas spatial filters are linear combinations of the eigenvectors themselves, and as such should be regarded mostly as patterns of independent spatial dimensions.
3. For details about the estimation of SIMs see among others, Haynes and Fotheringham (1984), and Sen and Smith (1995).

REFERENCES

Anselin L. (1988) *Spatial Econometrics: Methods and Models*. Kluwer Academic Publishers: Dordrecht Boston

Anselin L. (2007) Spatial Econometrics in RSUE: Retrospect and Prospect. *Regional Science and Urban Economics* 37 (4), 450–6

Anselin L., R.J.G.M. Florax and S.J. Rey (eds) (2004) *Advances in Spatial Econometrics*. Springer: Berlin, Heidelberg, New York

Bockstael N.E. (1996) Economics and Ecological Modeling: The Importance of a Spatial Perspective. *American Journal of Agricultural Economics* 78 (5), 1168–80

Chun Y., R. Bivand and M. Tiefelsdorf (2005) Using Open Source Data Analysis Environments for Prototyping Modelling Implementations for Spatial Data: Weights in R. Paper presented at the Geo-Computational Meeting, Ann Arbor, MI, August 1–3

Getis A. (1990) Screening for Spatial Dependence in Regression Analysis. *Papers of the Regional Science Association* 69, 69–81

Getis A. (1995) Spatial Filtering in a Regression Framework: Examples Using Data on Urban Crime, Regional Inequality, and Government Expenditures. In L. Anselin and R.J.G.M. Florax (eds), *New Directions in Spatial Econometrics*. Springer: Heidelberg, pp. 172–85

Getis A. and J. Aldstadt (2004) Constructing the Spatial Weights Matrix Using a Local Statistic. *Geographical Analysis* 36 (2), 90–104

Getis A. and D.A. Griffith (2002) Comparative Spatial Filtering in Regression Analysis. *Geographical Analysis* 34 (2), 130–40

Griffith D.A. (1981) Towards a Theory of Spatial Statistics: A Rejoinder. *Geographical Analysis* 13, 91–3

Griffith D.A. (1988) *Advanced Spatial Statistics*. Kluwer Academic Publishers: Dordrecht

Griffith D.A. (1996) Spatial Autocorrelation and Eigenfunctions of the Geographic Weights Matrix Accompanying Geo-Referenced Data. *The Canadian Geographer* 40, 351–67

Griffith D.A. (2003) *Spatial Autocorrelation and Spatial Filtering: Gaining Understanding through Theory and Scientific Visualisation*. Springer: Berlin, New York

Griffith D.A. (2004) A Spatial Filtering Specification for the Autologistic Model. *Environment and Planning A* 36 (10), 1791–811

Haining R. (1991) Bivariate Correlation and Spatial Data. *Geographical Analysis* 23, 210–27

Haynes K.E. and A.S. Fotheringham (1984) *Gravity and Spatial Interaction Models*. Sage Publications: Beverly Hills

Kosfeld R. and C. Dreger (2006) Thresholds for Employment and Unemployment. A Spatial Analysis of German Regional Labour Markets 1999-2000. *Papers in Regional Science* 85 (4), 523–42

Lacombe D.J. (2004) Does Econometric Methodology Matter? An Analysis of Public Policy Using Spatial Econometric Techniques. *Geographical Analysis* 36 (2), 105–18

Oden N.L. (1984) Assessing the Significance of a Spatial Correlogram. *Geographical Analysis* 16, 1–16

Rogerson P.A. and D.A. Plane (1984) Modeling Temporal Change in Flow Matrices. *Papers in Regional Science* 54(1), 147–64.

Sen A. and T.E. Smith (1995) *Gravity Models of Spatial Interaction Behavior*. Springer: Heidelberg and New York

Tiefelsdorf M. (2000) *Modelling Spatial Processes – The Identification of Spatial Relationships in Regression Residuals by Means of Moran's I*. Springer-Verlag: Berlin

Tiefelsdorf M. and B. Boots (1995) The Exact Distribution of Moran's *I*. *Environment and Planning A* 27, 985–99

Tiefelsdorf M. and D.A. Griffith (2007) Semiparametric Filtering of Spatial Autocorrelation: The Eigenvector Approach. *Environment and Planning A* 39 (5), 1193–221

Tiefelsdorf M., D.A. Griffith and B.N. Boots (1999) A Variance Stabilizing Coding Scheme for Spatial Link Matrices. *Environment and Planning A* 31, 165–80

Weinhold D. (2002) The Importance of Trade and Geography in the Pattern of Spatial Dependence of Growth Rates. *Review of Development Economics* 6 (3), 369–82

PART FOUR

Spatial Mobility

12. The role of computing in urban travel forecasting: how transportation planning practice shaped software, and software impacted transportation planning practice

David Boyce and Hillel Bar-Gera

INTRODUCTION

Digital computers, and programmes written for their use, had a profound effect on the development and evolution of travel forecasting methods from the very beginning of the field. Urban transportation planning, and travel forecasting specifically, was an early civilian application of computers, first in the United States and in the United Kingdom, and today throughout the world. This chapter traces the 50-year evolution of travel forecasting applications of computing from the first conceptualisation of a package of computer programmes in 1955 to modern software systems now used by all practitioners. The emphasis in this review is on computer programme packages, batteries, or software systems, whatever they were called in their own time, developed to perform travel forecasts. Software for supporting analyses and parameter estimation was sometimes incorporated into these systems, but other times not. By and large, this related software is ignored in this review. Also outside our scope are software systems for micro-, meso- and macro-simulation of traffic and transit systems.

The chapter is organized in the following way. First, a brief overview of the computing environment that evolved over the period 1951–2007 is offered. This overview includes an analysis of the characteristics of computers in terms of computational speed and memory. Second, the first governmental programmes in the US and UK to undertake the development of computer programmes for travel forecasting are reviewed, sometimes in conjunction with innovations in specific transportation planning agencies. Third, the evolution of government programmes into an international private software development niche that characterises the field today is traced, and its current status is described. A discussion of the future direction of travel forecasting and software for travel forecasting concludes the chapter.

OVERVIEW OF THE COMPUTING ENVIRONMENT

During the 50 years examined in this review, computing hardware and travel forecasting software have progressed to such an extent that today's novice may not be able to imagine the conditions experienced at the outset of this field. In this section, a brief overview of the development of computing hardware and software is offered, including a description of the early use of electronic accounting machines and first generation computers in the 1950s. Much of the following description was gleaned from Wikipedia, www.wikipedia.org and da Cruz, www.columbia.edu/acis/history.

Mainframe Computing

From the beginning of the use of computers in the late 1950s until well into the 1980s, travel forecasting models in the US and elsewhere were solved on mainframe computers, predominantly produced by IBM. These included the first generation (vacuum tube) computers, such as the IBM 704 and 709 during the late 1950s, the second generation (transistor) computers, such as the IBM 7090 and 7094, from the late 1950s well into the 1960s, and the third generation (integrated circuit) computers known as the IBM System/360 and System/370 from the 1960s onwards. Often smaller computers such as the IBM 650, 1401 and 1620 were used for secondary computations, or to prepare tape inputs from punched cards for input to a mainframe computer. Operating systems for first and second generation IBM computers, such as IBSYS and OS/360, evolved in parallel with the hardware.

Prior to and during the first years of the use of mainframe computers, the processing and analysis of the vast data files collected were undertaken with electromechanical accounting machines produced by IBM. One of the first transportation planning studies to use this equipment was the Detroit Metropolitan Area Traffic Study (1955, 1956), which was remarkable in many methodological respects, including its use of computational methods of the early 1950s. At the time of the Detroit study, the IBM 407 Accounting Machine was the last and best of the all-electromechanical IBM accounting machines, previously known as tabulators. The IBM 407 read a deck of 80-column punch cards on an integrated card reader, accumulated totals, subtotals, or other simple statistics in counters made of gears, and printed the results on an integrated 132-column line printer. Speed of operation was 100 to 150 cards per minute. As with all IBM punch-card equipment, except the key punch and sorter, a control panel was wired to specify the details of operation: what card columns to read and what to do with them, and how to format the report. Although the IBM 407 was really just a big adding machine, creative use could be made of the control panel programme. These machines were large and heavy, as shown in Figure 12.1, requiring floors sufficiently strong to support their weight (da Cruz, www.columbia.edu/acis/history).

The standard punch card, invented by Herman Hollerith, was first used for vital statistics tabulation by the New York City Board of Health and several states. After this trial use, punch cards were adopted for use in the 1890 census. After Hollerith had perfected his first series of electromechanical punch-card machines, including a punch, a tabulating machine to accumulate statistics from the information punched on cards and a sorting machine, he founded the Tabulating Machine Corporation. As with later high-tech start-ups, it had a somewhat rocky start until an experienced manager entered the scene. Thomas Watson, who previously worked for the National Cash Register Company, took over. One of Watson's moves was to change the name of the company to International Business Machines. Within a few decades, his company had expanded to the point that the Federal government sued IBM for anti-trust violations (Jones, www.cs.uiowa.edu/~jones/cards/history.html).

The experience accumulated by the Detroit Study was transferred to the next major transportation study, the Chicago Area Transportation Study (1959, 1960, 1962) through the migration of its study director, J. Douglas Carroll, Jr., and several staff members, from Detroit to Chicago in 1955. The Chicago study coincided with the availability of the first generation of mainframe computers, in particular the IBM 704. Building on its experience with the use of electromechanical accounting machines in the Detroit study, the Chicago staff moved aggressively to incorporate this new resource into their work programme.

The IBM 704 Computer was the first mass-produced computer with core memory and floating-point arithmetic; see Figure 12.2. From 1955 to 1960, 123 IBM 704s were sold. The IBM 700 series were binary, as opposed to decimal, vacuum-tube logic computers with 32,000 words of 36-bit length. The IBM 709 succeeded the 704, adding overlapped input/output, indirect addressing and decimal instructions. The IBM 7090 was a 709 with transistor, rather than vacuum-tube, logic. The 7040 and 7094 were scaled-down and scaled-up variations of the 7090. The 36-bit 700- and 7000-series were IBM's scientific computers from 1952 until the introduction of the 32-bit system/360 in 1964 (da Cruz, www.columbia.edu/acis/history).

Source: da Cruz, www.columbia.edu/acis/history

Figure 12.1 IBM 407 Accounting Machine

Source: da Cruz, www.columbia.edu/acis/history

Figure 12.2 IBM 704 Computer

The IBM System/360 (S/360) was a third generation, mainframe computer family introduced in 1964. It was the first family of computers that provided a suite of compatible designs at different prices. Computers used for travel forecasting were often the largest models available, such as the S/360 65. The S/360 introduced the 8-bit byte, byte-addressable memory, as opposed to word-addressable memory, and 32-bit words. The IBM System/370 (S/370) was a mainframe family announced in 1970, which maintained backward compatibility with the S/360. New architectural features distinct from the S/360 range included: standard dual-processor capability; full support for virtual memory; and 128-bit floating point arithmetic. S/370 computers were replaced by subsequent IBM mainframe models in the early 1990s; however, the transition to minicomputers and microcomputers effectively ended the dominance of the mainframe for travel forecasting.

In addition to IBM, seven smaller competitors in the US produced mainframe computers. Most noteworthy among these in travel forecasting was Control Data Corporation (CDC), whose chief designer was Seymour Cray, later the designer of the Cray supercomputer in the 1970s. The CDC 6600 was a mainframe computer first delivered in 1964. The CDC 3000 and 6600 series gained some market share during the late 1960s, especially in universities, later succeeded by the CDC CYBER series after 1970. IBM and CDC mainframe computers suitable for travel forecasting were initially far too expensive to be owned and operated by planning agencies, or most state and local governments. Computer time was purchased from service bureau (facility established to provide computer services), large corporate computing facilities, leading universities or federal research agencies, such as the (US) National Bureau of Standards. Execution of programmes was often performed by programmemers physically visiting the service bureau, as remote terminals were often not available.

Transition to Minicomputers and Microcomputers

During the 1970s and 1980s the computing hardware choices available to engineering consulting firms and universities, and in a few cases public agencies, began to broaden. Minicomputers were introduced, most successfully in the US by Digital Equipment Corporation (DEC) with its PDP and VAX computers with VBS or UNIX operating systems. These minicomputers were physically smaller and much less expensive, so that smaller consulting firms and university departments could aspire to acquire them. Software applications were either custom made, or highly specialised. As described in more detail below, travel forecasting software for use by practitioners began to be developed for these systems, as well as for university research.

As computer processor technology evolved, the next development stage was the Apple II microcomputer, offered by Apple in 1977 using the CP/M operating system. Then IBM introduced its Personal Computer (IBM PC) in 1981; see Figure 12.3. The CP/M (Control Programme for Microcomputers) operating system first used by these computers was soon eclipsed by the MS-DOS (disk operating system) of Microsoft. The IBM design team first decided to build its machine with "off-the-shelf" parts from a variety of different original equipment manufacturers and countries. Second, they decided on an open architecture so that other manufacturers could produce and sell peripheral components and compatible software. In this manner IBM determined the standards for personal computers, which other manufacturers soon followed, and the IBM PC-compatible computer was soon established as the standard for office computing.

This seemingly accidental standardization allowed software producers to develop and market software which would operate on hardware systems produced by many firms. In addition to general purpose software for spreadsheet analysis, statistics and word processing, specialised travel forecasting software began to be produced, as detailed below.

Figure 12.3a. Apple Lisa, 1983, an improved version of the original Apple II, 1977
Source: http://www.computerhistory.org/

Figure 12.3b. IBM PC (model 5150), 1982
Source: http://en.wikipedia.org/wiki/IBM_PC

Paralleling the development of personal computers was the engineering workstation, a higher-performance technical computing desktop microcomputer designed primarily to be used by one person at a time, but also accessed remotely by other users as necessary. At first, workstations offered higher performance than found in a personal computer, especially with respect to graphics, processing power and multitasking ability. In contrast to PCs, the operating systems were some "flavour" of the UNIX operating system, an earlier innovation of Bell Laboratories in the 1960s. In the early 1990s, as UNIX workstations were considerably more powerful than PCs, it was unclear which computing platform should be used for travel forecasting. Further technical development and price competition from PC producers gradually tipped the balance in favour of the evolving PC, relegating engineering workstations to more specialised engineering and research tasks, as well as servers for larger scale computing systems.

Comparisons of Computing Speed, Memory and Cost

How can we compare the performance of computers used in the early period of travel forecasting with those in use today? Computer scientists interested in this question have performed tests during the past 35 years. Perhaps the most noteworthy is J.J. Dongarra; despite his ongoing efforts, comparisons are not perfect. One metric for comparisons is the number of floating point operations per second (flops) for solving a standard system of linear equations; one difficulty seems to be that the problem used earlier is no longer demanding enough.

Figure 12.4 shows two data series for this measure for computers introduced between 1951 and 2007. The data series from Strohmaier et al. (2005) shows megaflops for several computers introduced between 1951 and 1982. The source and basis for their performance is not given. The trendline shown was proposed by the authors. The second and more detailed series is based on measurements performed by Dongarra (2009) for a standard system of linear equations. The points shown are but a small selection of the measurements performed to date. The trend established by early mainframe computers and extended by Cray supercomputers and more exotic machines continues to this day. The speed of the original IBM PC was similar to the IBM 704. Subsequent advances in microcomputers during the past 25 years have surpassed three gigaflops. Memory should also be compared, but few data are available. The early IBM mainframes had 32K words of memory, each word consisting of 36 bits, or approximately equivalent to 256 kilobytes (KB). The IBM S/360 and S/370 had up to 512KB of memory, where a byte consisted of 8 bits, as is true today. The maximum memory available during the evolution of mini-computers and engineering workstations was also typically 512KB. In the last several years the memory of PCs has expanded, first to 1,024KB (1 megabyte (MB)) and more recently to 2 gigabytes (GB) per 32-bit processor. As 64-bit PCs become available, memory limits are expanding to 4 to 16 GB.

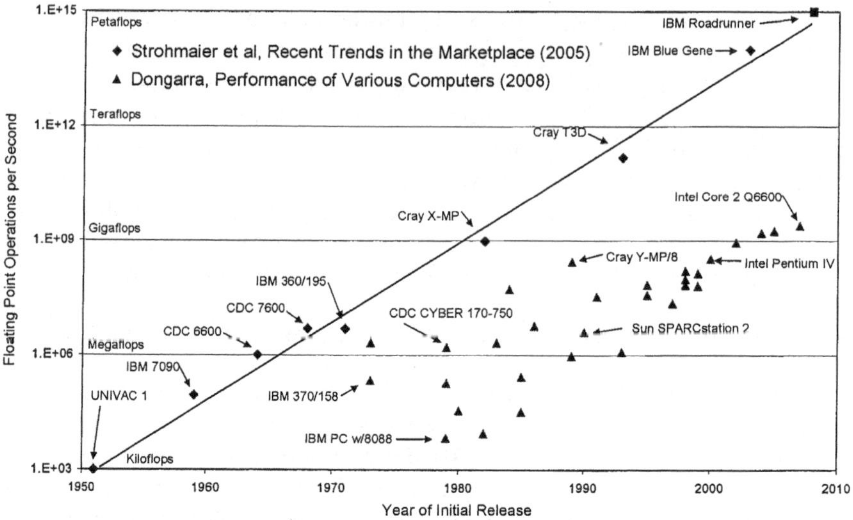

Figure 12.4 Performance of mainframes, minicomputers and microcomputers, 1951–2008

Before the advent of minicomputers, and especially microcomputers, commercial mainframe computing time could cost as much as $20,000 per CPU hour. Estimation of a logit model on a mainframe could easily cost several thousand dollars. Solving travel forecasting models was cost-prohibitive unless one was affiliated with a university or a public agency. Using microcomputers to perform analytical analyses and deliver the results in a form useful to clients was seen as a compelling business opportunity. To reap the benefits of PC technology, practitioners would clearly require extensive software for myriad applications. With the innovation and ongoing development of the IBM PC, software developers believed that the cost of computing would be nominal, once the cost of computer hardware was amortized.

TRAVEL FORECASTING PROGRAMME DEVELOPMENT IN THE US

Programmes for First and Second Generation Computers, 1958 – 1967

In 1955, the (US) Bureau of Public Roads (BPR) first undertook the development of computer programmes to solve: (1) trip distribution models of the gravity type; and (2) traffic assignment models (Brokke, 1967). Programmes for calibrating and testing a gravity model were initially written for the IBM 704 and later modified for use on the IBM 7090/94 and other computers of that era (US Department of Commerce, 1963b, p. ii). Programmeming for a "battery" of assignment programs was initiated for the IBM 704 in 1958 under contract to the General Electric Computer Department, Phoenix, Arizona. "This project produced a battery of high-speed computer programmes that would assign nondirectional interzonal traffic movements", including an option of using a diversion or all-or-nothing assignment. These programmes were extended in 1960 to permit directional assignments as well as turn penalties and prohibitions (US Department of Commerce, 1964, p. I-3). Following the transition from first generation to second generation machines during 1961–1962, BPR incorporated its major programmes into a "system or library of programs". Two early manuals distributed by the US Department of Commerce (1963a, 1964) describe these models and computer programmes.

Mertz (1961) reported on the findings of a survey of agencies solving traffic assignments with computers ranging from the IBM 650 to IBM 704, including the cost of programme preparation and use, and efforts to share programmes. For traffic assignments he reported computer use charges of $1,000 to 2,000 per assignment for a large mainframe computer. (At the time of the survey, the salary of an entry level civil engineer was about $500 per month.) Service bureau arrangements, in some cases including programme development, were implemented with IBM and with the General Electric Company for an IBM 704. In Canada, Irwin and von Cube (1962) implemented a relatively detailed travel forecasting model with feedback, which was applied in metropolitan Toronto; evidently, this work was performed on an IBM 650 and later on an IBM 7070 (Mertz, 1961, p. 100).

Alan M. Voorhees (1955), an early advocate of travel forecasting models, formed Alan M. Voorhees and Associates (AMV) in 1961 at the time when many urban transportation planning studies were being initiated based on the experience with the Detroit and Chicago studies described above, and later in response to the Federal-Aid Highway Act of 1962. Voorhees, together with Walter Hansen and Thomas Deen, led the development of a computational capability built upon the BPR battery described above. Computer time was rented from service bureaus as needed and used mostly at night when the staff would have the entire computer to themselves.

In 1963, the Transportation Planning Computer Programme Exchange Group, somewhat later known as T-PEG, was formed by the US Bureau of Public Roads, urban transportation studies and consultants to facilitate the exchange of technical information concerning computer programmes used in urban travel forecasting. A newsletter was issued beginning with Volume 1, Number 1 dated July 15, 1963, and continued through Volume 2, Number 4, dated October 1, 1965. No newsletters were issued in 1966, but in 1967, semi-annual meetings were initiated, which evidently continued through June

1973, when T-PEG ceased to meet. Newsletters were issued from each meeting. This collection of newsletters provides insights into the activities of early computer programmemers in the travel forecasting field in the US, Canada, the UK, Australia and elsewhere. All together, the newsletters comprise 270 pages; an electronic, searchable version of the newsletter can be accessed at http://books.northwestern.edu/viewer.html?id=inu:inu-mntb-0005779868-bk.

As BPR, the early urban transportation studies, and then the consultants began to extend these computer programmes for travel forecasting for second-generation mainframes in the late 1950s, IBM quickly emerged as the dominant computer platform. Control Data Corporation (CDC), which was emerging as one of IBM's principal competitors for its mainframe business, undertook to compete by creating its own travel forecasting package, T/P. (For a history of CDC and its competitive strategies, see Worthy, (1995).

Development of T/P (for TRANsportation PLANning or TRANPLAN) was initiated by CDC in the early 1960s as a service bureau application for the CDC 3600 mainframe introduced in 1962 in order to compete with travel forecasting programmes available for the IBM 7090/94 (CDC, 1965). CDC's *Urban Affairs Newsletter* (CDC, 1969, Sept., p. 3) described TRANPLAN as "an integrated system of computer programmes which not only performs the data processing functions traditionally employed by transportation planners, but also contains several advantages over previously implemented systems. The system developed for the 3600 computer by Control Data Corporation, in association with a number of leading transportation planning consultants, is a powerful tool for the transportation planning process."

The "previously implemented systems" to which the newsletter refers were BPR's battery of trip distribution and assignment programmes. Brokke (1967) noted that CDC "has developed at their own expense a battery of CDC 3600 programmes that are capable of handling larger networks than can be accommodated on the 7090/94". Although the names of the transportation consultants involved in the development of TRANPLAN were not divulged, the principal consultant in the development of TRANPLAN was AMV beginning in early 1965. Richard Bunyan was the project leader and Raif Kulunk wrote CDC's tree builder, after Bunyan showed him how the algorithm authored earlier by General Electric worked. TRANPLAN was delivered by AMV to CDC during 1966–1967, whereupon AMV itself became a primary user of TRANPLAN. The CDC newsletters issued during 1969 reported applications of TRANPLAN in several urban transportation studies in the midwestern and western US.

In early 1965 Robert Dial joined AMV at age 27 after working for two years for the Puget Sound Regional Transportation Study (PSRTS), followed by the completion of an M.S. degree in civil engineering at the University of

Washington. Dial's undergraduate major in mathematics was the basis for his first job as a statistician at PSRTS. There he became "enthralled" by the modelling and computational challenges of this emerging field, and gained experience with the use of early computers. Working under Edgar Horwood, a geographical information systems pioneer, Dial studied transportation engineering and programmemed the IBM 709 to implement a rudimentary GIS design for his thesis. At AMV, Dial's first assignment was to lead three experienced programmemers in creating a transit assignment package for the state-wide network of Connecticut with 2,000 zones and 32,000 links. (These statements are based on notes by Robert Dial.)

The Urban Mass Transportation Act of 1964, as amended in 1966, assigned responsibility for transit planning to the US Department of Housing and Urban Development (US HUD) (Weiner, 1997, Chapters 4 and 5). Charles Graves, a transit planner also from PSRTS, believed that US HUD should develop a programme battery for public transit systems comparable to BPR's highway battery. A contract was awarded to AMV for development of computer programmes for modal split, transit assignment and related peripheral functions for the IBM 7090/94. Hansen headed the project, and Bunyan and Dial were the initial programmeming staff. Bunyan prepared a process diagram, comparable to BPR's highway package, which consisted of seven modules, as summarized in Dial and Bunyan (1968): network description; transit path builder; OD travel times; mode split model calibration; mode split model application; load trip table; and report generator. Descriptions of these modules, whose names began with *AV* for Alan Voorhees, are given in US HUD (1966, 1967, 1968).

Robert Dial's contributions to the representation of transit networks and route finding in transit networks stem from this project. The programmes were written in FORTRAN except for the tree builders written in assembly language for the IBM 7090 to achieve practical running times (Dial, 1967). Dial also wrote a tree builder in FORTRAN to test his algorithm's design. Taken together with the programme batteries developed by BPR, the AMV's transit planning programmes developed for HUD constituted a relatively complete battery of programmes for urban travel forecasting.

Upon completion of the HUD package, AMV received a terse letter from CDC demanding all of the HUD source code under the Freedom of Information Act. AMV promptly complied; at that time any findings accomplished at the taxpayer's expense were considered to be in the public domain. When CDC received the source code, their programmemers noticed that the tree builder was written in assembler language. In response to a complaint, Dial offered CDC his FORTRAN test code. Without this small programme to guide them, CDC might have never developed a transit planning capability or achieved a substantial use of the programme.

In the same *Urban Affairs Newsletter* cited above, another system was described, acknowledging AMV's and HUD's contribution: "HUDTRAN is a series of computer programmes, available for the CDC 6600, which is used for planning public transportation systems. The programmes were designed and written by Alan M. Voorhees and Associates, Inc. (AMV), under contract to the Office of the Secretary, Department of Housing and Urban Development. Conversion of the programmes to the CDC 6600 was done by Data Services personnel to complement 3600 TRANPLAN."

In the December 1969 newsletter, the release of TRANPLAN II was announced for July 1970 for the CDC 6600 and 3300 computers, which were recently introduced at seven CDC Data Centres in a nationwide network called CYBERNET Service, accessible by voice-grade or wideband telecommunication lines. TRANPLAN II was described as offering: a comprehensive, integrated set of transportation planning functions; application of key advances within the computer industry; problem-oriented command and control language; and the capability to insert or change individual modules. The basic functions of TRANPLAN II were to include: data capture; highway and transit network builder; highway and transit path builder; highway and transit network loader; gravity and Fratar trip distribution; modal choice; file utilities; report generator; and graphic displays. The capability to implement the programmes from a remote terminal was emphasized. The September 1970 newsletter stated that TRANPLAN II was scheduled for release in January 1971. As the archived newsletters end with the January 1971 issue, the eventual disposition of TRANPLAN II is unknown; evidently, the effort was abandoned near the end of 1970.

Programmes for Third Generation Computers, 1968 – 1980

In 1964, BPR realized that its programme library would need to be adapted to third generation computers, the IBM System/360 announced by IBM in 1964 (US DOT, 1972a, pp. I-8, I-9). According to Brokke (1967), BPR's S/360 computer installation replaced three IBM second generation machines at roughly the same cost, and handled a problem twice as large at the same speed and one-third the rental cost. The computer programmes implemented for the IBM S/360 by BPR and its consultant, Peat, Marwick, Livingston & Co., were designated the Urban Transportation Planning Battery for Trip Distribution and Traffic Assignment (US DOT, 1969).

By 1968, AMV had completed transferring its HUD modal split and transit assignment programmes to the IBM S/360, thereby completing a travel forecasting capability for IBM's third generation computer system. By now, AMV had its first in-house computer, a leased IBM S/360 Model 40 with 128KB of memory, four tape drives and one disk, half the memory and speed of the first IBM PC. The Ontario Department of Transportation (ODT) contracted with AMV in 1967 to write a new highway modelling package, including matrix manipulation tools; the code development undertaken by three ODT programmemers under the direction of Robert Dial included: a highway network builder, a highway network path builder, a path skimmer and a gravity trip distribution model. Programmes developed or modified during the ODT project had the prefix *OV* for Ontario/Voorhees. The resulting programme package was named TRIPS (TRansportation Improvements Programmeming System). TRIPS became a foundation of much work performed by AMV, but was not sold commercially. The HUD package was also a key element of TRIPS.

In 1966, the US Department of Transportation was formed. Following a transition period, the Bureau of Public Roads, formerly in the US Department of Commerce, effectively became the Federal Highway Administration (FHWA). The responsibility for mass transportation in US HUD was reconsidered following the authorisation of US DOT, and the Urban Mass Transportation Administration (UMTA) was created in US DOT in 1968. Computer programme development activities in the US HUD were transferred to UMTA. (Much later, UMTA was renamed the Federal Transit Administration.) FHWA continued the development and dissemination of its travel forecasting programmes for the IBM S/360 and subsequently the IBM S/370. Later, the FHWA battery came to be known as PLANPAC/BACKPAC. PLANPAC was a core set of programmes used in travel forecasting, and BACKPAC contained programmes related to data and utilities; see, for example, US DOT (1974).

In 1968, Robert Dial returned to Seattle and pursued a Ph.D. at the University of Washington under Edgar Horwood, while continuing to work part-time for AMV. His thesis research resulted in his probabilistic multipath assignment model (Dial, 1971). Following completion of his degree in 1970, Dial joined UMTA's Office of Research, Development and Demonstration, where he led the development of the UMTA Transportation Planning System, abbreviated UTPS (US DOT, 1972b). AMV elected to donate TRIPS to UMTA, thereby providing UMTA with the source code, and sparing itself the cost of maintaining TRIPS. The initial version of UTPS mainly involved changing the *AV* and *OV* prefixes of TRIPS programmes to *U*. The first release of UTPS was just TRIPS in different clothing, but then development of TRIPS and UTPS diverged.

Subsequently, FHWA joined with UMTA in the further development and enhancement of an integrated system of programmes for road and transit planning, renamed Urban Transportation Planning System (also UTPS) for the IBM S/370. UTPS consisted of eight principal modules incorporating those in TRIPS, but added INET and also UROAD, a capacity-restrained assignment module that included user-equilibrium assignment. Descriptions of the development of UTPS are found in Dial (1976), Dial and Quillian (1980), and Dial et al. (1980). In addition to government staff, consultants to UMTA and FHWA participated in the development of UTPS programmes, documentation and training manuals (e.g., US DOT, 1977).

Robert Dial led the development of UTPS through the 1970s, as well as engaging in research activities at UMTA. By 1981, computer technology was rapidly evolving, and the days of the mainframes of that era were numbered. Minicomputers (e.g., Digital Equipment Corporation's PDP series) were approaching the speed and memory of mainframes at substantially lower costs. Microcomputers were entering the market, including the Apple II in 1977. Although Dial did not write any code while at UMTA, the opportunity to experiment with these new machines proved irresistible. In 1981, Dial joined the University of Texas at Austin as a tenured professor; following his departure, no further development of UTPS occurred.

COMPUTER PROGRAM DEVELOPMENT IN THE UK

Wilbur Smith and Associates, a prominent traffic engineering consulting firm based in New Haven, Connecticut, also developed its own travel forecasting programmes. In the early 1960s, the firm collaborated with a British firm, Freeman Fox and Partners, to undertake elements of the first London Transportation Study. A joint consultancy formed for this purpose, Freeman Fox Wilbur Smith and Associates, played an important role in other transportation studies in the UK and elsewhere. One result of this collaboration by the end of the 1960s was Traffic Analysis Programmes (TAP), which incorporated models for allocating travel to various modes of travel and public transport routes as well as translations of codes for the IBM 7090/94 obtained from BPR (Wigan, 1977; Wootten, 2004).

During the mid-1960s the UK Ministry of Transport's Mathematical Advisory Unit (MAU) undertook the development of a travel forecasting model for the SELNEC Transportation Study based on Manchester; see Wilson et al. (1969). MAU was responsible "for the development, calibration and testing of a model of transport flows" for the region. To implement its model computationally, the Ministry of Transport purchased rights to the TAP package in mid-1967, and performed testing of the model on a CDC 3300 computer in London. To facilitate the application of the model in Manchester, arrangements were made for the use of a CDC 6600 computer, also located in London, but accessed via a CDC 200 terminal over telephone lines. According to "Computing Procedures" (SELNEC, 1973), the computer used for running a model on the scale of SELNEC must be extremely powerful in terms of speed, memory size and storage facilities. "The CDC 6600 offers an extremely fast 'turn round' time – usually of the order of 30 minutes for a large job – and a competitive price/performance ratio."

In 1968, the UK Ministry of Transport contracted with AMV to undertake a land use/transportation study called the Tyne Wear Plan. The contract reflected the fact that British consultants were predominately highway engineers, whereas AMV had a reputation in transit and land use planning. AMV brought TRIPS to the UK for this study, and AMV Ltd. was formed. This development led to new life for TRIPS, which had been subsumed under UMTA's development of UTPS in the US.

TRANSITION TO MINICOMPUTERS AND MICROCOMPUTERS

The evolution of computing hardware from mainframes to minicomputers to microcomputers began in the 1970s. Several transportation planning agencies, universities and consultants began to develop and offer programmes for travel forecasting, learning from the development of FHWA's PLANPAC and UMTA/FHWA's UTPS experience. About 1980, these consultants complained to the Federal Government that US DOT was developing software in competition with the private sector. Such complaints may have received a receptive hearing by the Reagan Administration, which took office in January, 1981. In any event, as noted above, UMTA and FHWA ceased further development of UTPS and PLANPAC, although maintaining it for several years. FHWA also terminated development of its Quick Response System (QRS), and gave the rights to its developer, Alan Horowitz at the University of Wisconsin-Milwaukee.

This change in policy clearly had a huge impact on the evolution of the travel forecasting software sector. In the eyes of some, it also contributed to the inability of US DOT to advance the state of practice in travel forecasting methods. Given the emerging role of private software developers, FHWA ceased recommending to planning agencies and consultants which software system they should use. Various software systems being developed and sold to state DOTs and MPOs had different capabilities and qualities. Users of these systems had modest abilities to evaluate the qualities of the systems, which tended to produce different results. Moreover, standards and protocols for software did not emerge, as they sometimes did in other software sectors. To facilitate exchange of information about this emerging software, UMTA and FHWA compiled a *Microcomputers in Transportation Information Source Book*. The first issue was dated August 1982, and the last evidently appeared in June 1987; e.g., see US DOT (1984). A new series called *PCs in*

Transportation Software Directory appeared in July 1988, issued by the University of Kansas Transportation Centre, and continued for several years (Anderson, 1988). Another series launched during this period was the University of Florida McTrans microcomputer software catalogue and sales, which continues to the present.

Although these source books and directories issued during the 1980s cannot be regarded as comprehensive, they do offer a window into the activities of the period. The *Software and Source Book* issued in 1984 shows that several relatively comprehensive software systems were under development, as listed below in the order in the publication (US DOT, 1984).

Name	Developer	Computing Environment
MicroTrans	TRANSWARE Systems	IBM PC-DOS
TMODEL	Robert Shull, Professional Solutions	Apple,IBM PC-DOS, etc.
MINUTP	Larry Seiders, Comsis Corp.	BM PC-DOS
MTPS	San Diego Systems, Inc.	Apple, IBM PC
TGDA	Wayne Kittleson, CH2M Hill	Apple
UMOT	Mobility Systems, Inc.	Apple
TP-System	Kenneth Roberts & Associates, Inc.	Altos 68000-UNIX
TRANPLAN	James Fennessy, DKS Associates	IBM PC
MicroTRIPS	PRC Voorhees/MVA Systematica	IBM PC
EMME/2	Michael Florian, INRO	Pixel 100/AP-UNIX
MOTORS	Bob Lewis, M. M. Dillon Ltd.	IBM PC

By the 1987 edition of the *Software and Source Book*, the following had been added:

Name	Developer	Computing Environment
UTPS	Mark Hallenbeck, Univ. of Washington	Not Stated
(microcomputer version of the mainframe version of UTPS)		
QRS II	Alan Horowitz, AJH Associates	IBM PC-DOS

In addition, two systems were being offered in Europe:

Name	Developer	Computing Environment
VISUM/VISEM	T. Schwerdfeger, PTV AG, Germany	HP 1000-UNIX
SATURN	Dirck Van Vliet, University of Leeds, UK	IBM PC-DOS

Early editions of the *Source Book* also described software under development at M.I.T. by Marvin Manheim, and at the University of Texas, Austin by Robert Dial.

A review by Ferguson et al. (1992) provides another useful benchmark in the development of travel forecasting software. By this point in time, software was primarily oriented to the IBM PC-compatible with the 80286/80386/80486 series CPU, with 640 KB memory and at least 60 MB disk storage and a graphics monitor (VGA/SVGA) operating under MS-DOS. Some vendors used engineering workstations as their software development platform, and also offered systems for one or more UNIX-based computers. In their review, the authors identified eight vendors offering software suited to both road and transit analysis with a four-step procedure capability. In alphabetical order, the systems were: EMME/2; MINUTP; QRS II; System II; TMODEL2; TransCAD; TRANPLAN; and TRIPS. To these should be added PTV's VISUM/VESEM offered in Western Europe and Atkins-ITS's SATURN. (SYSTEM II was a product briefly offered by JHK and Associates in the early 1990s.)

These software systems varied substantially in size of model (numbers of zones, nodes and links), graphics and network editing capabilities (limited graphics to a full GIS), interactive capabilities (batch, menu-driven, mouse) and price (nominal to quite expensive). All vendors offered training programmes and technical support. Thus, only ten years after the advent of the IBM PC, a highly specialised software market niche had developed, limited primarily by the initially small but rapidly expanding capabilities of microcomputers. While large planning organizations continued to rely to a considerable extent on mainframe computers and PLANPAC and UTPS software, the conversion to commercial software systems for the PC, as well as UNIX-based computers, was clearly the trend for the future.

OVERVIEW AND STATUS OF TRAVEL FORECASTING SOFTWARE DEVELOPERS AND PRODUCTS

Software evolved for minicomputers and microcomputers in three ways. First, mainframe programmes were rewritten and downsized for minis and micros. The authors of these early systems based on mainframe programmes were often the same programmers, practitioners and consultants who wrote applications for the mainframes. Second, as academic research on travel forecasting methods began to catch up with practice beginning in the late 1960s, academic research-based software developers initiated software systems that were distinctly different from their practice-based counterparts. Third, several software systems were created by practitioners based on their experience and interest in applying the emerging minicomputers and microcomputers from the late 1970s.

In the first group there were three principal development streams, which have now merged into one firm, Citilabs:

1. TRANPLAN, based on the pioneering efforts of AMV to create TRIPS, which formed the basis of CDC's TRANPLAN;
2. MICROTRIPS, which also evolved from the original TRIPS of AMV;
3. MINUTP, based in part on its developer's earlier experience with BPR and PLANPAC.

An unrelated product was TRACKS, an early software product of a New Zealand consultant, Gabites Porter, which initially was a mainframe application.

The second group consists of six academic research-related firms that began at about the same time in the 1980s:

1. INRO, developer of EMME/2, based on research at the Centre for Research on Transport, University of Montreal, Canada (Florian, 1999, 2008);
2. PTV AG, developer of VISUM, based on research at the University of Karlsruhe, Germany;
3. Atkins-ITS, developer of SATURN, based on research at the Institute for Transport Studies, University of Leeds, UK (Van Vliet, 1982);
4. AJH Associates, developer of QRS II, based on research at the University of Wisconsin-Milwaukee;
5. MCT, developer of ESTRAUS, based on research at the Department of Transport Engineering, Catholic University of Chile, Santiago, Chile (de Cea et al., 2005);
6. Caliper Corporation, developer of TransCAD, built on academic research of that era (Slavin, 2004).

The third group consists of several smaller software developers, sometimes just one person working alone. These include TMODEL and UFOSNET, developed in the northwestern US, JICA STRADA, developed in Japan, and OMNITRANS, developed in the Netherlands.

HOW DEVELOPERS, PRACTITIONERS AND RESEARCHERS SHAPE SOFTWARE AND ITS EVOLUTION

From the history of travel forecasting software development described above, we may observe that software and its use have changed markedly over periods in the past. Its evolution is strongly influenced by different actors in this market. In this final section of the chapter, we examine how the various actors and their interactions shape the software available to our field and its application. From this examination, we seek to raise some questions about how our field may improve its effectiveness in the future.

For the purpose of this discussion, we divide travel forecasting activities into three types:

1. Practice in support of transportation planning decisions, including region-specific model preparation;
2. Software development, as generic tools for practical use in any region;
3. Research, ranging from questions of direct practical implications to development of new methods to conceptual/fundamental/theoretical studies.

It is convenient to consider three corresponding groups of actors: practitioners, software developers and researchers, recognizing that some individuals play roles in more than one group, and that all three activities occur in the private sector, the public sector and the academic sector. We shall examine the influence of these groups one by one.

Practitioners

Practitioners are a heterogeneous lot. Their understanding of the software they apply ranges from very advanced to almost nil. Some view software as a black box, and do not desire an understanding of its properties. Others become proficient in the use of a specific software system, but do not fully understand why it behaves as it does. Still others are highly informed about the properties of the models and algorithms, as well as their advantages and disadvantages.

Choices of practitioners among the available software options are influenced by many legitimate criteria, including: available resources; analysis needs; heritage and the great difficulty of converting from one software system to another, including personnel training; convenience and usability; and finally their own knowledge and ability to benefit (or not) from advanced features. As a result, travel forecasting practice varies widely by agencies and consultants, each local practice differing from others.

There are several advantages to this situation. Advanced practitioners dealing with particularly challenging applications use significant resources to explore new frontiers and advance the state of the practice and often the state of the art as well. At the other extreme, regions with limited resources can rely on modestly skilled personnel using low-end software tools to support their decisions. While the quality of forecasts obtained in this way may not be admirable, they may still be sufficient for the needed analyses, and are probably an improvement to the option of making decisions without any forecasts at all. In view of the statement that "all models are wrong, some are useful", attributed to the prominent statistician George E.P. Box, there seem to be a very wide range of cases in which transportation models are useful.

The downside of this disparity in the abilities of practitioners is the risk of flawed decisions and their consequences. Choice of software is but one of many choices that practitioners make. Being unaware of the implications of some of these choices, practitioners occasionally attempt to perform a certain analysis with an inappropriate tool. This problem is aggravated by the fact that subtle issues may lead to meaningless results, which are considered by the uninformed practitioner as solid truth, since they were produced by a widely used software tool.

Many practitioners work for government agencies at the local, regional and national levels. Some have special responsibilities for policy related to software, such as administration of grants for methodological research and government programmes for software development. Generally, it seems that governments prefer not to interfere with the private market, especially when the firms serving that market compete internationally. Nevertheless, some governmental decisions do affect the market. Therefore, officials should take care that their actions do no unintended harm. At the same time, they may positively affect the advancement of the field by rewarding bidders that use better software, for example.

Software Developers

As described above, at present and in the foreseeable future most practical travel forecasting is and will be performed using software developed by commercial developers. Software development is a very tricky business, since the investment in its "production" is practically zero; most investment is devoted to "development" leading to a modified product, whose value is unknown.

In a free market environment, developers are obliged to give serious consideration to the desires of their users. Therefore, there is a natural tendency of vendors to focus on advances that will be appreciated by the majority of practitioners. For example, all practitioners can comprehend usability improvements, like graphical interfaces; however, only some of them will notice methodological improvements, like consistency between model steps, convergence, plausible route flows, and so on.

In general, software developers maintain high levels of in-house expertise in travel forecasting, and make continuous efforts to be informed about research advances in the field. Decisions concerning which new methods to implement in a commercial product, and when, are quite challenging, especially in view of the risk of practitioner appreciation, as described above. Nevertheless, gradual improvement in software methods may be observed; for example, recently nearly all developers introduced traffic assignment codes able to achieve more precise solutions, as compared with the Frank-Wolfe algorithm that dominated the market only ten years ago. The computing resources needed for the application of these new traffic assignment methods in practical contexts, particularly in RAM, became available only recently, providing another example for the connection between the evolution of computing technology and the progress in travel forecasting methods and practices.

Researchers

In the first stage of innovation of a method, researchers typically consider a stylized and simplified problem, possibly focusing on a very specific aspect of the transportation system, which may require a substantial amount of specialised knowledge. The second stage may be to enhance the new method to represent more realistic conditions, to deal with large-scale problems, and to integrate it into the entire process of practical travel forecasting. This process can be slow and gradual, requiring both specialised understanding of the specific aspect under consideration, as well as general knowledge of prevailing practices. At this second stage, it is often necessary to implement the method in a computer code. In most cases, this is done with a "research code", which is rarely useful for real practice. The transition to an implementation in a practical software tool is a barrier that few innovations cross.

To justify the investment in an implementation usable by the wider practicing community, researchers can explore the significance of the new method for practical modelling and decision making. Given the subtle nature of some improvements in methods, getting the message across and making the knowledge available, accessible and understood can be a fairly challenging research task. However, only after these last stages are accomplished is the new method or model likely to become useful

Travel forecasting researchers have a wide range of interests, aspirations, capabilities and resources available, and their influence on software development and its usage varies greatly. As a group, contributions of researchers can be found in all stages of innovation development, but their distribution between the stages is not necessarily best. This situation may be due to the structure of academic rewards, the lack of interaction with the practitioner community, or for other reasons. Therefore, while the impact of researchers on software development is noticeable, it does not fully reflect the invested efforts.

What May Be Expected and What Should Be Done?

Advances in computing technology have been, and probably will continue to be, rapid. The present market structure of travel forecasting software has proven to be reasonably good in forcing developers to keep up with this rapid technology race, and offer significant improvements in usability and productivity to users. Practitioners have been quite effective in utilizing the advances in computing hardware and software to develop models of increasing scale, complexity and sophistication. It is natural to expect continued effectiveness in these respects.

Methodological improvement has been less effective, as methods often remain unchanged despite research achievements. General questions, including how new methods should be adopted, when and how old methods should be declared obsolete, and how the variety of methods should be maintained by the travel forecasting community, are beyond the scope of the present discussion. Since our focus is on computing, we narrow the question to the availability of new methods in practical software tools. As indicated above, new methods do appear in the market, but the whole process from research innovation to software improvements to practical application is often ineffective.

Practitioners appear as active players in many methodological debates. Yet many take a fairly passive point of view regarding software development, perhaps trusting software developers to keep up with the state of the art. Understanding the structure of this small niche market, and their potential influence on it, may lead some practitioners to change their attitudes. They may make additional efforts to follow research advances independently, to stay on the guard for methodological problems in their work, and most importantly to demand solutions from software developers.

Software developers in certain cases act as leaders of the travel forecasting community, developing solutions that only a few appreciate upfront, but all enjoy in retrospect. Whether the motivation for such action is the long-term benefit of a better market position, or whether they simply want to do the right thing, the result is very positive. In other cases, the same developers may express a different point of view, as if they prefer to follow the herd, and wait for market demand before acting. The choice between these two approaches has a critical impact on the effectiveness of improvements.

Researchers that choose to study transportation are in most cases inspired by practical problems, and view potential practical impacts as a positive benefit. Funding opportunities in this field are an additional motivation to take impacts on practice into consideration. Yet many research efforts end without any practical impact. In part, this result is due to the risky nature of research, or to other actors, but researchers can improve their own effectiveness as well. They can maintain closer connections with practitioners and their concerns. They can become more knowledgeable about the entire travel forecasting procedure. Perhaps most importantly, more researchers should follow innovations through all their stages, keeping in mind that finding a correct answer is only part of the job; convincing others that it is so is not less important.

REFERENCES

Anderson II, L.D. (ed.) (1988), 'PCs in transportation software', *Directory, Transportation Centre*, Lawrence, KS, US: University of Kansas.

Brokke, G.E. (1967), "Urban transportation planning computer system", American Association of State Highway Officials Conference, St. Paul, MN, US; Traffic Assignment and Peripheral Programmes, pp. 32–68.

Chicago Area Transportation Study (1959, 1960, 1962) *Survey Findings*, Volume I, *Data Projections*, Volume II, and *Transportation Plan*, Volume III, Chicago, IL, US.

Control Data Corporation (1965), "T/P Transportation Planning System for the Control Data 3600 Computer, Users' Manual", Application Programme No. 7, Data Centres Division, Minneapolis, Minisota, MN, US.

Control Data Corporation (1969–1970), "A Data Services Newsletter Covering the Application of Computer Sciences to Urban Affairs", *Urban Affairs*, Minneapolis, MN, US: Data Services Marketing Staff.

de Cea, J., J.E. Fernandez, V. Dekock and A. Soto (2005), "Solving network equilibrium problems on multimodal urban transportation networks with multiple user classes", *Transport Reviews*, **25**, 293–317.

Detroit Metropolitan Area Traffic Study (1955, 1956) Data Summary and Interpretation, Part I, and Future Traffic and a Long Range Expressway Plan, Part II, Detroit, MI, US.

Dial, R.B. (1967), "Transit pathfinder algorithm", *Highway Research Record*, 205, Highway Research Board, 67–85.

Dial, R.B. (1971), "A probabilistic multipath assignment model which obviates path enumeration", *Transportation Research*, **5**, 83–111.

Dial, R.B. (1976), "The Urban Transportation Planning System: UTPS philosophy and function", *Transportation Research Record* 619, Transportation Research Board, 43–48.

Dial, R.B. and R.E. Bunyan (1968), "Public transit planning system", *Socio-Economic Planning Sciences*, **1**, 345–362.

Dial, R.B., D. Levinsohn and G.S. Rutherford (1980), "Integrated transit-network model (INET): A new urban transportation planning system programme", *Transportation Research Record*, 761, Transportation Research Board, 33–40.

Dial, R.B. and L.F. Quillian (1980), "Introduction to aggregate data analysis by using UTPS: UMATRIX", *Transportation Research Record*, 771, Transportation Research Board, 17–22.

Dongarra, J.J. (2009), "Performance of Various Computers Using Standard Linear Equations Software, CS-89-85", Electrical Engineering and Computer Science Department, University of Tennessee, Knoxville, TN, US.

Ferguson, E., C. Ross and M. Meyer (1992), "PC software for urban transportation planning", *Journal of the American Planning Association*, **58**, 238–243.

Florian, M. (1999), "Untangling traffic congestion", *ORMS Today*, **26** (2).

Florian, M. (2008), "Models and software for urban and regional transportation planning", The CRT, INFOR, Information Systems and Operational Research, 46, 29–50.

Irwin, N.A. and H.G. von Cube (1962), "Capacity restraints in multi-travel mode assignment programmes", *Bulletin*, 347, Highway Research Board, 258–289.

Mertz, W.L. (1961), "Review and evaluation of electronic computer traffic assignment programs", *Bulletin*, 297, Highway Research Board, 94–105.

SELNEC Transportation Study (1973), "Computing Procedures", Technical Working Paper No. 11, Manchester, UK.

Slavin, H.L. (2004), "The role of GIS in land use and transport planning", Chapter 19 in *Handbook of Transport Geography and Spatial Systems*, D.A. Hensher, K.J Button, K.E Haynes, and P. Stopher (eds), Elsevier, The Netherlands, 329–356.

Strohmaier, E., J.J. Dongarra, H.W. Meuer and H.D. Simon (2005), "Recent trends in the marketplace of high performance computing", *Parallel Computing*, **31**, 261–273.

US Department of Commerce (1963a), *Calibrating and Testing a Gravity Model for Any Size Urban Area*, Bureau of Public Roads, Washington, DC.

US Department of Commerce (1963b) *Newsletter of the Transportation Planning Computer Program Exchange Group*, Bureau of Public Roads, Washington, DC, volumes 1 and 2, plus semi-annual newsletters issued through 1973; searchable electronic version available at: http://books.northwestern.edu/viewer.html?id=inu:inu-mntb-0005779868-bk.S

US Department of Commerce (1964), *Traffic Assignment Manual*, Bureau of Public Roads, Washington, DC.

US Department of Housing and Urban Development (US HUD) (1966), *Factors Influencing Transit*, Technical Report 1, and *Computer Program Specifications*, Technical Report 2, Urban Mass Transit Planning Project, Alan M. Voorhees and Associates, McLean, VA.

US Department of Housing and Urban Development (1967), *IBM 7090/94 Computer Programs General Information Manual, Volume I, IBM 7090/94 Computer Programs Users' Reference Manual, Volume II, Technical Report 3, Modal Split Simulation Model, Technical Report 4* and *Recommendations for Urban Mass*

Transportation Research, Technical Report 5, Urban Mass Transit Planning Project, Alan M. Voorhees and Associates, McLean, VA.
US Department of Housing and Urban Development (1968), *IBM System/360 Computer Programs, General Information Manual, Volume I, User's Manual, Volume II, Technical Report 6*, Urban Mass Transit Planning Project, Alan M. Voorhees and Associates, McLean, VA.
US Department of Transportation (US DOT) (1969), *Urban Planning System 360, Trip Distribution and Peripheral Programs,* and *Traffic Assignment and Peripheral Programs,* Federal Highway Administration, Washington, DC.
US Department of Transportation (1972a), *Urban Transportation Planning, General Information*, Federal Highway Administration, Washington, DC; updated and reissued periodically.
US Department of Transportation (1972b), *U.M.T.A. Transportation Planning System – Reference Manual*, Urban Mass Transportation Administration, Washington, DC.
US Department of Transportation (1973), *Traffic Assignment, Methods, Applications, Products*, Federal Highway Administration and Comsis Corporation, Washington, DC.
US Department of Transportation (1974), *Urban Transportation Planning Program Documentation*, Federal Highway Administration, Washington, DC.
US Department of Transportation (1977), *User-Oriented Materials for UTPS, An Introduction to Urban Travel Demand Forecasting*, Federal Highway Administration, Urban Mass Transportation Administration, Washington, DC.
US Department of Transportation (1984), *Microcomputers in Transportation, Software and Source Book*, Urban Mass Transportation Administration, Washington, DC; updated annually through 1987.
Van Vliet, D. (1982), 'SATURN – A modern assignment model', *Traffic Engineering and Control*, **23**, 578–581.
Voorhees, A.M. (1955), 'A general theory of traffic movement', *Proceedings*, Institute of Traffic Engineers, New Haven, CT, USA.
Weiner, E. (1997), *Urban Transportation Planning in the United States, An Historical Overview*, Fifth Edition, DOT-T-97-24, US Department of Transportation, Washington, DC. http://tmip.fhwa.dot.gov/resources/clearinghouse/158; also available from Springer, Berlin.
Wigan, M.R. (ed.) (1977), *New Techniques for Transport Systems Analysis*, Australian Road Research Board, Vermont, Victoria, Australia.
Wilson, A.G, A.F. Hawkins, G.J. Hill and D.J. Wagon (1969), 'Calibration and testing of the SELNEC Transport Model', *Regional Studies*, **3**, 337–350.
Wootten, H.J. (2004), 'Traffic forecasting and the appraisal of road schemes', Chapter 7 in *The Motorway Achievement*, Volume 1, *The British Motorway System: Visualisation, Policy and Administration*, P. Baldwin and R. Baldwin (eds), Thomas Telford, London, UK.
Worthy, J.C. (1995), 'Control Data Corporation: The Norris era', *Journal of the History of Computing*, **17**, 47–53.

13. Sustainable transportation development and travel behavior

Yoram Shiftan and Daniel Shefer

INTRODUCTION

The challenges facing transportation planners have grown continuously over the years owing to mounting problems of congestion, concerns with environmental degradation and global warming, enhanced awareness of safety, and the increasing complexity of travel behavior patterns associated with modern life.

Modern life has brought about more travel, more leisure time, and more engagement in out-of-home, non-work activities. Modern life, especially in more recent years, has also witnessed rapid population and economic growth in many urban areas and the decentralization of residential, commercial, and work places. It has also seen more women entering the labor market. Accompanying these changes has been the relaxation of some constraints, such as the need to commute at fixed hours, thus providing more degrees of freedom of travel. Because of the significantly increased alternative activities and travel patterns from which households can now choose, travel patterns have become more complicated. In addition, the total number of trips has increased, trip chaining is more frequent, and traffic peaks are becoming smoother. All this has resulted in making the analysis of travel behavior more complex.

Understanding travel behavior is a first necessary step in the design of various policies towards sustainable transportation development that will on the one hand support economic growth and well-being of the population, and on the other hand will minimize transport externalities. Road users create externalities, often referred to as social costs that are imposed on other road (and non-road) users and, therefore, are not considered by their generators. Newbery (1988a, 1988b) classified these social costs into four main types:

1. Congestion costs –These are considered the classic transport externalities. Imposed on all other road users by any vehicle entering already congested traffic, these costs reveal themselves primarily in increased travel time (or cost) but also in increased vehicle operating costs (because of excessive fuel consumption and wear and tear on the vehicle).
2. Road damage costs – Newbery (1988a, 1988b, 1989) treated road damage costs quite extensively in a series of published articles. He defined road damage externalities as the increased vehicle operating cost imposed on subsequent vehicles driving on an increasingly damaged road. Road damage costs are not trivial, particularly in developing countries, and therefore they should be taken into account in the cost-allocation calculus of road-user charges.
3. Environmental pollution costs – These are externalities imposed primarily on non-road users in the form of air pollution, noise, and obstructed vision (US Federal Highway Authority, 1982; Quinet, 1989; and ECMT, 1990). For example, Monzon and Guerrero (2004) estimated the annual health and social costs of transport-related air pollution in Madrid at 357 million Euros.
4. Accident externalities – These refer to the cost of road accidents which appears to be appreciably high, much higher than often thought (see, for example, Jones-Lee et al., 1985; O' Reilly et al., 1992; and Department of Transport, 1978a). Lemp and Kockelman (2008) estimated that the most significant externalities are associated with crashes and congestion. The US National Safety Council in 1994 assessed the economic costs of motor-vehicle accidents, including deaths and disabling injuries, at $176.5 billion (1995, p. 78). In Israel, the annual cost of accidents was estimated at 4 billion USD, representing 2.5% of the Israeli GDP or $450 per capita (Cohen, 2004 and 2006). In both these cases, the costs include fatalities and injuries, pain and suffering, property damage, traffic delays, and administrative costs.

The significance of accident externalities and the inadequate knowledge concerning their generation prompted Newbery (1987) to address these issues in the following terms:

> The present state of knowledge concerning accident externalities is very unsatisfactory, but on calculation suggests that these costs are so large that all classes of road users are being undercharged by the present system of road taxes. Further research on this topic would therefore seem to merit high priority.

These externalities are especially pertinent in metropolitan areas where the infrastructure networks are most intensively used and development densities are high. The phenomenon of transport externalities was analysed in seminal work by such eminent classical economists as Pigou (1920) and Knight (1924) and later Walters (1961) and Vickery (1969), who also suggested ways to correct these market failures.

Increased externalities from motor vehicles called for the development of new policy and planning objectives toward sustainable transportation. These policies, varying from auto-restraint measures, such as congestion pricing and various parking arrangements, through emission-control programmes to a wide array of urban design and land-use policies are aimed at reducing the greenhouse effect, improving air quality and public health, reducing congestion and road accidents, and enhancing the overall welfare and quality of life of inhabitants in urban areas. The World Bank (1996) listed as one of its most important objectives increasing the social sustainability of transport by making poverty reduction an integral part of national and local transport strategies.

The growing complexity of travel patterns and the need to estimate changes in travel behavior in response to new policies made transportation planning more challenging and called for a better understanding of travel behavior. This means responding to such issues as how travel behavior is affected by new information and communication technologies (ICT), how land use and growth management affect travel behavior, how travelers respond to auto-restraint policies, how much travel is induced as a result of a new infrastructure, and how the various policies affect equity issues. An understanding of such effects is essential for a better design of new policies. In this regard, travel behavior lies at the core of procedures for analysing and evaluating transportation-related measures aimed at improving urban mobility, environmental quality, safety, and at achieving a wide variety of social objectives. Policy analysis and planning, then, rely on travel behavior studies and travel-demand models.

Travel behavior changed over time in parallel with changes in transportation systems, planning goals, and travel behavior. In the 1960s, the development of the American interstate highway system led to the development of an operational macro-level system of models providing planners and engineers with travel demand forecasting tools. This system (also known as the four-step process) which depends on statistical tools enables the forecast of various travel behavior elements in an aggregate manner, given a certain spatial distribution of land use and the socio-economic characteristics of the population. In the late 1960s, additional emphasis was placed on economic approaches, this motivated by the need not only to better understand travel behavior but also to derive a value of behavioral travel time saving for cost–benefit analysis. In the 1970s, the economic theory of the allocation and valuation of time had developed toward an econometric approach in providing a framework for modeling travel behavior. This approach was based on random utility modeling (RUM). Thereafter, a growing interest in and the need to better understand the motivation of travel behavior prompted the involvement of psychology, among other social sciences, which questioned whether the RUM paradigm could accurately represent the set of decisions that travelers use. A rules-based paradigm was developed as an alternative method. In both approaches, the econometric and the social science, there was a move toward an activity-based framework, in which an effort was made to understand travel behavior as derived from activity participation and time-use behavior.

Policy-making today requires more sophisticated tools, and these have been developed in terms of advanced models, most of them activity-based, that go into different levels of detail on various scales: spatial, temporal, and social. Together with the development of activity-based models, various specific models have been estimated and implemented in support of sustainable policy analysis, some of them as auxiliary to activity-based models and some of them as stand-alone models. Scenario analysis has also been widely used to analyse sustainable transportation policies. The purpose of this chapter is to highlight and emphasize the important role of understanding travel behavior and the complex relations between the various travel externalities to the development of sustainable transport policies. Sustainable transportation also means a safe one with a minimum number of casualties, therefore we first (in the next section) introduce the complex relations between congestion, speed variance, vehicle mix and road safety. We then (in the third section) present our view of the required actions for the transition toward sustainable transportation-planning practice and the role of better understanding of travel behavior in this process. The fourth section provides a few examples where modeling and evaluation have contributed to understanding the potential effect of various sustainable transport policies. The fifth section provides, with some examples, an overview of activity-based models as improved tools to understand travel behavior and explain their contribution to travel behavior in support of sustainable transportation planning. Finally, the sixth section suggests some future research and implementation recommendations.

CONGESTION, SPEED VARIANCE, VEHICLE MIX, AND ROAD SAFETY

Among the factors influencing the number of fatalities on highways are the following: speed, speed differences, and traffic composition. The lower speeds caused by congestion will lead to a lower number of fatal accidents. As a result, we expect a parabolic relationship between density and fatal accidents on highways. When densities increase, there will first be a positive relationship because of the increase in the number of cars in the system. However, when density becomes so high that speeds are influenced negatively, the number of accidents will decrease. The conclusion is that a positive impact exists in addition to the negative impact of congestion in terms of time losses, since fatalities are reduced. Some supporting evidence is found in a number of countries where relatively low numbers of fatalities are observed during the morning peak hours (see Shefer and Rietveld, 1997).

Road Fatalities and Density

Based on the hypothesis that the rate of road fatalities is closely related to traffic density and speed (flow), a gradual increase in cars while the level of the road capacity is held constant follows the curve described in Figure 13.1 This is certainly a possibility if we postulate that travel speed decreases with the increase in density stemming from congestion, thus causing a reduction in the number of road fatalities (see the movement on the curve in Figure 13.1 from point B to point A). It should be emphasized that we are specifically and intentionally limiting our analysis to the case of fatalities. (It is conceivable that with the increase in density, the number of road accidents may still continue to increase; however, the reduction in travel speed will cause these accidents to result only in injuries or in physical damage to the motor vehicles involved.)

A social objective with respect to road fatalities is the minimization of fatalities. Such an objective could be attained in an unrealistic situation, such as imposing a ban on the use of vehicles or, alternatively, creating such a high level of congestion that travel speed is reduced to zero – a gridlock situation (see Shefer, 1994). (Both of these extreme conditions are described by the far left and far right, respectively, of the curve in Figure 13.1.)

Road Fatalities and Speed Variance

In recent years, a heated debate has been waged by several scholars over the effect of speed variance, rather than average speed, on the rate of highway fatalities. The links between speed variance and accident rates was first studied by Solomon (1964), who argued that the relationship between car

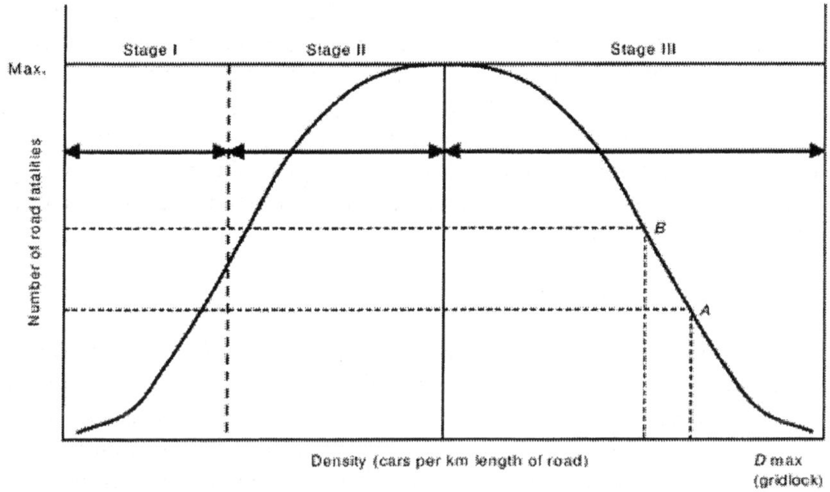

Figure 13.1 Hypothetical road fatalities: rate–density function

accidents and the speed of travel could be described by a U-shaped curve. Variations around the average speed of travel, whether above or below it, are associated with an increase in the rate of car accidents. The background of this result is that the rate of car accidents is closely linked to the number of overtakings, and overtakings are closely related to speed differences. The introduction of a minimum speed limit (in addition to a maximum limit) would decrease total costs. This assertion concurs with Hauer's conclusion that "the indiscriminative public crusade against speeding should be replaced by a balanced approach emphasizing the dangers of both fast and slow driving" (Hauer, 1971, p. 7). Solomon's findings were subsequently corroborated by several other studies, including that of Lave (1985), which sparked a debate in the *American Economic Review* (see Levy and Asch, 1989; Fowles and Loeb, 1989; and Snyder, 1989). Further reviews are given by Fildes and Lee (1993) and Finch et al. (1994). The higher the speed differences the higher is the probability that a driver will be involved in an accident. The question in the present context is, to what extent are these speed-difference related costs internal or external? This question is important, since differences that lead to external costs would be a source of unnecessarily high levels of social costs of transport (for more on this issue, see Rietveld and Shefer, 1998).

Speed variance positively affects the level of road safety, particularly the road-fatalities rate. When speed variance is held constant and the average speed of travel increases, the level of road safety decreases. Likewise, when travel speed is held constant and the speed variance increases, the level of road safety also decreases. These suppositions have far-reaching implications for public policies on road safety. For example, the debate over whether to increase speed limits should be discussed jointly with policies aimed at reducing speed variance so as to achieve any improvement in road safety. This raises the issue of introducing minimum speeds in conjunction with maximum speed limits. (A more detailed analysis of the role of average speed and variances in speed as determinants of accidents can be found in Shefer and Rietveld, 1997.)

Road Fatalities and Vehicle Mix

A similar argument may be presented with respect to traffic composition, or vehicle mix. As the share of heavy vehicles (trucks) in the traffic on highways increases, the level of road safety is more adversely affected. Given a certain fixed level of speed, an increase in the traffic composition reduces the level of road safety; by the same token, holding traffic composition constant and increasing speed also reduces the level of road safety. This provides another good reason for separating heavy vehicles and private cars as has been implemented on particular roads in metropolitan areas in several countries, notably the United States and the Netherlands. Another positive effect of such a separation is that it more easily enables the imposition of different speed limits according to type of vehicle (car versus truck). After separation, different speed limits do not have an adverse effect on speed variance. Especially relevant for urban areas, too, is the barring of vehicles with hazardous freight from particular routes.

The joint effect of speed variance and traffic composition is apparent. If one variable is held constant and the remaining variable increases, then the level of road safety is expected to decrease. Again, the ramifications of this contention for public policies on road safety are far reaching; i.e., the level of road safety is jointly determined by speed, speed variance and traffic composition. Legal speed limits both within and outside urban areas will have an effect on both average speed and speed variance and, thus, positively affect safety on roads (for an analysis, see Rietveld and Shefer, 1998).

The mixture effects mean that different types of vehicle use the same roads. This leads to high potential risks, especially for non-motorized road users (mainly bicycles and pedestrians). Separation of road-user types (through, e.g., the construction of bicycle lanes but also the barring of vehicles with hazardous freight from certain routes) will substantially improve safety. On roads where only motorized traffic is allowed, mixture continues to play a role (cars, trucks, motorcycles). The mixture effect and the speed variance effect are to some extent related, but not always. There still may be a speed variance effect with completely homogeneous road-users (no mixture effect) (for more on this issue, see Shefer and Rietveld, 1997; Rietveld and Shefer, 1998). This effect is especially relevant when accounting for accidents on expressways. Officials should consider homogenizing speeds, not only maximum speeds but also minimum speeds.

In addition to the costs of congestion in the form of time losses and (possibly) higher levels of pollution, there is a potential gain that is often overlooked: congestion leads to lower numbers of fatalities. The reason for this is that, with lower speeds, the probability of fatal accidents will be reduced. Congestion will also indirectly affect safety, because the longer travel times by car lead to a shift in modal choice toward public transport. This has a favorable effect on safety because safety levels are higher in public transport than in private cars.

Congestion and Cost–Benefit Analysis (CBA)

Congestion has relevance for cost–benefit analyses because it demonstrates that external effects are interdependent. Ignoring such interdependencies may lead to sub-optimal results. In cost–benefit terms, it shows that reducing congestion not only has beneficial effects (in terms of time gains and emissions rates), but also negative effects in terms of an increase in the number of fatalities. This does not imply, of course, that congestion should just be allowed to continue for the sake of traffic safety. There are most probably more cost-effective ways to improve safety.

A reduction in the level of road congestion and in road fatalities represents two conflicting objectives. Thus, public policies, such as improving road capacity in order to reduce congestion, could inadvertently produce an increase in the number of road fatalities, as well as in total emissions if more cars use the road as a result of the improved capacity. By the same token, policies that yield the largest reduction in road accidents are not necessarily the best or most socially desirable policies. The most effective policy is the one that yields the highest net social benefits. The optimal level of a programme is reached when the marginal social costs of the programme equals the marginal social benefits. Every intervention policy entails some direct and indirect costs; thus, it is imperative to perform a comprehensive cost–benefit analysis for each alternative intervention policy. These analyses will enable decision-makers to select the most cost-effective policy aimed at reducing congestion or reducing the number of road accidents with fatalities.

The foregoing discussion leads to the unavoidable and critical question: what is the socially optimal combined level of congestion, fatalities, and emissions? Although decision-makers may continue to avoid this sensitive, even highly explosive, issue, it is clear that a public programme is justified only when the total social benefit exceeds the total social cost.

Congestion increases air pollution in urban areas, but at the same time, as discussed earlier, it may reduce the number of road fatalities. Thus, the optimal level of congestion must be determined while simultaneously taking into account all direct and indirect social costs and benefits. The optimal level of congestion will be found at the point where the social marginal cost is equated to the social marginal benefit. At this point, the total social net benefit will be maximized. A more appropriate social objective is one that tries to minimize road fatalities, subject to a certain agreed (acceptable) level of mobility (accessibility) or some maximum acceptance cost entailed by the reduction in opportunities for travel.

The trade-off that exists among congestion, safety and pollution costs is especially important for sustainable transport planning in metropolitan areas, where these problems are of paramount importance. As mentioned, this will also have implications for the need to develop infrastructure for other transport modes.

SUSTAINABLE TRANSPORTATION PLANNING

Better understanding of travel behavior, in particular of drivers' responses to various policies instruments, improved evaluation tools, better institutional and planning organizational forms, and greater public participation in the planning process can contribute to the successful implementation of sustainable transportation planning. These issues are discussed in the following subsections.

Modeling Tools

Most metropolitan planning organizations (MPO) use the traditional aggregate four-step model as their main travel demand model for investment programmes and policy analysis. These models, which were developed in the United States in the 1950s and 1960s for planning and evaluating the interstate highway system programme, were mainly intended to analyse road infrastructure. Although these models have improved with the years to better analyse transit alternatives, they still have many limitations when it comes to analysing policies toward sustainable transportation (Shiftan and Suhrbier, 2002; Shiftan, 2000, 2008), mainly because our understanding of the impacts of such measures lags behind our understanding of the impacts of new transportation-infrastructures projects. Much has been written on the problem of uncertainty in travel demand models (see, for example, Skamris and Flybjerg-Bent 1997; Walmsley and Pickett, 1992; Pickrell, 1990; and Stopher, 1993). This problem is magnified in the evaluation of transportation policies. Our inability to correctly predict travel behavioral responses to such policies and, as a consequence, to evaluate their benefits results in more barriers to their implementation. Policy-makers understand the criticism of these models and question analyses based on them. This lack of trust in the potential impact of such policies may discourage policy-makers from implementing them. There is also some evidence that these models overestimate highway project benefits and underestimate transit and auto restraint policy benefits (Shiftan et al., 2008).

Evaluation Tools

Better evaluation tools constitute an important element to support the implementation of various transportation policies by showing policy-makers and the public the full benefits and costs of such policies which is not a simple matter as already discussed in the section on Congestion and Cost–Benefit Analysis. The World Bank (1996) indicated the need to understand the mechanism by which environmental and ecological impacts emerge and the values that society places on them as still another obstacle in the way of sustainable transportation. However, to date, no adequate framework for appraisal, funding, implementation, and monitoring of integrated transport policies is available (Hatzopoulou and Miller, 2006, 2008). Hatzopoulou and Miller (2008) propose assessing the extent of institutional integration in the appraisal, funding, and implementation of transport policies and stress the need for cross-sector evaluation of transport policies.

For the evaluation procedure to better account for and promote sustainable transportation policies, it should include wider impacts and benefits of transport improvements (see, for example, Graham 2005, 2006; Vickerman, 2007). New road-safety and environmental impact analyses should be integrated in the evaluation procedure, since ignoring these impacts may bias the evaluation in favor of highway projects. Other impacts include equity considerations, as well as accessibility and level-of-service indices. The World Bank (1996) also pointed out that full accounting for external effects and rigorous economic appraisal procedures were crucial for sustainable transportation development.

The Metropolitan Planning Organization (MPO)

When discussing different barriers to the implementation of sustainable transport policy measures, Banister (2002) points to the institutional/political structure as one such barrier. Significant progress has been made in using regional strategies, but tension continues between issues of efficiency and privatization, on the one hand, and equity and the provision of public good, on the other (Haynes et al., 2005). In the context of metropolitan transportation in the United States, interest in the regional operation of metropolitan transportation facilities and the provision of transportation services in metropolitan areas has raised issues of coordination among jurisdictions to facilitate, oversee, and monitor the implementation of new technologies in the transportation area (Haynes et al., 2005). In this context, the World Bank (1996) listed another challenge to the transition toward sustainable transportation: this is the relegation of public sector transport functions to the level at which they are most effectively performed, with the corresponding transfer of technical and financial assistance and the separation of a system's operations as much as possible from direct political influence.

Shiftan et al. (2003) showed that agreement exists among policy-makers, academicians, and practitioners in Israel as to the efficacy of establishing a metropolitan transportation planning organization to help overcome barriers toward sustainable transportation in the Tel-Aviv Metropolitan area, the principal and largest metropolitan area in Israel.

Public Participation

Public participation is an important element in the transition toward sustainable transportation development for two reasons. First, by involving the public in the planning process, it may be possible to better assure the public's support of policies and thus to eliminate potential objections. Second, sustainable transportation development should reflect the public interest. Involving the public in the planning process can assure that policies are designed and implemented with the public's consent and welfare in mind. The World Bank (1996) has also advised engaging non-governmental local community organizations in the planning, supply, and management of local transport services.

EVALUATING ALTERATIVE POLICY MEASURES – CASE STUDIES

Various policy measures can be used toward more sustainable transportation development including improved alternatives to the private car, auto restrain policies such congestion pricing and various parking policies, land use policies and information communication technologies. This section provides a few examples where modeling and evaluation have contributed to understanding the potential effects of such policies.

Employer-Provided Transportation

Shefer et al. (1999) examined the effect of employer-provided transportation by three large Israeli employers in the Haifa Bay Area in reducing private auto use. Israeli employers have been providing transportation services for employees for many years. Although this policy was initially designed as a social policy, it could also be effectively used for achieving environmental goals, as has been done in other countries.

A survey was designed for this purpose and distributed among employees of the three companies including both revealed preference (RP) and stated preferences (SP). Logit mode choice models were estimated based on the SP data derived from the effect of individual choices using three different scenarios on vanpool use and emissions:

Scenario I: No vanpool services. The objective of this scenario is to estimate the contribution of current vanpool programmes to air quality.

Scenario II: Enhanced vanpool services, mostly by improved frequency.

Scenario III: Radical changes in the metropolitan transportation system, providing significant travel time advantage to vanpools over private cars.

The results of Scenario 1 show that current vanpool programmes are estimated to reduce weekly car trips from 70% to 57% and, therefore, save about 140,000 km of travel by private car annually. The improved frequency of the vanpools in Scenario II significantly improves vanpool share from its current 20% to about 40% and reduces the car share from its current 50% to about 43%. VKT by car would decrease to 27%, and kilometers traveled by bus would increase up to 50%. With more significant investments in improving vanpool services and their travel time compared to cars (Scenario III), the car-use rate would drop to lower than 40% and vanpool trips would exceed 40%. The kilometers traveled by cars would decrease to 60% of the current rate, and the kilometers traveled by buses would increase by 67%.

The annual emissions of cars and buses were estimated by multiplying the annual kilometers traveled by their respective emissions factors. The results show that when vanpools are offered as a travel mode alternative to about 7,000 workers in the city of Haifa, CO_2 annual emissions are reduced by approximately 100 tons (Current vs. Scenario I), HC emissions by 7 tons, and NOx emissions by more than 5 tons. The improved vanpools frequency (Scenario II) could lead to significant reductions in the emissions generated by work-trips. Annual CO_2 emissions would be reduced by 50 tons, and both HC and NOx emissions by approximately 3 tons. Further improvement of vanpool services, by providing them with a time advantage using HOVs (Scenario III), could reduce annual CO_2 emissions produced by work trips to the three workplaces by 75 tons, HC emissions by 5 tons, and NOx emissions by about 4.5 tons.

Fuel Tax and Partial Congestion Tolls

Shiftan et al. (2000) compared the effectiveness of a partial (second best) congestion toll and a fuel tax as alternatives to full (first best) congestion pricing, in which every driver is charged the difference between the marginal cost he or she imposes on the system and the average cost the driver pays in terms of travel time. Such full-congestion pricing has been shown to be the best approach to minimizing total system travel time; however, the implementation of such a toll is in practice not feasible. For this reason, we simulated drivers' behavior in the city of Haifa under different scenarios. Total traffic demand was assumed to be constant; i.e., only the impact of changes in route-choice behavior was evaluated. Obviously such pricing scenarios can also affect mode, time, and destination choices, as well. The simulation was performed using EMME2 software and the equilibrium-assignment procedure.

The following scenarios were simulated:

- Existing condition.
- Optimal congestion pricing.
- Selective congestion pricing (three scenarios, in each of which various segments of the main congested roads are tolled).
- Selective congestion pricing on all segments, with V/C>1.25.
- Fuel tax (constant fee cost per km).

Except for optimal congestion pricing, all the other scenarios were simulated using Wardrop's first Principle of User Equilibrium reflecting individual behavior. For the optimal congestion pricing scenario, Wardrop's second Principle of System Equilibrium was used, in which the average (or total) travel time on the system is minimized, reflecting the situation that can be achieved with optimal congestion pricing as discussed above.

Table 13.1 shows the results of the different scenarios. As can be seen, the full congestion pricing of MC-AC will minimize the system's total travel time and reduce average trip time by 3%, which can accumulate to a significant saving in system travel time. However, selective congestion tolls, a fuel tax, and a fixed toll per km will all increase total system travel time compared not only to optimal congestion-pricing scenarios but also to existing conditions. This implies that such second-best solutions will not always improve congestion and must be analysed carefully before implementation. However, the reader should take note that this examination took into account only route-choice behavior and assumed no mode or destination changes. Obviously such pricing mechanisms may have other long-term effects.

Table 13.1 Results of the different scenarios examined

Scenario	Total travel time (hours)	Average travel time (minutes)
Existing cond.	29,555.3	22.26
Opt. cong. pric.	28,700.9	21.61
Selective tolls 1	29,860.0	22.48
Selective tolls 2	30,159.0	22.71
Selective tolls 3	30,177.0	22.73
Avg. per km	32,551.4	24.51
Toll on v/c > 1.25	30,085.0	22.66

Transport Control Measures and the Location of Businesses in Central Business Districts

Shiftan et al. (2004) studied the effect of various transport control measures on the location decision of small businesses in the midtown business district of Haifa. They looked at three policies implemented in this area: parking restrictions, pedestrian mall (pedestrianization of one of the main streets), and a bus lane and tried to evaluate their impact on the dispersal of businesses away from the center. An understanding of the responses of visitors and business to auto-restraint policies will aid planners in designing effective policies and in finding the preferred mix and balance among the various policies to be implemented. On the one hand, such a mix should encourage travelers to shift from private vehicles to public transportation or to travel during less congested times; on the other hand, they should not force travelers to switch to other destinations so as eventually to cause the relocation of businesses from the CBD.

A survey designed for this purpose was conducted among shop and office owners currently operating in the midtown business area, as well as among those who had operated there in the past but have since moved to another location. The results indicated that among the policies that were investigated, shortage of parking spaces and parking fees ranked as the most important factors causing businesses to leave the CBD. The bus lane had only a marginal effect, and the pedestrian mall influenced only 10% of the businesses, principally businesses already located on the pedestrian mall. Overall, the results show that the implementation of auto-restraint policies in the CBD caused businesses to a great extent to move away from the CBD.

Businesses currently operating in the CBD were also asked about their potential response to the application of a number of hypothetical auto-restraint policies. The results showed that a scenario excluding private cars from the CBD caused most of the businesses to declare that they would either move out or consider moving out. The two scenarios of eliminating parking on the street on which the business was located and of creating a pedestrian mall on that street elicited a similar response. Shops were somewhat less affected by a scenario of pedestrian mall than were offices. The scenario of reducing the number of parking spaces in the CBD elicited a response similar to the doubling of parking fees there. The scenario that had the smallest effect on business relocation was the dedication of traffic lanes to public transport on all of the main streets of the CBD. No significant differences in responses to the different scenarios were detected between offices and shops.

Among all the auto-restraint policies employed in the CBD, it appears that restrictions on the supply of parking caused the largest number of businesses to leave the area. Parking fees elicited somewhat less sweeping responses, and it appears that this policy has a smaller impact on the location decision of businesses than does the reduction in parking supply. The two hypothetical scenarios of "reducing the number of available parking spaces" and "doubling the parking fee" produced almost identical responses; however, parking supply was indicated as a more important location factor and a stronger reason for leaving the Hadar CBD.

ACTIVITY-BASED MODELING AND SUSTAINABLE TRANSPORTATION DEVELOPMENT

Activity-based modelling treats travel as being derived from the demand for personal activities. Travel choices, therefore, become part of a broader activity-scheduling process, based on modeling the demand for activities rather than merely trips. The explicit modeling of activities and the consequent tours and trips enable a more credible analysis of responses to policies and their effect on traffic and air quality, thus, can better support the evaluation and implementation of such policies.

Travel Behavior

Ben-Akiva et al. (1996) proposed a practical activity-based modeling approach in which a comprehensive travel demand-modeling framework captured the mobility, activity, and travel decisions of individuals and households. A corresponding system of models can be used for planning and policy analysis. Since then, researchers have continuously tried to improve our understanding of activity participation and travel behavior, calling for a need to analyse, as well, the context of the activities: the why, when, with whom, and how long, in addition to the sequence of those activities (see Bhat and Koppleman, 1999; Goulias et al., 2004). Required, too, is a detailed understanding of how households and individuals acquire and assimilate information about opportunities for activity participation and travel options, how this information is used to determine time allocation for activities and travel, and whether the attributes of activity episodes are determined jointly or sequentially (Bhat and Koppelman, 1999).

Some researchers have claimed that for a good understanding of activity participation, there is need to understand the evolution of activity schedules from intentions to final outcomes over the period of a week (Lee and McNally, 2003). Our understanding of the scheduling process, furthermore, can be advanced by observing household scheduling behavior in realistic planning conditions (Doherty and Miller, 2000).

Activity-based models improve our ability to analyse sustainable transportation policy measures mainly because of their ability to predict how a person's entire daily activity pattern may change as a result of a specific measure taking individuals' time and space constraints into consideration. For example, assume that newly imposed congestion pricing causes a commuter to change his or her mode from drive alone to transit; because the person no longer drives to work, there may be other adjustments to the commuter's daily activity schedule, such as being unable to stop on the way back home to buy groceries. Upon returning home, the person may then take the car and drive to some other store. A view of the person's detailed daily activity pattern may reveal that it is not really feasible for this commuter to shift to transit and only an activity-based model can deal with these types of responses. Land-use policies can affect all activity and travel decisions; therefore, only an activity-based model can be sensitive to such measures through its ability to consider adjustments in one's daily activity pattern. However, relatively few studies have applied the activity-based framework to investigating the relationships between urban forms and travel behavior (Zhang, 2005). The effects of ICT on daily activity pattern and travel behavior are more far reaching than impacting specific trips, and in a similar fashion, only an activity-based model can capture these effects. One of the main contributions of one such activity-based model, the Portland model, is its ability to distinguish between in-home and away-from-home activities and to make the trade-off between work and any other activity at home or away from home as was demonstrated by Shiftan and Suhrbier (2002).

The importance of activity-based modeling to better analyse various policies also lies in its ability to account for latent demand or, more generally, in its sensitivity to level-of-service variables at all dimensions of travel choice, including trip generation. This is a critical issue in policy-making in regard to transportation investment. The benefits from a new highway can be significantly biased by not taking into account induced demand as travel speeds will be much lower in such cases. It has been shown that this overestimation of benefits outweighs the benefits to new users, who are not considered by the traditional models (Williams and Yamashita, 1992; Shiftan et al., 2008). The advantage of activity-based models lies in their integrated approach, including log-sum variables that bring level-of-service variables up the structure to the daily activity pattern model. Thus, the decision to participate in various activities may be a function of the accessibility of opportunities.

Recent developments in activity-based models have placed emphasis on aspects of behavioral realism that are important for the analysis of sustainable transportation policies. In the Tel Aviv Model, for example, emphasis was placed on such policies as parking supply and fees and congestion pricing. For this purpose, a special parking survey was conducted to collect data on parking demand and supply, as was a stated preference survey of potential travellers' responses to such policies. These data were used to estimate parking search time, walking time to one's destination, and parking fees at each destination zone, all of which are important determinants of travellers' choice behavior. A detailed time-of-day model was also estimated to support the analysis of these various policies by time of day.

In San Francisco, the supply, fees, and availability of parking were developed from a variety of sources, including parking surveys, a stated preference survey, parcel data, and aerial photographs of on-street parking. The San Francisco model also developed a policy variable to measure the potential impacts of an improved pedestrian system and of expected growth, which would likely influence future travel demand (Outwater and Charlton, 2006).

Air Quality

Activity-based models provide better outputs of vehicle-miles traveled and vehicle-hours traveled, which are needed as inputs for emission modeling and air-quality analysis and which enable a better prediction of changes in these variables. The prediction of trips as part of a tour and of tours as part of a daily activity pattern can also identify whether a trip is a cold or a hot start, which is very important for emissions analysis. Activity-based models can predict the time and location of all starts and trips at much better resolution then traditional models, as well as identify the vehicle class and model that makes each trip. Integration with dynamic traffic-assignment models, which are now evolving (Lin et al., 2008), will provide much better input for emissions analysis by enabling the development of driving profiles. Activity-based models therefore provide input for emissions and air-quality analysis that is simply not provided with the traditional four-step models (Shiftan, 2000).

Road Safety

Like better input for emissions modeling in terms of vehicle-miles traveled and vehicle-hour traveled, this input is also important for better analyses of accidents and road fatalities. Activity-based models can provide additional important input for accident analysis that is not provided by traditional models. This input includes accumulated hours of travel, which can indicate driver fatigue, as well as other details concerning the person's activity that have been shown to affect the probabilities of being involved in accident (Elias, et. al, 2010).

Application to Policy Analysis

Few authors have tried to use the activity-based approach to analyse the potential effects of transport policies. Recker and Parimi (1999) used an activity-based approach with the Portland data as a case study to show the potential of transportation demand managements (TDM) to reduce vehicle emissions. They estimated an upper bound for such emission reductions by posing the research question as: "Given a set of activities, location, and various constraints, if all individuals were to act to minimize CO_2 emissions by trip chaining and ridesharing in the most efficient way possible, what activity patterns would result, and how would they differ in CO_2 emissions from their observed patterns?" Kitamura et al. (1995) used a dynamic and integrated microsimulation forecasting approach to test a few TDM in the Washington DC area (see also Pendyala et al., 1997). Kitamura (1997) also provided a review of studies in which activity-based models have been applied to demand forecasting and policy analysis comparing structural equation model systems and micro-simulation model systems. Shiftan and Suhrbier (2002) used the Portland activity-based model to analyse the potential impact of various sustainable transportation policies including pricing of automobile use (rising parking prices in the city center and peak hour toll for single occupancy vehicle drivers), telecommunication incentives, transit improvements and a combination of the three. They showed that the evaluated pricing policies have a large impact in reducing travel to the central business district, especially during the morning peak period. Overall regional travel, though, is not significantly impacted. Telecommuting reduced the number of work tours but there is a higher reduction in transit tours than auto tours and part of this reduction is offset by increase of other tours. The proposed comprehensive improvements to bus transit produce the desired decreases in auto travel, although the number of tours occurring by walk and bicycle also are decreased. Approximately one-third of the increase in bus travel is forecast to take place for non-work trip making. The synergistic effects estimated for the particular mix of policies analysed in this Portland testing were slightly negative rather than positive as normally hypothesized. Once a shift in individual transportation behavior is induced, additional transportation actions that are directed towards the same end would have a decreasing marginal effectiveness. Reductions in vehicle miles of travel and emissions for the combination of the three transportation policies may be less than expected given the nature of the TDM policies being analysed. Emissions during the morning peak period for the Portland analysis region are estimated to be reduced by approximately 3%.

Shiftan (2008) also explained and illustrated the potential of these models in analysing various land-use policies. Using again the Portland case study he showed the potential effects of a package of various land use and transport policies encouraging people to move the urban growth area. The results showed that also 6% of the suburban population would move to the urban growth area, increasing the urban center population by 16% and the rest of the urban growth area by 13%. Under such a scenario the impacts on regional travel would be marginal. Overall, there is an increase in the number of tours by all modes and even a slight increase in auto tours (2.5%) as a result of improved accessibility. Average trip length decreased by 9% but the combined effects of shorter trips and more trips is an overall increase of 1.4% in VMT.

CONCLUSIONS AND FUTURE RESEARCH DIRECTIONS

Transportation and travel have many positive characteristics both for the individual user and for society as a whole. This explains why the transport sector, for several decades now, has shown unprecedented growth. At the same time, transportation has undesirable side-effects. The almost unlimited demand for transportation cannot be facilitated on an equal basis by the construction of new infrastructures, which induce more traffic and continues to leads to congestion. Other concerns relate mainly to safety and emissions. Governments and other stakeholders are generally aware that policy measures are needed in order to find a balance between accessibility, safety and sustainability, taking into account the various effects of alternative transportation policies. This is an enormous challenge, and the question arises as to how one finds the right balance among economic growth, congestion, safety, and pollution in order to maximize social welfare.

Travel behavior lies at the core of procedures for the analysis and evaluation of transportation-related policies that are aimed at improving urban mobility, environmental quality, road safety, and social objectives. Analysis and planning of a sustainable transportation policy is the main objective of travel behavior studies and travel-demand models. Travel behavior has advanced over the years, and the new generation of travel-demand models, activity-based models, is now making the transition from theory to practice and is starting to assist decision-makers in policy analyses toward sustainable transportation. This chapter has shown the importance of improved behavioral realism in travel demand models in order to better understand travelers' responses to policy measures and better analyse their benefits and costs.

More research is required to further improve behavioral and travel demand models, our evaluation tools, and the design of policy measures that best achieve this balance (and trade-off) among economic growth, congestion, safety, and pollution. Activity-based models have been advanced significantly in the past decade, but practical activity-based models have only recently become operational. This is the time to test these models, use them for policy analysis, and make the transition toward a wider use of activity-based models. To make this transition, future research should point to the balance and trade-off between behavioral realism and complexity so that they can be widely used by planning agencies.

Evaluation tools should be improved to better estimate the various externalities, in particular emissions, pollution, and road safety. To find the right balance, however, better understanding is also required of the impact of transport projects on economic growth and regional development, as well as of such issues as the benefits of improving travel time reliability and the effects of private financing of public infrastructure. To this end, comprehensive risk analysis, system analysis of urban safety impacts, and the new activity-based models should be incorporated into the evaluation procedure.

From a policy point of view, one of the main challenges is how to improve and accelerate the transition toward sustainable transportation planning. While there is some agreement on desired policy measures toward sustainable transportation development, their implementation is lagging.

Better insight into transition management is required to assist policy-makers and stakeholders to speed up implementation.

REFERENCES

Ben-Akiva, M., J.L. Bowman and D. Gopinath (1996). Travel demand model system for the information ear. *Transportation*, 23, pp. 241–166.

Bhat C.R. and F.S. Koppelman (1999). A retrospective and prospective survey of time-use research. *Transportation*, 26, pp. 119–139.

Banister, D. (2002). Overcoming barriers to implementation. Draft paper for presentation at the STELLA Focus Group 5 on Institutions, Regulations, and Markets in Transportation Meeting, Brussels, Belgium, April 26–27.

Cohen, S. (2004). *Road Accidents in Israel – Characteristics and Damage to the National Economy*. Matat – Transportation Planning Center Ltd., Israel Ministry of Transport (Hebrew).

Cohen, S. (2006). *Road Accidents in Israel – Characteristics and Damage to the National Economy – 2006 Update*. Matat – Transportation Planning Center Ltd., Israel Ministry of Transport (Hebrew).

Department of Transport (1978a) COBA 9, London: Assessments Policy Methods Division.

Doherty, S. and E. Miller (2000). A computerized household activity scheduling survey. *Transportation*, 27(1), pp. 1–23.

ECMT (1990) *Transport Policy and the Environment*. Paris: The European Conference of Ministers of Transport.

Elias, W., T. Toledo, and Y. Shiftan (2010), The effect of daily activity patterns on crash involvement, *Accidents Analysis and Prevention*, 42, pp 1682–1688.

Fildes, B.N. and S.J. Lee (1993) *Road Environment, Behaviour, Speed Limits, Enforcement and Crashes*. New South Wales: Federal Office of Road Safety.

Finch, D.J., P. Kuompfner, C.R. Lockwood and G. Maycock (1994) *Speed, Speed Limits and Accidents*. Crowthorne, Berkshire: Transport Research Laboratory, Report 58.

Fowles, R. and P.D. Loeb (1989) Speeding coordination and the 55-mph limit: comment, *American Economic Review*, 79 (September), pp. 916–921.

Gomez-Ibanez, J.A. and G.R. Fauth (1980). Downtown auto restraint policies, *Journal of Transport Economics & Policy*, 14, pp. 133–153.

Goulias, K.G., T. Kim and M.L. Patten (2004). On activity type classification and issues related to the with whom and for whom questions of an activity diary: Preliminary findings and pattern classification. Paper presented at the Conference on Progress in Activity-Based Analysis, May 28–31, 2004, Vaeshastel Castle, Maastricht, the Netherlands.

Graham, D. (2005). Wider economic benefits of transportation improvements: Link between agglomeration and productivity. Stage 1 Report. London: Department for Transport.

Graham, D. (2006). Wider economic benefits of transportation improvements: Link between agglomeration and productivity. Stage 2 Report. London: Department for Transport.

Hatzopoulou, M. and E.J. Miller (2006). Indicators of sustainable transport for integrated policy appraisal: Linking sustainable transport plans with decision making in the Greater Toronto Area (GTA). In: *Proceedings of the New Scholars Conference on Sustainable Transportation*, Bloomington, IN (CD).

Hatzopoulou, M. and E.J. Miller (2008). Institutional integration for sustainable transportation policy in Canada. *Transport Policy*, 15(3), pp. 149–162.

Hauer, E. (1971). Accidents, overtaking, and speed control, *Accident Analysis and Prevention*, 3, pp. 1–13.

Haynes, K.E., J.L. Gifford and D. Pelletiere (2005). Sustainable transportation institutions and regional evolution: Global and local perspectives. *Journal of Transport Geography*, 13(3), pp. 207–221.

Jones-Lee, N.W., M. Hammerton, and P.R. Phillips (1985) The value of safety: results of a national sample survey, *Economic Journal*, 95, pp. 49–72.

Kitamura, R. (1997). Applications of models of activity behavior for activity based demand forecasting. In: *Activity-Based Travel Forecasting Conference.* Proceedings of the Activity-Based Travel Forecasting Conference in New Orleans, June, 1996. Washington, DC: US Department of Transportation.

Knight, F.H. (1924) Some Fallacies in the Interpretation of Social Cost, Quarterly *Journal of Economics*, 38, pp. 582–606.

Lave, C.A. (1985) Speeding coordination and the 55-mph limit, *American Economic Review*, 75 (December), pp. 1159–1164.

Lemp, J.D. and K.M. Kockelman (2008). Quantifying the external costs of vehicle use: Evidence from America's top-selling light-duty models. *Transportation Research Part D: Transport and Environment*, 13(8), pp. 491–504.

Lee, M.S. and M.G. McNally (2003). On the structure of weekly activity/travel patterns. *Transportation Research Part A*, 37, pp. 823–839.

Levy, D.T. and P. Asch (1989) Speeding coordination and the 55-mph limit: comment, *American Economic Review*, 79 (September), pp. 913–915.

Lin, D.-Y., N. Eluru, S.T. Waller, and C.R. Bhat (2008). Integration of activity-based modeling and dynamic traffic assignment. *Transportation Research Record*: *Journal of the Transportation Research Board*, No. 2076. Washington DC, pp. 52–61.

Monzon, A. and M.-J. Guerrero (2004). Valuation of social and health effects of transport-related air pollution in Madrid (Spain). *Science of the Total Environment*, 334–335, pp. 427–434.

Newbery, D.M. (1987) Road User Charges in Britain, Center for Economic Policy Research. Discussion Paper 174, (April).

Newbery, D.M. (1988a) Road user charges in Britain, The Economic Journal (Conference Papers), 88 (Supplement), pp. 161–176.

Newbery, D.M. (1988b) Road damage externalities and road user charges, *Econometrica*, 56, pp. 295–316.

Newbery, D.M. (1989) Cost Recover y from Optimally Designed Roads, *Econometrica* , 56 (May), pp. 165–185.

O' Reilly, D., G. Loomes, and S.P. Philip, et al. (1992) The value of road safety: UK research on injury valuation. Unpublished papers prepared and delivered at the 6th WCTR, Lyon, France, 29 June–3 July.

Outwater, M.L. and B. Charlton (2006). The San Francisco model in practice: validation, testing, and application. Paper presented at the Innovations in Travel Modeling 2006 Conference, Austin, TX, May 21–23.

Pendyala, R.M., R. Kitamura, C. Chen, and E.I. Pas (1997). An activity-based microsimulation analysis of transportation control measures. *Transport Policy*, 4(3), pp. 184–192.

Pickrell, D.H. (1990). *Urban Rail Transit Projects: Forecasts Versus Actual Ridership and Costs*. US Department of Transportation, Report No. DOT-T-01-04.

Pickrell, D.H. (1992) A desire named streetcar: fantasy and fact in rail transit planning. American Planning Association. *Journal of the American Planning Association*, (58), 158–176.

Pigou, A.C. (1920) *The Economics of Welfare*. London: MacMillan.

Quinet, E. (1989) The Social Cost of Land Transport. OECD, Paris (draft).

Recker W.W. and A. Parimi (1999). Development of a microscopic activity-based framework for analysing the potential impacts of transportation control measures on vehicle emissions. *Transportation Research Part D*, 4(6), pp. 357–378.

Rietveld, P. and D. Shefer (1998). Speed choice, speed variance, and speed limits. *Journal of Transport Economics and Policy*, 32, Part 2, pp. 187–202.

Shefer, D. (1994). Congestion, air pollution, and road fatalities in urban areas. *Accident Analysis and Prevention*, 26(4), pp. 501–509.

Shefer, D. and P. Rietveld (1997). Congestion and safety on highways: Towards an analytical model. *Urban Studies*, 34(4), pp. 679–692.

Shefer, D., Y. Shiftan and M. Rotstein (1999). The effect of employer-provided transportation on the emission of pollution from motor vehicles. The 39th Congress of the European Regional Science Association, Dublin, August.

Shiftan, Y. (2000). The advantage of activity-based modeling for air quality purposes: Theory versus practice and future needs. *Innovations*, 13(1), pp. 95–110.

Shiftan, Y. (2008). The use of activity-based modeling to analyse the effect of land-use policies on travel behavior. *Annals of Regional Science*, 42(1), pp. 79–97.

Shiftan, Y. and J. Suhrbier (2002). The analysis of travel and emission impacts of travel demand management strategies using activity-based models. *Transportation*, 29(2), pp. 145–168.

Shiftan, Y., S. Kaplan, and S. Hakkert, (2003). Scenario building as a tool for planning a sustainable transportation system. *Transportation Research D: Transport and Environment*, 8, pp. 323–342.

Shiftan, Y, N. Sharaby and I. Solomon (2008). Transport project appraisal in Israel, *Transportation Research Record*, 2079, pp. 136–145.

Shiftan, Y., D. Shefer and G. Avraham (2004). Impact of auto-restraint policies on the location of businesses in city centres: The case of Hadar, Haifa, *European Spatial Research and Policy*, 11(2), pp. 119–132.

Shiftan, Y., D. Shefer, A. Parber, S. Bekhor and H. Aviram (2000). Comparison of fuel tax and partial congestion toll as alternatives to complete congestion tolls. The 6th World Congress of the Regional Science Association International, Lugano, Switzerland, May.

Skamris, M. and K. Flyvbgerg-Bent (1997). Inaccuracy of traffic forecasts and cost estimates on large transport projects. *Transport Policy*, 7, 4(3), 141–146.

Snyder, D. (1989) Speeding coordination and the 55-mph limit: comment, *American Economic Review*, 79 (September), pp. 922–925.

Solomon, D. (1964) Accidents on Main Rural Highways Related to Speed, Drivers and Vihicle, Washington DC: U S Department of Commerce.

Stopher, P.R. (1993). Deficiencies of travel-forecasting methods relative to mobile emissions. *ASCE Journal of Transportation Engineering*, 119(5), pp.723-741.

US Federal Highway Authority (1982) Final Report on the Federal Highway Cost Allocation Study , Washington, DC.

US National Safety Council (1995) Accident Facts.

Vickerman, R. (2007). Recent evolution of research into the wider economic benefits of transport infrastructure improvements. Discussion Paper No. 2007-9, Canterbury, UK: OECD Joint Transport Research Centre.

Vickery, W.S. (1969) Congestion Theory and Transport investment, *American Economic Review*, Vol. 59, pp. 251–260.

Wafa, E. (2008). The effect of activity patterns on road accidents: Case studies of bypassed towns, Ph.D. Thesis, The Technion-Israel Institute of Technology, Haifa.

Walmsley, D.A. and M.W. Pickett (1992). The cost and patronage of rapid transit systems compared with forecasts. Research Report 352, Crowthorne, UK: Transport Research Laboratory.

Walters, A.A. (1961) The Theory and Measurement of Private and Social Cost of Highway Congestion, *Econometrica*, 29, pp. 676-699.

Williams, H.C.W.L. and Y. Yamashita (1992). Travel demand forecasts and the evaluation of highway schemes under congested conditions. *Journal of Transport Economics and Policy*, 26(3), pp. 261–282.

World Bank (1996). *Sustainable Transport: Priorities for Policy Reform.* Washington, DC: The World Bank.

Zhang, M. (2005). Exploring the relationships between urban form and nonwork travel through time use analysis. *Landscape and Urban Planning*, 73, pp. 244–261.

14. Reinventing the wheel: planning the rail network to meet the mobility needs of the 21st century

Moshe Givoni and David Banister

INTRODUCTION

The sustainable mobility agenda (Banister, 2008) places a key role on the railways, as this mode provides an efficient form of transport and it encourages a modal switch. With the overriding policy concern about the environmental agenda, rail is considered to have an advantage over other modes of transport (on long-distance, inter-city journeys), and promotion of rail transport has become an important element in achieving a (more) sustainable transport system. Indeed, rail is experiencing a renaissance, not only in terms of its appeal to policy makers and transport planners, but more important, in terms of its appeal to travelers. Demand for rail travel is now at record levels.

The above argument might suggest the way forward, namely further promotion of rail transport through pro-rail policies and investments in the rail network. But there are also major challenges facing rail transport as a mode for passenger transport. The car is still dominant for many shorter trips, and it seems to provide a more attractive option for travelers on most journeys. And over longer distances the growth in low-cost air travel means that rail is now having to compete much more effectively with these alternatives. Yet, as can be seen in Table 14.1, the demand for rail is still increasing despite the growth in the popularity of the alternatives, and this reflects the general growth in travel distances.

If the success of the railways and development of the rail network, especially the High-Speed Train (HST) network, simply leads to increased mobility, to increase in trip distances (especially commuting trips) and further sprawl of cities, then promoting rail transport can be counterproductive for sustainable transport. Technological innovation and social change have also had a major impact on travel demand, social activities and the ways in which people view the world. These new possibilities also present the railways with opportunities and challenges, and transport policy makers with a real dilemma – how should the railway take a larger share of the current demand for transport without inducing new demand?

This chapter aims to discuss some of the different aspects of the challenges and opportunities facing rail transport as well as the implications these have for transport policy in the pursuit of sustainability. The basic argument here is that if the railways are to really play a major role in meeting the mobility needs in the 21st century and if they are to really contribute to sustainability, it is not enough to simply expand the rail network and facilitate demand for more travel. Rather, a different type of rail system is required, namely one that accommodates more of the existing levels of demand – the wheel needs to be reinvented.

This chapter briefly reviews trends over the last 35 years of passenger rail travel in the EU,[1] and it then raises a series of issues for discussion. The heavy investment in HST has helped the rail renaissance, but it has also encouraged longer and faster journeys that in turn may not lead to a more sustainable transport system. The sustainability of rail travel is also questioned, particularly if most investment is in speed. From an environmental perspective, it is argued that the railways are not green, but probably greener than other modes. Comments are made on the difficult issue of travel time and travel time savings, as well as the notion that travel time is wasted time. Arguments are put forward for better integration between rail and other services so that the total journey is considered, not just the main part of it. This explores the potential for rail stations playing a much more prominent role as places (or nodes) on the network and the need for better access to stations. Finally, the benefits of rail travel are contrasted with the income levels of the users of rail services, and whether investment in rail reduces or increases the equity of the transport system. A synthesis of the above issues, at the end of the chapter, provides some guidelines for a sustainable development of the rail network.

PASSENGER RAIL USE OVER TIME

Passenger travel by rail, at least in Europe, is on the rise. Some refer to recent years as the second railway age (Banister and Hall, 1993). In the UK the number of rail passenger-km is at its peak, over 48 billion passenger-km in 2007, the highest ever for the UK (Figure 14.1). In Europe, demand for rail transport has increased in recent years, although it is not yet close to its past peak (Table 14.1).

The key message from Table 14.1 is the very different patterns of growth in the EU countries joining up to 1995 (EU15), as compared with the substantial decline in the new EU countries (EU12). Over the last 36 years, there has been a uniform growth across all EU15 countries (apart from Greece), reflecting low levels of rail use in the 1970s, and the probability of an increase in travel distances since that time (there is evidence for this in the UK in last 10 or so years, represented in the growing gap between the two curves in Figure 14.1). The greatest growth in passenger rail use has been in Ireland (+217%) from a low base level, but many of the major EU15 countries have shown dramatic increases in passenger kilometres traveled, including France (+92%), the Netherlands (+84%) and Sweden (+109%). There is also a group of countries where the increase has been over 50% (Denmark, Finland, Spain and the UK).

Table 14.1 European rail, 1970–2006 (1000 mio pkm)

	1970	1980	1990	1995	2000	2001
EU27	326.4	366.0	400.7	347.7	368.3	369.4
EU15	220.2	246.9	268.9	273.3	306.9	310.7
EU12	106.2	119.1	131.8	74.4	61.4	58.7

	2002	2003	2004	2005	2006
EU27	362.2	358.4	363.5	373.8	384.0
EU15	308.4	306.5	312.8	324.5	334.2
EU12	53.8	51.9	50.7	49.3	49.9

Notes: EU15: Countries joining the EU up to 1995. EU12: Countries joining the EU between 1995 and 2007; EU 27: current members of the EU.

Source: DGTREN (2008)

A completely different picture is given with the new EU12 countries, where the decline in rail use has been of a similar magnitude. Estonia (-75%), Latvia (-73%) and Lithuania (-81%) have recorded the highest falls in passenger rail use, but most of the other countries have seen levels reduced by more than 50% over the 36 year period – Slovak Republic -66%, Romania -55%, Poland -51% and Bulgaria -61%. The Czech Republic has had a 48% reduction over a shorter period (1990–2006), and even Hungary and Slovenia have reported a 41% and a 43% reduction respectively.

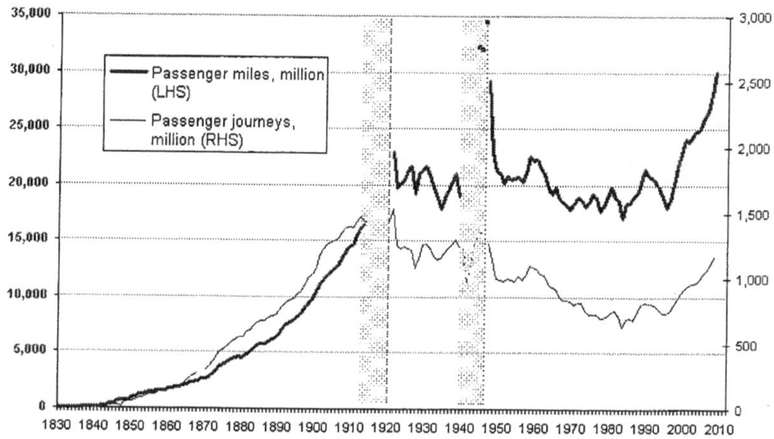

Source: ATOC (2008)

Figure 14.1 Development of passenger rail in the UK, 1830–2007

The explanation for these two very different, yet internally consistent, patterns is difficult. It might relate to investment in rail services, the quality and frequency of services offered, the growth in car ownership, levels of expendable income, and the location of rail services with respect to the population (i.e. rail proximity). One possible explanation can be interpreted from the graph of rail passenger journeys in the UK (Figure 14.1). Here it can be seen that there has been a steady decline in rail passenger journeys and passenger-km made over the period 1945–1980 (about 35 years). The low point just after 1980 is about 50% of the peak figure in 1945 for the number of journeys. But since that time, most of the loss in passenger journeys has been regained in the period 1980–2006 (26 years), and all the loss in passenger-km has been more than regained. The EU15 countries could all be located on this second part of the curve, whilst the EU12 countries are still in the earlier phase of declining use of rail by passengers. There is a clear policy message here for the EU12 countries, and an opportunity to revive rail use, and this chapter now discusses these possibilities in more detail.

The relative weight given to the railways in transport policy, especially European (CEC, 2001) and in countries like the UK (DfT, 2007b), and the constant increase in rail use in recent decades, might give the impression that this is the main passenger mode of transport, but this is not the case. The railways, still, play a relatively small role in meeting overall demand for transport, although they do play a key role on some specific markets and corridors. This is evident in Table 14.2, where the rail share in the overall transport of people has remained relatively stable between 1995 and 2006, at just above 6% of passenger-km, smaller than the role bus and coach play in European passenger transport.

Table 14.2 The position of rail in overall travel in the EU27

	Passenger Cars	Powered Two-wheelers	Bus & Coach	Railway	Tram & Metro	Air	Sea	Total
	1000 mio passenger-km							
2006	4,602	154	523	384	84	547	40	6,333
2000	4,283	136	514	368	77	456	42	5,876
1995	3,855	123	501	348	71	335	44	5,277
	Modal split %							
2006	72.7	2.4	8.3	6.1	1.3	8.6	0.6	100
2000	72.9	2.3	8.8	6.3	1.3	7.8	0.7	100
1995	73.0	2.3	9.5	6.6	1.3	6.3	0.8	100

Source: DGTREN (2008)

Furthermore, despite the growth in rail use, especially since the 1990s, their share of passenger transport is in decline. While rail transport is not a substitute for many of the journeys by car and bus, and certainly not an alternative to journeys by sea, it is clear from T14.2 that some of the reduction in rail's share can be attributed to the increased use of air transport. For many journeys these two modes are substitutes. In comparison to the US, the European railways are doing well. In the US, rail transport carried only 0.5% of the passenger-km in 2004 (the figure also includes light and commuter rail), the same as in 1990 (DGTREN, 2008). The differences from Europe are attributed mainly to the dominance of freight transport on the US rail network compared to that of passenger transport in Europe.

The main factors contributing to the observed increase in rail use in Europe can be associated with the development of the HST, road congestion and (rail) regulatory reforms, which result in introducing competition for the provision of rail services and changes to the way the industry is operated and managed. The above factors are likely to continue to influence demand for rail transport, and together with continued economic growth (which is likely to lead to further growth in demand for transport) rail use can be expected to increase. It is not clear, however, whether rail transport, in Europe and elsewhere, can at the same time increase its share across all modes of transport. From a sustainable transport perspective this is the really important question and what transport policy should aim to increase – railways' share rather than railways' use.

THE HIGH-SPEED TRAIN (HST)

The 'comeback' of the railways is very often attributed to the development of the HST. Although rapidly growing, the HST still does not account for the majority of rail travel across the EU, only 23% of the EU27 passenger-km in 2006.[2] However, in some countries, much of the growth in rail travel has taken place in HST. In France, HST services in 2006 accounted for 57.1% of all rail passenger-km. The figures for Spain, Germany and Sweden are 38.3%, 27.4% and 25.9% respectively. In Italy the figure is slightly lower at 19.2% and in Belgium it is now 10.4% (DGTREN, 2008).

The main reason for the development of the HST was the need to increase capacity through higher speeds, and not to reduce travel (Givoni, 2006). Nevertheless, the reduction in travel time the HST offers is certainly a major contributor to the increase in rail use. This increase can be attributed to shift of demand from other modes, as rail becomes more competitive, or simply an increase in demand for transport – whether it is the former or the latter has important implications for sustainable transport.

In many respects HST can be considered as an ideal mode of transport for long-distance land journeys of up to 1000km. It is fast and (relatively) environmental friendly and thus can offer improved accessibility at lower environmental cost. In addition, the development of the HST can lead to further effects such as economic development around railway stations, although the evidence on this is not conclusive and mainly points to the need for other conditions to be met for the HST to contribute to (local) economic growth (Banister and Berechman, 2000).

The two main benefits HST services can provide, reduction in journey time and (as a result) higher accessibility (which might lead to economic development), can result in undesirable effects – longer journeys and lower accessibility, as explained below.

There is growing evidence that the average daily travel time budget is constant across the population (Metz, 2008). The implication of this is that as the speed of services increases, and travel time is cut, travelers do not necessarily use the saved time for activities other than travel, but instead choose to carry out their activities further away. Thus, the introduction of HST services can lead to more travel. There are suggestions, for example, that the use of tele-working might reduce the number of commuting trips, but those which are still taken are much longer, resulting in an overall increase in travel (Mokhtarian et al., 2004). Thus, HST can encourage infrequent commuting from places further away from the workplace and in general a longer commute. Furthermore, there is evidence that the introduction of HST services, and better and faster service, results in the generation of new demand (Givoni, 2006). The introduction of HST services might be counterproductive to sustainable transport, even when it promotes rail use.

An important element in achieving a substantial reduction in travel time by HST services is the spreading of stops at hundreds of km intervals, since an HST service running with frequent stops would not achieve a high average speed (and shorter travel time). This results in many cities and regions being bypassed by the HST. For these areas, the HST is not different from an aircraft, and the inauguration of HST services will not change the level of accessibility. More to the contrary, since the HST often replaces or supplements an existing conventional railway service, these bypassed cities along the HST line will see substantial reduction in rail services because they will lose the through traffic to the HST and local demand would not support the same level of service. In turn, this reduction in rail accessibility can contribute to economic decline.

Another reason the HST is promoted by policy makers is its ability to compete with, and thus substitute, the aircraft (CEC, 2001). Such a modal shift is expected to reduce the environmental impact from the transport operation. Comparison of HST and aircraft operations' impact on the environment confirms that substituting an aircraft seat with an HST seat will result in environmental gains (Givoni, 2007). However, this advantage of the HST does not necessarily translate to environmental gains in practice. Givoni and Banister (2006) found that the introduction of HST services between London and Paris did not result in a substantial reduction of the number of flights on the route. Since a route like London–Paris is essential for carriers like British Airways and Air-France for their Hub and Spoke operation, and in their competition against each other, they continue to offer a very high frequency service on this route (in the order of 60 daily flights one-way between the cities) even when the HST operator captured over 70% of the market, and when evidence suggest that such a route is not profitable for the airlines.

Cooperation between air and HST can take place if the airport is situated on the rail and HST networks[3] allowing airlines to retain connecting traffic from short-haul routes to their more profitable long-haul services, thus giving up flights without compromising their Hub and Spoke operation. This is the case in airports like Frankfurt, CDG and Schiphol, where the above conditions exist.

For the HST to reduce air transport impact on the environment two conditions must be met. First, the HST service must be integrated with other aircraft services provided by the airline and thus it must complement it rather than compete with it. Second, the airline (or airport) should not use the freed runway slots for other flights. With the development of the HST network and successful integration of air and rail services (for example at CDG) the first condition is now being met at a number of airports. However, given current levels of demand for air transport services and the lack of runway capacity to meet it, it is unlikely that the second condition is met. This would require some form of demand management policies to be introduced at airports.

The HST have been very successful in competing with the 'traditional' airlines, especially on routes between major cities like London–Paris and Paris–Brussels, leading to the removal of flights on the latter. However, the growth in low-cost carriers' (LCC) operation in Europe has brought new forms of competition that have proved more problematical for HST to compete or cooperate with. The different characteristics of HST and LCC services, including the price advantage of the latter, reduce the level of mode substitution. The lower fares offered by LCCs might be more than compensated for by the cost of accessing the airports from which the LCCs operate (usually small airports far from the centre of the main cities, where the HST stations are located), but it is not clear to what degree passengers account for these costs when choosing their mode of transport.

In the UK, the development of a HST network (along the main four rail corridors) is again on the agenda (Greengauge 21, http://www.greengauge21.net) although there are no concrete plans for its construction (DfT, 2007b). The debate on the future of HST in the UK highlights all the advantages of the HST mentioned above – namely faster travel time, increased accessibility, economic development, mode substitution and environmental benefits – while the shortcomings mentioned above are ignored. In the context of sustainable transport, which is central to UK Government transport policy, the risk that an HST network would lead to more travel is apparent. There is a risk that such a network would bring London and Manchester within commuting distance of each other. The risk of lower accessibility to cities and regions not on the network is apparent. Furthermore, current plans do not put Heathrow airport as a central point on the HST network. This limits the possibility for mode substitution and the potential benefits from HST. To some extent the above argument also applies to further development of the European HST network and the introduction of HST in other parts of the world, most notably China and Korea.

The implications of the above are that the development of the HST must be carefully considered to ensure the benefits are fully realised, and the possibility for adverse effects are limited. Integration between the HST network and that the conventional rail network seems crucial, as only this will ensure the accessibility benefits from HST services are spread across the entire rail network. For this, the location of the HST stations must consider the entire rail network and not only local demand from cities. In addition, the HST stations must be designed to a relatively seamless transfer between HST services and other rail (or air) services.

ARE RAILWAYS GREEN OR JUST GREENER?

Rail transport is generally considered as being green, and this is its main appeal for policy makers. However, it is hard to see how rail transport can be regarded as 'green' if its energy is sourced from diesel fuel or electricity which is generated mainly by coal, oil and natural gas. It can be regarded as 'green' if it uses electricity generated by renewable sources or nuclear power, but with a few exceptions (notably France) this is not the case. Yet, there is evidence that rail transport is more energy efficient than other modes of transport used for long-distance travel and thus can at least be regarded as 'greener'. Whether green or just greener, different types of trains and different types of rail services will have substantially different environmental impact and these differences must be recognised to minimise the impact of rail transport on the environment and maximise the environmental benefits from mode substitution, namely road and air transport.

Rail transport's impact on climate change is mainly affected by the sources of energy used to power the train. Two main categories can be identified, diesel and electric traction. The contribution of the latter to CO_2 emissions and climate change depends on the mix of sources used to generate the electricity. In the UK, electricity is currently (2006) mainly generated using natural gas (39%), coal (34%), nuclear power (20%), renewable sources (e.g. wind, 5%) and others (2%), and this mix results in the CO_2 intensity of electricity of 554 g/kWh (DTI, 2007). In 2020, with the implementation of policies described in the UK Energy White Paper (DTI, 2007) it is expected that the CO_2 intensity of electricity will fall to 415 g/kWh, a 25% reduction. This reduction is associated mainly with a 10% reduction in the share of coal in the electricity generation mix, a 4% increase in the use of natural gas and a 4% increase in the share of renewable energy to generate electricity.

Considering the current electricity generation mix in the UK, electric train operation results in 54 gCO_2/passenger-km compared with 69 gCO_2/passenger-km for diesel trains, a difference of more than 20%. The differences are not associated with differences in the load factor for each type of train as these were similar (Givoni et al., 2009). However, when accounting for the cost of completing the electrification of the UK rail network (estimated at between £550K and £650K per single track km; The Railway Forum, 2008) it does not appear to be cost-effective considering the

CO_2 reduction this can bring (at more than £3,000 per tonne of CO_2 saved and including future years). Other options might prove to be cost-effective, including running fewer diesel trains under existing electrified sections, small-scale electrification to allow connection of existing electrified routes, and selective electrification of additional routes.

Another important difference between the impact of diesel and electric train operations on the environment is their contribution to local air pollution (LAP). Electric trains have virtually zero impact on LAP at point of use (i.e. alongside the rail network) unlike the operation of diesel trains which results in the emission of sulphur dioxide, particulate matter, nitrogen oxides and ozone. This difference is especially important in highly populated urban areas. LAP also occurs at point of electricity generation, but this is usually away from densely populated areas resulting in a relatively low impact. Thus electric trains can be said to have an advantage over diesel trains in terms of both climate change and LAP impacts.

Train are usually high-capacity vehicles with about 100 to 400 seats. HST will have even higher capacity, as the Eurostar trains operating between the UK and Europe have a capacity of over 750 seats, and some Japanese HSTs can accommodate 1000 seats. The energy efficiency and environmental impact of trains depends very much on the load factor or occupancy rate. In the UK it is assumed that the average occupancy rate is 40% for intercity trains and 30% for other services. Thus, there is a potential for improving the environmental performance of rail, when considered in passenger-km terms and especially when compared to other modes, by increasing the average load factor. Most airlines would achieve an average load factor of 70% while some LCC can achieve 80%, to a large extent filling seats through the creative use of the pricing mechanism. Rail can follow such practices and it is important to balance demand between peak and off-peak travel, as well as aiming to balance demand between the two directions of travel on specific routes (Rietveld, 2002). This imbalance in the demand for rail transport can result in a marginal environmental burden during the peak that is much higher than is usually thought, whereas it is almost zero at the off-peak period, due to the need to add services when demand increases at the peak, while it can be accommodated on existing capacity during the off-peak period (Rietveld, 2002).

The operational characteristics of the rail service are also important in determining the environmental impact of rail transport. It is expected that energy consumption and therefore environmental impact will increase with speed, yet this is partly offset in the case of HST by aerodynamic design. More important is the number of stops along the route, whether at stations or at junctions, as the acceleration requires more energy which is not compensated for when slowing down.[4] The spacing of intermediate stops along the route is probably the one factor affecting energy consumption that operators have most control over. Here a trade-off between exploiting commercial opportunities, picking up more passengers by stopping at more stations, and fuel and energy savings must be struck. However, increasing the number of stops also has commercial penalties as more stops increase travel time for passengers, making the railways a less attractive option (Givoni et al., 2009).

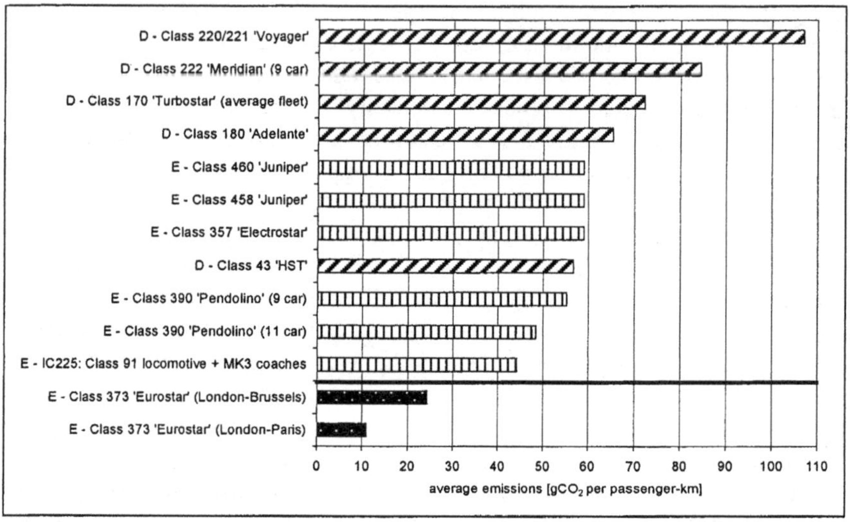

Notes:
1. 'D' stands for diesel traction, also indicated by diagonal pattern; 'E' for electric traction, indicated by vertical pattern. 'Eurostar' figures indicated by black pattern with white dots.
2. The analysis assumes passenger load factors of 0.4 for UK intercity trains ('Adelante', 'HST', 'Pendolino', 'IC225') and 0.3 for other UK trains ('Voyager', 'Meridian', 'Turbostar', 'Electrostar', 'Juniper').
3. Class 373 'Eurostar' emissions values taken directly from Eurostar (2008).

Source: From Givoni et al. (2009) based on RSSB (2007), DfT (2007b), and Eurostar (2008).

Figure 14.2 Average emissions of CO_2 for a number of typical UK trains (in gCO_2/passenger-km)

Figure 14.2 illustrates the above by showing how different the CO_2 emission levels per passenger-km can be for different types of trains operating on the UK rail network. In determining emission levels, it appears that route characteristics, and especially the number of stops on the route and the distance between them, are more important than the speed at which trains are running at. Figure 14.2 suggests that the faster trains are responsible for lower CO_2 emission levels per passenger-km.[5] Thus, although the faster Eurostar trains use more electricity per seat-km (in the region of 0.05 kWh/seat-km), the combination of higher load factors, aerodynamic design and lower CO_2 content of the electricity supplied on non-UK route segments (France, Belgium) yields significantly lower average emissions per passenger-km then other trains. Surprisingly, the most polluting (diesel) train in Figure 14.2 is probably the newest train, that entered service after 2000. Its high level of emissions is hard to explain.

Naturally, the environmental aspects must be considered together with other considerations when developing and promoting rail transport. Yet, careful examination of the environmental impact of rail transport is important to provide a service that not only meets demand but also considers the most effective way to doing so from commercial and environmental perspectives. For example, countries in which large parts of the network are not electrified are advised to take measures to reduce emissions from diesel trains, electrify where the carbon content of electricity is low (e.g. France, Norway), and invest the money in carbon abatement options in other transport sectors and the wider economy (Givoni et al., 2009).

RAIL TIME – IS IT ALL WASTED TIME?

The notion underlying transport planning is that travel time is wasted time and therefore any changes to the transport network which can reduce travel time are beneficial – to passengers and to the economy. Indeed, reduction of travel time is considered to be the main rational for expanding the HST network and services, even if originally the HST was 'invented' to increase capacity rather than shorten travel time. However, this notion has been questioned recently. First, Mokhtarian and Salomon (2001) found evidence that, in some circumstances, travelers might not suffer any 'loss' from the need to travel. One of these circumstances is when travel time can be used effectively, or at least partly, to engage in other activities when traveling (e.g. read, rest). Rail travel offers such opportunities. Second, the evidence that the average daily travel time is constant over time (Metz, 2008) suggests that when travel time is cut on one journey, for example by a faster train service, travelers (in the long run) use this time to travel longer distances or make more journeys.

In Britain, the average trip time for rail (and including trips on the London Underground) is 71 minutes (Lyons et al., 2007) and this provides ample opportunities for passengers to 'use' travel time to do 'something worthwhile'. Table 14.3 (from Lyons et al., 2007: 110) clearly demonstrates this. There are many issues arising from the information presented on how rail passengers use travel time, but what is important here is that rail time is not necessarily a (complete) waste of time.

Table 14.3 Comparison, by journey purpose and direction of travel, of the percent of travelers undertaking activities for some time during the train journey and (shown in brackets) for most of the time

Activity	Journey purpose							
Figures in percent	All		Commute		Business		Leisure	
	Out	Return	Out	Return	Out	Return	Out	Return
Reading for leisure	53(33)	56(35)	62(42)	62(42)	43(22)	53(29)	46(27)	51(31)
Window gazing/ people watching	56(18)	58(18)	49(12)	48(11)	51(12)	58(14)	66(27)	68(27)
Working/ studying	26(13)	25(12)	27(13)	29(13)	55(35)	48(27)	13(6)	12(5)
Talking to other passenger s	16(6)	14(5)	11(4)	10(3)	14(5)	11(4)	23(9)	21(8)
Sleeping/ snoozing	13(3)	19(4)	16(5)	23(5)	11(2)	17(3)	9(1)	14(3)
Listening to music/ radio	9(3)	9(3)	12(4)	13(5)	4(1)	5(1)	7(3)	6(2)
Not answered	1(11)	1(10)	1(9)	1(9)	1(10)	1(10)	1(12)	1(11)

Source: Lyons et al. (2007; 110)

The possibility to use travel time effectively is very much enhanced in the 'information age' where developments in ICT release many activities from their ties to a specific location. The mobile phone and laptop computer are probably the most notable examples. Although most rail passengers (T14.3) do not engage in activities requiring the use of ICT, the possibility to use it is still important, and might influence mode choice (many of the passengers surveyed noted that they have brought 'work' with them for the trip but did not use it, the same with mobile phones; Lyons et al., 2007).

If travel time can be used in some way, speed might not be so important any more. In addition, there is increasing evidence that passengers put a lot of weight on valuing a service on travel time reliability. Passengers want to arrive on time! This time is the service advertised time, the time passengers planned to arrive at the destination station. Many studies found that passengers will trade travel reduction with travel time certainty and that passengers place a value on the service reliability (Bates et al., 2001). However, the adverse effects of an unreliable service can nowadays be more easily mitigated by the use of ICT (e.g. mobile phones) allowing people to rearrange their schedule while on the move. Furthermore, Urry (2007) claims that development in ICT technologies, and specifically the development and widespread use of mobile phones (and text messaging), changes our perception of time and has made scheduling of activities more flexible. In this case, it might be that both speed and reliability are less important than they are perceived to be. Lyons and Urry (2005) suggests the following (p. 263):

> The boundaries between travel time and activity time are increasingly blurred. Specifically, many people are using travel time itself to undertake activities. The 'cost' to the individual of travel time is reduced as travel time is converted into activity time. In turn, less of the individual's travel time budget is used, enabling more travel or encouraging greater use of modes that may enable en-route activities to be undertaken.

This has implications for the attractiveness of rail, as this time can be used for work, leisure and other networking activities. The key conclusions here are that travel time is not uniform in terms of its costs or of its potential value to the traveler, and provided that a sufficient level of comfort is provided and there is the necessary support available, then many activities can be undertaken. So travel is no longer only about speed – it is about comfort as well and providing the opportunity to use time (more) effectively. These positive attributes of rail travel need to be countered by the sustainable mobility dimensions that might indicate that there is an incentive to travel more and live even further away from work places and other activities.

Rail has potentially two advantages over other modes, as it can provide shorter travel time and as it provides an opportunity to use this time more effectively. Yet, these advantages might only lead to longer journey distances. In the context of increased congestion on the road network and slower car journeys, rail transport becomes relatively faster even without the expensive investments in increasing the train speed. It would then be logical to conclude that rail operators ought to make their services more comfortable, to reduce the disutility of travel time and improve the conditions that allow this time to be used, rather than an overriding concern with higher speeds. This in turn would also probably make rail services more reliable, thus meeting all the passenger requirements: faster (than other modes), reliable and a comfortable journey.

RAILWAYS IN INTEGRATED TRANSPORT – THE DUAL ROLE OF STATIONS

The increase in rail use, the development of the HST and the promotion of rail by policy makers marks a comeback for the railway station, which is recapturing its central position in the urban landscape. From a wider policy context of promoting rail transport, there are mainly two issues at play, which might prove difficult to reconcile. One of the roles of the station is as a node on the transport network and the other role is as a 'place' in the urban socio-economic landscape.

A railway journey is almost always part of a journey 'chain' that includes a journey to, and later from, the railway station by different modes of transport. The integration of these components is essential to achieve a continuous travel, door-to-door when using rail, and in order to make the railway a viable and attractive alternative to the car. The accessibility of the railway station is crucial. Research shows that improving access to railway stations can be one way to increase rail use,[6] and this option is relatively cheap and easy to implement in comparison with improvements to the actual rail journey (e.g. higher speed or frequency). Good access to railway stations requires the station to be designed for a relatively quick and seamless transfer between several different access modes and the train (Givoni and Rietveld, 2007; Brons et al., 2009).

Car parks at railway stations can be seen as one way to divert at least part of the journey from car use to rail use. But these could also encourage car use at the expense of other modes to get to the station, and car parks are expensive to build and hard to accommodate at city centre stations. Public transport accessibility to the stations is therefore crucial, but also these facilities need to compete for the land in and the around the station, land that might be more profitable for other uses (see below). At large stations; even the need to accommodate parking facilities for bicycles is a challenge, as the Dutch experience reveals. Some Dutch railway stations have cycle parking facilities with a capacity of several thousand places.

Table 14.4 reflects current patterns of access to the UK rail network, giving high levels of access by walking (cycling is only a small part of that group). This suggests that for many journeys, the proximity of the rail station to the home or activity place is a key determinant of whether that service will be used, perhaps also indicating the importance of network density and rail catchment areas. In the Netherlands, cycling and walking are the main modes (together with public transport) used to access railway stations. At the home end station, access to and egress from the station is 38% by bicycle, 27% by public transport and 20% by walking. At the activity end station, almost half of the journeys (47%) to/from the station are by walking followed by public transport (35%). Bicycles are only used on 10% of the journeys to/from the station at the activity end (Givoni and Rietveld, 2007).

Table 14.4 Main mode used in the UK to travel to station of origin

	Percentage
Walk/cycle	55
Bus/coach	10
Car	20
Underground/light rail/metros/trams	15
Total	100

Notes: From the 2007 UK National Rail Travel Survey. The NRTS has about 436,000 records and the data have been expanded to represent approximately 2.7 million rail trips on an average weekday.

Source: DfT (2007a)

The data from the Netherlands highlight the role non-motorised modes can play in the accessibility of the rail network. The high share of walking at the activity end is probably related to the success of the Dutch in concentrating commercial development and employment centres close to and around railway stations. Public transport can supplement non-motorised modes to reach the station from further apart, where the distances are too large for cycling. However, in promoting public transport use to railway stations to reduce car use, an adverse effect can be a decline in bicycle use, as happened in the Netherlands when free public transport was given to students (Givoni and Rietveld, 2007).

The central place of railway stations in the urban landscape, very much due to their good accessibility (by rail and other modes) at the city, regional and often national and international levels (if HST services are provided), also makes them attractive for activities other than traveling. Most notably, the high accessibility of major railway stations makes them an attractive location for firms and businesses, and a very attractive employment centre. This attraction results in high demand for land and property close to rail stations, and this turn may lead to economic development (Pels and Rietveld, 2007). It marks an important opportunity for rail operation since the rail industry can capitalise on these, where it owns part of the land around the station and has building rights above it, and thus improve its financial viability. In the same way that major airports are profitable due to their commercial activities, which are cross-subsidising their aviation activities and used as a resource to finance infrastructure expansion, railway stations can improve their financial viability and the service they provide. Yet, the increase in land values around stations can lead to a conflict (as in the case with airports) between the transport and non-transport role of railway stations. If land around the station is developed as a work place and most workers arrive by train to it, the incentive to provide good access (by public transport) to the station from different parts of the city is reduced, especially when access to the station is seen as outside the control and responsibility of the railways, and is provided by other operators.

While more stations might be desirable in one way, as they increase the accessibility to the rail network especially by reducing the distance to the station, making non-motorised modes more attractive, they are expensive to provide (especially in urban areas). Additional stations also increases travel time for some passengers, and this reduces the overall attractiveness of rail. Given the cost of rail infrastructure (including stations) and the time lead it requires, the key to improving the accessibility of the rail network and making rail travel more attractive is very much in providing better access to railway stations, and by a range of modes (including the private car). This can enhance both the role of the station as a node on the transport network and as a place. However, in this process conflicts between these two roles will have to be resolved.

WHO IS BENEFITING FROM BETTER RAIL TRANSPORT?

Making rail travel more attractive can potentially have environmental and economic benefits, as illustrated above. The concept of sustainability, however, also includes a third dimension – the social aspect. This dimension, with respect to transport, focuses on accessibility, the possibility for everyone (without distinction of age, sex, race and income) to have reasonable access to a range of activities. Equity considerations are not an issue for rail operators, where rail services are run on a commercial basis, which is standard practice. For policy makers, rail operation presents a challenge in this respect. Equity issues with respect to rail operation arise mainly in relation to who uses rail transport, or more precisely who can afford rail transport, and where rail services should be provided.

Most rail travelers come from high income groups (Table 14.5), and the growth in the quality and speed of rail services has helped those travelers move further and faster. This presents decision makers with an interesting dilemma. Investment in high speed rail may disproportionately benefit the higher income travelers, as they are the main users of rail services. Investment in local rail services and bus services may correspondingly help low income travelers. There are also important issues here relating to whether there is a subsidy for the service or whether the traveler pays the full costs of travel. Again there are distributional outcomes.

Table 14.5 Rail travel in the UK by household income

	Percentage
Below £7,000	7
£7,000 – £12,500	8
£12,501 – £17,500	9
£17,501 – £35,000	30
£35,501 – £50,000	23
£50,501 – £75,000	15
More than £75,700	9
Total	100

Notes: From the 2007 UK National Rail Travel Survey. The NRTS has about 436,000 records and the data have been expanded to represent approximately 2.7 million rail trips on an average weekday. The UK household median income is about £26,000 (2006).

Source: DfT (2007a)

On a spatial level, intuition might suggest carrying out investments in the rail network in areas where currently the level of rail service is relatively low and the access to it relatively poor – in the periphery. However, Brons et al. (2009) showed for the Netherlands that when the aim is to increase rail use investments should actually be directed towards areas where the level of service (in terms of the rail service and the access to it) is already relatively high and to the most populated areas – the urban centres.[7] The policy implication of this conclusion is that investment should not be directed towards improving the rail service at the periphery of the network but (further) improving it at its centre. In other words, investments in rail infrastructure must be very selective and targeted at the main urban agglomerations.

Conversely, in those areas which are reached by the rail network but where the level of rail service is relatively low (generally the peripheral parts of the network), it is more effective to improve the access to the station rather than the rail service itself. The Dutch rail network already covers most of the country, so improving the current level of service might be more important than further expanding the rail network. This would imply that in the peripheral areas of the Netherlands, transport services may be supplied by modes other than rail and to increase accessibility from these areas to the rest of the country these modes would be used to provide better access to the existing railway stations on the network. The conflict between efficiency and equity is apparent here.

SYNTHESIS – THE ROLE OF THE RAILWAY IN SUSTAINABLE TRANSPORT

This chapter presents the main dilemmas with regard to developing passenger rail transport in a way that will contribute to sustainable transport. There are many opportunities for further increases in the use of rail transport, and more important, an opportunity for a real change in modal shift – where rail will carry a larger proportion of passenger travel. The chapter also provides a health warning and indicates the danger that some of the advantages of rail transport, and especially shorter travel times through faster service, could simply lead to an increase in demand for travel and to an increase in the distance traveled. Thus promotion of rail transport might jeopardise the efforts to move towards a sustainable transport system. Another dimension of the dilemma faced by those promoting sustainable transport through better rail services is that on environmental grounds all forms of rail travel may be better than long distance travel by car or by air (at least this is currently the situation in Britain), but rail travel is not good for the environment. Furthermore, rail (and especially long-distance high-speed services) is used mainly by very specific segments of the population (those on a relatively high income). This conclusion also raises questions concerning the appropriateness of rail transport as a champion of sustainable transport. This is not to write off the railways' role in sustainable transport, on the contrary, but to emphasise that for this role to be meaningful in the promotion of and investment in rail transport must be considered carefully.

Returning to the questions raised at the beginning of this chapter, there is one key unresolved issue if rail is to contribute to the new sustainable mobility paradigm. The renaissance of rail seems to be well advanced, at least in the EU15 countries, and similar trends will probably occur in the new EU12 countries over the next 10–20 years, but the growth rates are still lower than for all other modes (apart from bus and coach). The railways' modal share is gradually shrinking or at best being maintained while the travel market is continuing to expand. This expansion is unsustainable, particularly as that expansion is taking place in the most energy-consumptive forms of transport. Rail is not making a contribution to taking a larger modal share, it is only making a modest contribution to taking on some of the growth in travel. For rail to make a real contribution to the sustainable transport system, it must capture a larger share of the journeys, and this means that people must switch to rail from other modes. This does not seem to be happening, except in the HST market, where there is some evidence of switching from air and car.

The main message emerging from this chapter is that unless the overall levels of travel are reduced and rail takes a much greater share of that market, there seems little opportunity for rail (in particular the HST) to contribute to sustainable mobility. Another important message relates to integration. This can be seen as having two dimensions. First, since HST lines are only viable (from a socio-economic and environmental respects) on a relatively few routes, integration between the HST network and the conventional rail network must be achieved. This can take place by adopting an HST technology where the trains can also run on the conventional network (the TGV model) and by designing the HST stations to allow fast and seamless transfer between HST services and other (national and regional) services. The concept of short- and long-haul air travel networks which are integrated at the hub can be adopted for integrating HST services with other rail services. In air transport, the distinction between these two networks is conceptual without any noticeable differences at the airport. Second, the rail network (HST and conventional) needs to be fully integrated with the rest of the transport network. The air transport network on one hand, and the urban/local transport network on the other, should work together. The main barrier for such integration is likely to be with respect to institutional barriers, and the very much uni-modal focus within the institutions governing transport network development and operation. Since investment in rail infrastructure and changes to the physical structure of the rail network is expensive and requires a long lead time, investments might be better directed to the other elements of the rail journey, including those outside the train, but this depends on removing the institutional barriers mentioned above.

There is a segment of the population for which travel by rail is the first option (those passengers will often choose their work and living place next to a railway station) and there is also a, presumably large, segment of the population for which the car is the first option while the train is not an option at all. For those two segments of the population, improvements in rail transport might not change much their travel behavior and mode choice. A third, probably large, segment of the population uses rail transport infrequently, and increasing the use of rail transport for this segment of the population is probably the main challenge to increasing the railways' share in passenger transport.

NOTES

1. The analysis does not include Switzerland, which in terms of rail supply and patronage is doing better than any other European country with a share for the railways of 15.3% of land passenger transport. The figure for the EU27 countries and EU15 countries was only 6.9% in 2006 (DGTREN 2008). It is important also to note that predominately rail transport in Switzerland is electric powered and the electricity is to a large extent generated from renewable sources (see section 4).
2. That is, 89.9m passenger-km in 2006, up from 58.8 (16% of rail passenger-km) in 2000 and only 15.2m passenger-km in 1990 (DGTREN, 2008).
3. Equally important is that the airport railway station be easily accessed from the aircraft gates and the transfer between the modes will be facilitated by, for example, hand-free transfer of baggage, through ticketing, coordinated time-table, etc. (see Givoni and Banister, 2007).
4. For electric traction, the rail industry is pursuing energy savings through the introduction and further roll-out of regenerative braking, i.e. using the electric motors as generators that convert kinetic energy released while braking back into electrical energy. Electric trains which have already been equipped with such a system have shown energy savings of up to 20%, but it is not expected that regenerative braking will compensate for the energy needed to accelerate (see Givoni et al., 2009).
5. For example, maximum speeds of the Class 357 Electrostar, Class 390 Pendolino and Class 373 Eurostar trains are 161, 225 and 300 kph respectively. Class 43 HST (diesel) has a maximum speed of 201 kph.
6. Based on the Dutch Railway customer satisfaction survey Brons and Rietveld (2009) show that the total quality score for a railway journey by a railway passenger in the Netherlands is determined 41% by the rail trip itself, 21% by the access and egress and 25% by the station experience.
7. These two qualities, large population and high level of rail service, are of course highly correlated.

REFERENCES

ATOC (2008), The billion passenger railway – lessons from the past: prospects for the future, Report by the Association of Train Operating Companies. London, UK.

Banister D. (2008), The sustainable mobility paradigm, *Transport Policy*, **15**, 73–80.

Banister D. and Berechman J. (2000), *Transport Investment and Economic Development*, UCL Press, London.

Banister D. and Hall P. (1993), The second railway age, *Built Environment*, **19 (3/4)**, November, 157–162.

Bates J., Polak J., Jones P. and Cook A. (2001), The valuation of reliability for personal travel, *Transportation Research Part E*, **37**, 191–229.

Brons M. and Rietveld P. (2009), Improving the quality of the door-to-door rail journey: a customer-oriented approach, *Built Environment*, **35 (1)**, 122–135.

Brons M., Givoni M. and Rietveld P. (2009), Access to railway stations and its potential in increasing rail use, *Transportation Research Part A*, **43**, 136–149.
CEC, Commission of the European Communities (2001), *White Paper – European Transport Policy for 2010: Time to Decide*. Commission of the European Communities, COM(2001)370, Brussels, September, [http://europa.eu.int /comm/off/white/ index_en.htm (15/2/2002)].
DfT, Department of Transport (2007a), *National Rail Travel Survey: Provisional 2007 Report*, Published by the Department of Transport and Transport Scotland, March.
DfT, Department of Transport (2007b), *Delivering a Sustainable Railway*, White Paper CM 7176 and supporting documents. Department for Transport, July.
DGTREN, Directorate General for Energy and Transport (2008), *EU Energy and Transport in Figures: Statistical Sourcebook 2007–2008*, Office for Official Publications for the European Commission, Luxembourg.
DTI (2007), *Energy White Paper – Meeting The Energy Challenge*, Department for Trade and Industry, The Stationery Office, London.
Eurostar (2008), Greener than flying, based on research conducted for Eurostar by an independent consortium of Paul Watkiss Associates and AEA Technology Environment. Eurostar Group Ltd.
Givoni M. (2006), The development and impact of the modern high speed train, *Transport Reviews*, **26(5)**, 593–612.
Givoni M. (2007), Environmental benefits from mode substitution – comparison of the environmental impact from aircraft and high-speed train operation, *International Journal of Sustainable Transport*, **1(4)**, 209–230.
Givoni M. and Banister D. (2006), Airline and railway integration, *Transport Policy*, **13**, 386–397.
Givoni M. and Banister D. (2007), The role of the railways in the future of air transport, *Transport Planning and Technology*, **30(1)**, 95–112.
Givoni M. and Rietveld P. (2007), The access journey to the railway station and its role in passengers' satisfaction with rail travel, *Transport Policy*, **14**, 357–365.
Givoni M., Brand C. and Watkiss P. (2009), Are railways 'climate friendly'?, *Built Environment*, **35 (1)**, 70–86.
Lyons G. and Urry J. (2005), Travel time use in the information age, *Transportation Research Part A*, **39**, pp. 257–276.
Lyons G., Jain J., Holley D. (2007), The use of travel time by rail passengers in Great Britian, *Transportation Research Part A*, **41(1)**, 107–120.
Metz D. (2008), The myth of travel time savings, *Transport Reviews*, **28(3)**, 321–336.
Mokhtarian P. L. and Salomon I. (2001), How derived is the demand for travel? Some conceptual and measurement considerations, *Transportation Research Part A*, **35**, 695–719.
Mokhtarian P.L., Collantes G.O., Gertz C. (2004), Telecommuting, residential location, and commute-distance traveled: evidence from State of California employees, *Environment and Planning A* **36(10)**, 1877–1897.
Pels E. and Rietveld P. (2007), Railway stations and urban dynamics, *Environment and Planning A*, **39**, 2043–2047.
Rietveld P. (2002), Why railway passengers are more polluting in the peak than in the off-peak: environmental effects of capacity management by railway companies under conditions of fluctuating demand, *Transportation Research Part D*, **7(5)**, 347–356.
RSSB (2007), Improving the Efficiency of Traction Energy Use, Enotrac and Interfleet Technology for the Rail Safety and Standards Board, June.

The Railway Forum (2008), *Electrification.* Available at http://www.railwayforum.com/electrification.php, last accessed 15/07/2008.

Urry J. (2007), *Mobilites*, Polity Press, Cambridge.

15. Are urban transportation benefits absorbed fully by land values?

Haim Aviram and Daniel Shefer

INTRODUCTION

There is no doubt that transportation projects have a noticeable impact on land values. A number of theoretical and empirical studies verifying this assertion have been published (see, for example, Mohring, 1961; Mills, 1969; Gatz, 1975; Pines and Wiess, 1976, 1982; Wheaton, 1977; Arnott and Stiglitz, 1979, 1981; Evans, 1983; and more recently, Gatzlaff and Smith, 1993; McDonald and Osuji, 1995; Ryan 1999; Yiu and Wong, 2005; Chebreegziabiher, Pels and Rietveld, 2007). However, serious debate prevails over the nature of this impact. Some researchers claim that increased land values in the vicinity of the impacted areas of a transportation project are fully due to the benefits generated by the project. Others argue that the increase in land values in the vicinity of the project is often offset by a decrease elsewhere, and thus it is a "zero sum game": the impact is mainly an exchange of values between different parts of the urban area (see Mohring, 1961).

The method selected for a cost–benefit analysis of large-scale urban transportation projects may vary substantially according to the magnitude of the net impact of a project on the urban area. Classic methods of the economic evaluation of a transport project are restricted to dealing only with the transport sector and to focusing largely on travel-time impacts (Van der Tak and Ray, 1971; Nash and Preston, 1991), whereas more recent studies tend to include the spatial impacts of a project in assessing its overall effect (Mackett, 1991, 1993; Wegener, 1991; Blum, 1997; Cervero, 1997; McDonald and McMillen, 2004; Shefer and Aviram, 2005). Empirical studies that analysed the transportation–land use relationship found close interactions between them in a wide variety of projects (Dewees, 1976; Damm et al., 1980; Bajic, 1983; Hongbo and Mulley, 2006).

Improved transportation projects produce mixed impacts: on the one hand, they influence the geographic surroundings; on the other, they increase the number of trips. The latter impact is due to two primary reasons:

1. the reduction in transportation costs increases the quantity in demand since the demand curve slopes downwards;
2. the increase in land-use intensification (density) shifts the demand curve for trips to the right, thus increasing the number of trips.

Consequently, we should expect both of the following:

1. an increase in the demand for land and, therefore, an increase in land values;
2. an increase in the number of trips owing to diverted and induced traffic.

However, increased expenditures require additional resources. These resources derive from transport benefits. In the long term, we would expect all the benefits to be absorbed by land values because of a reduction in the location budget; otherwise, an unsatisfied demand for land will exist in the vicinity of the project. These contrasting results are the principal motivations for this chapter: "Are transportation benefits fully absorbed by land values?"

In the attempt to answer this question, the chapter will analyse the possible impact of urban transport improvement projects. Specifically, it will deal with two impacts of a transport project:

- The classic indirect impact of transport projects on land values and land area as presented by urban economists (Arnott and Stiglitz, 1981).
- The less frequently employed direct impact of a reduction in transport costs on the transport sector itself in the sense of induced traffic (Hills, 1996; Goodwin, 1996; Goodwin and Noland, 2003).

As a result of a reduction in transportation costs, the formal expression of all changes in household decisions can be derived from an analysis of the household budget if one disregards the household utility function. These changes, as well as others, depend on the tastes and preferences of the members of the household, and thus they belong to a broad "choice model." In reality, all outcomes are possible, but their magnitudes may vary according to the specific circumstances.

LITERATURE REVIEW

The classic theoretical model for evaluating the impact of transportation projects on land values is based on the consumer utility theory (Alonso, 1964). As a result, many studies treat residential land values, and relatively few studies discuss the impact of transportation projects on commercial or industrial land values. Aviram (2001) dealt with commercial and industrial land uses (see also, Ryan, 2005); Shefer (1973) analysed urban agglomerations (i.e., increasing returns to scale) in the manufacturing industry in the United States (see also, Graham 2006, 2007, 2008; and Venables, 2007). The two major arguments supporting the hypothesis that transportation and land values are tightly connected may be summarized as follows:

Each household has a "location budget" (Alonso, 1964; Herbert and Stevens, 1960; Alcaly, 1976), which consists of the cost of land (rent) and the cost of transportation. Any reduction in the cost of transportation for a specific household will in general be offset by an equal increase in the value of land or in the budget allocated for land (i.e., the quantity of land multiplied by its price). Otherwise, some households will bid for that specific parcel of land, which in terms of location budget will turn out to be cheaper. Increased demand will then push land prices upward, until the location budget reaches its former level.

Mohring (1961, 1965, 1993) rejects the hypothesis that land value captured is a net benefit of transportation improvement. He maintains that "increases in land value are not themselves net highway benefits... . Rather, they reflect an actual or potential transfer of benefits derived from the highways from one population group to another" (Mohring, 1961, p. 236) or from one location to another. According to Mohring, if a specific area is better served by transportation, it will become more attractive than other areas. As a result, although the value of land near the project increases, it will decline in other places. Hence, only a shift of values will be found, from "old" attractive areas to "new" ones.

Both considerations are clearly consistent with the framework of consumer utility theory. Hicks labelled these two outcomes "income effect" and "substitution effect". In reality, a logical conclusion is that both effects exist in parallel. The main difficulty is presented by the fact that, in contrast to other commodities, land is not foot-loose. Hence, if transportation benefits are absorbed in the short–term by other commodities than land, then no stable equilibrium will be reached in the urban land market. If land in the impacted area becomes more efficient than elsewhere, equilibrium will not be reached. This condition will hold until all benefits are absorbed by land values.

Utility theory is formulated mostly as a static model that defines an equilibrium condition. On the other hand, the urban area practically never reaches equilibrium (Getz, 1975); it is always in a dynamic state. Thus, every observation could be considered temporary and partial, and the analysis should not assume that full equilibrium is achieved in the urban land market. Throughout the analysis period, only portions of the impact of the transport project are detectable, and it is logical that visible impacts will be found in places where they can be absorbed quickly and easily. In other words, benefits will automatically be directed to places where weak land-use regulations exist or, in economic terms, where the market can easily absorb the reduction in transport costs because of the short-term high elasticities of either demand or supply. Such restrictions may be economic, legal, or social. This means that intensive development could take place on a relatively small area of vacant land, while only the price of land will increase in an already densely developed area. If restrictions exist on land uses, the consumption of other goods than land will rise, at least temporarily.

A number of studies (see, for example, Arnott and Stiglitz, 1981; Pines and Weiss, 1976, 1982; Wheaton 1977) discuss several specific results, such as increased land values or an increase in developable land areas. Nevertheless, they do not use a multiple choice set for the household, perhaps because of the problem that introducing choices would restrict and even impede the possibility of reaching equilibrium. On the other hand, researchers in the field of traffic analysis, transportation planning and transportation engineering do emphasize the importance and the considerable magnitude of "induced traffic" that results from new or improved transportation projects (Hills, 1996; Goodwin, 1996; Goodwin and Noland, 2003). In some cases, part of this newly generated traffic is diverted from competing routes or modes (buses, trains), or it owes to a different time of the day; however, there is no doubt that at least some of it is newly created traffic generated by the new, or the improved, transportation project.

In the section below, we develop a model based on the household budget constraint, but expand the section that corresponds to transportation in order to bring to the fore the impacts of transportation projects on both transportation and land value. Some empirical results will be presented at the end of the theoretical presentation to support the proposed theoretical model.

THE MODEL

Let us assume a monocentric urban structure, in the midst of which is a single CBD, surrounded by parcels of land occupied by homogeneous households.

One household is located on each parcel of land. The budget equation for a household may be written as:

$$Y = Q(x)\,R(x) + T(x)\,P(x) + Z(x) \tag{15.1}$$

Where:
Y is the disposable income of the household; $Q(x)$ is the quantity of land that the household consumes at a distance x from the centre of the city, the CBD; $R(x)$ is the monthly rent per unit of land; $T(x)$ is the total number of trips the household makes per month; and $P(x)$ is the cost per single trip from x; $Z(x)$ is the expenditure on other goods, measured in monetary terms.

The objective of the household is to optimize its utility function, U:

$$U = U(Q,T,Z) \tag{15.2}$$

The household wishes to maximize its utility, subject to budget constraints. Assuming that all parcels of land in the city are similar (i.e., $Q(x) = Q_0$) and that $P(x)$, the cost per trip, is linear with respect to distance (i.e., $P(x) = ax$), then it is possible to formulate an optimisation problem with a Lagrange multiplier:

$$L = U(Q,T,Z) - \lambda\,[Q_0R(x) + T(x)\,ax + Z(x) - Y] \tag{15.3}$$

Transportation improvement means a reduction in the cost per trip. Therefore, the optimisation has to be carried out with respect to the coefficient "a". We assume, first, that a reduction in transportation costs has no impact on the consumption of "other goods" and that the number of trips does not increase as a result of the reduction in the cost per trip (i.e., fixed demand – or perfectly inelastic demand). The first optimization condition, then, requires that

$$\partial L/\partial a = \partial U/\partial a - \lambda[Q_0\partial R/\partial a + T(x)\,x] \tag{15.4}$$

Assuming a constant marginal utility of money and that the impact of the improvement in transportation on utility is monetary and nothing else, then it is possible to state that

$$\partial U/\partial a = \gamma T(x)\,x \tag{15.5}$$

Where γ is a scalar representing the marginal utility of money.

Inserting equation (15.5) into equation (15.4) and rearranging the equation, we obtain the following:

$$\partial L/\partial a = \gamma T(x)\,x - \lambda[Q_0\partial R/\partial a + T(x)\,x] \tag{15.6}$$

Or, applying the optimisation condition $\partial L/\partial a = 0$:

$$[(\gamma - \lambda)/\lambda]T(x)\, x = Q_0 \partial R/\partial a \qquad (15.7)$$

Hence, the change in land value is proportional to the change in the transport cost. Note that under the assumption of a linear transportation cost function, the changes in land value in areas distant from the CBD will exceed changes in areas adjacent to the CBD.

If we relax the restriction on the number of trips (i.e., $T(x)$ may vary when trip costs change), then instead of equation (15.6), we obtain:

$$\partial L/\partial a = \gamma[T(x)\, x + ax\partial T/\partial a] - \lambda[\, Q_0 \partial R/\partial a + T(x)\, x + ax\partial T/\partial a] \qquad (15.8)$$

Solving the first-order optimization condition and setting the derivative equal to zero yield:

$$[(\gamma - \lambda)/\lambda]\, [T(x)\, x + ax\partial T/\partial a] = Q_0 \partial R/\partial a \qquad (15.9)$$

Equation (15.9) shows that the resulting change in land value is proportional to changes in transportation costs derived from two sources:

- The change (reduction) in trip cost for all trips carried out before the transportation project, plus
- The additional cost that is due to induced demand (generated traffic) caused by the reduction in trip costs.

The foregoing analysis is driven by the optimization of the household utility function, and hence it represents a steady-state equilibrium for the household. However, it is possible to show that even when no steady-state equilibrium exists because of the nature of the urban area, most results could still be derived from the household's budget constraints. According to utility theory, a household's reduced transportation costs will increase its disposable income and, hence, may impact on both land values and transportation. Accordingly, the model is based on a household's budget allocations without any optimization, and most of the basic assumptions used in the previous equilibrium model are removed. Note that the model refers to the division of household income, whether or not it is optimum.

The model is developed in two stages. In the first stage, the impact of transportation improvement will be discussed, maintaining the assumption of a monocentric city. In this phase, the restrictions of fixed demand and fixed land area consumed by the household are relaxed, and the sensitivities of these variables resulting from changes in transport costs are evaluated. In the second stage, the distance from the CBD is replaced by a general accessibility measure of household location, without any restriction on the direction or length of trips. Here the Origin–Destination trip matrix replaces the assumption of a monocentric urban structure. This last change permits an analysis of trip cost in any configuration of the urban structure.

Consider a household located in a monocentric city, where:

Y = household income;
Q = area occupied by a household;
R = unit price of land (land rent);
T = number of trips;
P = cost per trip;
Z = quantity of "other" goods and services consumed;
V = unit price for every unit of "other" goods.

The budget equation is as follows:

$$Y = QR + TP + ZV \tag{15.10}$$

We assume that a change in the household's income has no impact on the price of other goods, because of its negligible influence on market prices. All other variables, though, may be affected.

Differentiating equation (15.10) and setting the differential equal to 0 (that is, improved transportation) do not change the total income of the household:

$$\mathrm{d}Y = \mathrm{d}Q(R + \mathrm{d}R) + Q\,\mathrm{d}R + T\,\mathrm{d}P + \mathrm{d}T(P + \mathrm{d}P) + V\,\mathrm{d}Z \tag{15.11}$$

Setting $\mathrm{d}Y = 0$ and rearranging the right-hand side of Equation (15.11), we obtain the following equation:

$$- T\mathrm{d}P = Q\,\mathrm{d}R + \mathrm{d}Q(R + \mathrm{d}R) + \mathrm{d}T(P + \mathrm{d}P) + V\,\mathrm{d}Z \tag{15.12}$$

The left-hand side of equation (15.12) represents the savings from the improvement in transportation: $\mathrm{d}P$ (negative) is the reduction in transport cost per trip; it is multiplied by T, the number of trips made before the improvement. The result is presented by the dotted area in Figure 15.1a. The negative sign implies that it is increased income (saving), to be allocated among competing expenditure categories. The right-hand side of equation (15.12) is composed of four parts, each of which is open for consideration by the household.

Thus the household may respond as follows: $Q\,\mathrm{d}R$ is the impact of the change in the price of land on the total value of the parcel of land held by the household before the transportation improvement took place. As was pointed out, the sign of this argument could be positive or negative. It will be positive in places where the income impact of the improvement in transportation is dominant. The reduction in the price of transportation increases the household's disposable income for allocation among other goods. The example in Figure 15.1 shows the possible impacts of the change in income.

Figure 15.1a presents the situation in the transportation market. If there is a reduction in the overall cost of transportation, then the expenditure on existing trips decreases, releasing some resources and, hence, increasing the disposable income of the household for re-allocation. This causes an upward shift in the demand curve for land, as shown in Figure 15.1, resulting in an increase in the equilibrium price of land. Thus, the change in land prices, dR in equation (15.12), will be positive. Yet, in other places not positively impacted by the project, demand may decrease because of the substitution effect (negative cross-elasticity in demand between competitive zones). As shown in Figure 15.1, the change in land prices (dR) will be negative.

dQ(R + dR) is the increase in the quantity of land consumed by the household (after the improvement in transportation). Note that according to Figures 15.1, it is not clear whether the increase is positive. However, Alcaly (1976) presents the increase in the amount of land as the direct result of an overall reduction in the price of transportation. A similar result was obtained by Arnott and Stiglitz (1981): "Thus, reduction in unit transport costs leads to more land consumption and the same aggregate expenditure on transport costs and land rents" (p. 338).

We believe that a reduction in the household consumption of land is unreasonable, since physical construction takes place in adjacent areas almost immediately when the demand for land increases. In other locations, people do not abandon their places of residence. However, less intensive use of these other areas could happen, thereby representing a certain capital loss. Since a net reduction in total residential land consumption is not reasonable, it is quite clear that this argument should in general be positive.

dT(P + dP), depicted by the grey area (left of point b) in Figure 15.1a, is the cost of the increased quantity of transportation demanded. Assuming that the demand for transportation slopes downward, then dT must be positive, and hence this argument is always positive. Note that it refers to induced traffic, a factor almost always ignored in the evaluation of urban transportation projects. Newly generated traffic apparently results from the existence of negative price elasticity of the demand for travel (see Christensen, 1975; Lago and McEnroe, 1991). In terms of trip length, newly generated traffic is also a result of the observed phenomenon of urban sprawl (see Ewing, 1997; Gordon and Richardson, 1997). Induced traffic emanates from two completely different sources: diverted trips and new trips. Figure 15.2 presents a case of two competing road links: link A and link B. When the cost function of link A is reduced from AC0 to AC1, the volume of traffic increases from T_1 to T_2. This is induced traffic, diverted from link B to link A.

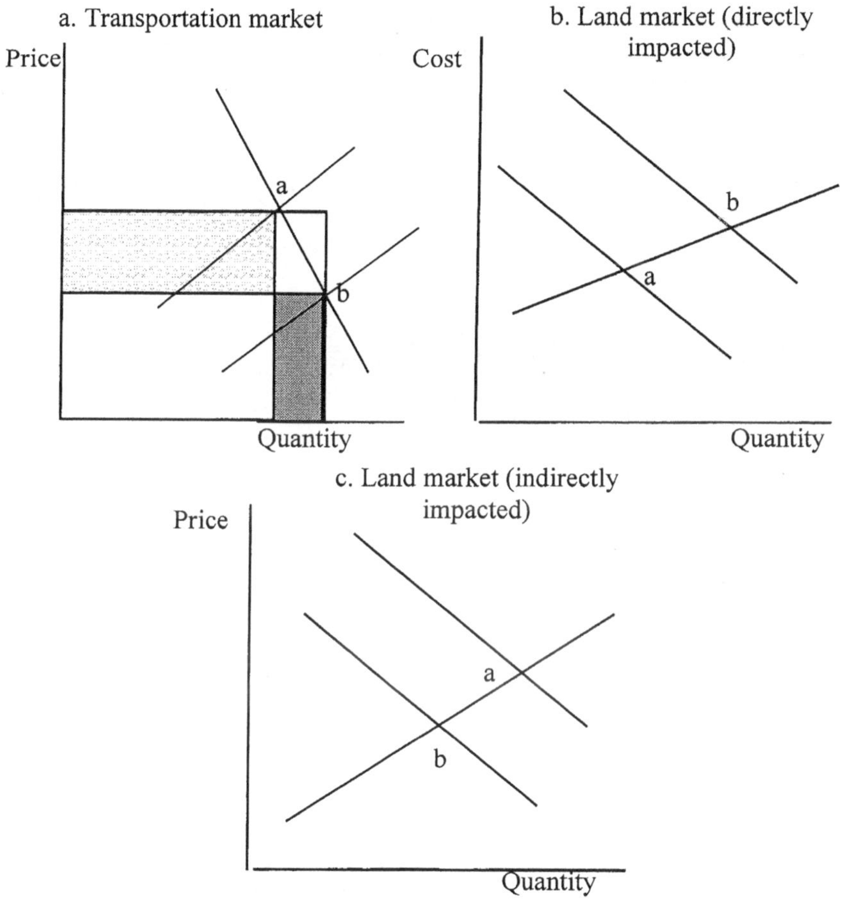

Figure 15.1 Transportation improvement impacts: transportation market, land market (directly impacted) and land market (indirectly impacted)

However, this type of diverted traffic is only one of the changes caused by the reduction in transportation costs. Note that the total volume of traffic is fixed in both links, A and B, and that the reduction in travel costs does not affect the total volume of traffic. This is not the normal situation. Transportation is a well-behaved commodity: "well behaved" in the sense that when its price declines, the quantity demanded increases. Hence, the description presented in Figure 15.3 is much more appropriate.

In Figure 15.3, the Y axis is not perpendicular to the X axis. This means that when the cost of transportation decreases, the total amount of traffic increases. In our example, the cost reduction on link A from P_0 to P_1 not only diverts the volume T_2 - T_1 from link B to link A, but also increases the demand by T_0 - T because of induced traffic, which had not existed at all before the improvement.

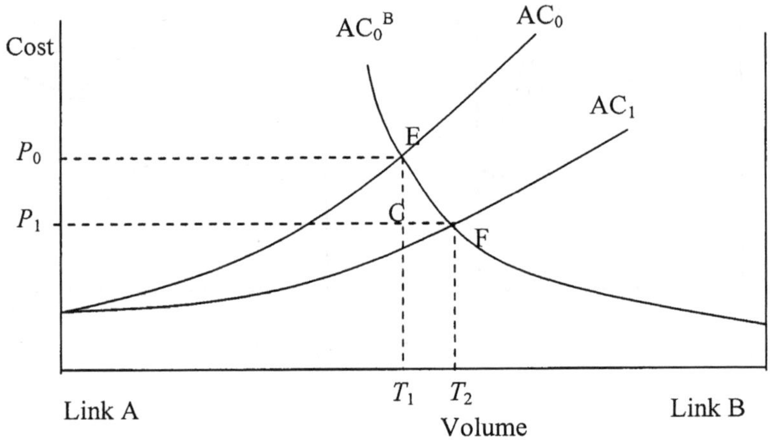

Figure 15.2 Two alternative roads

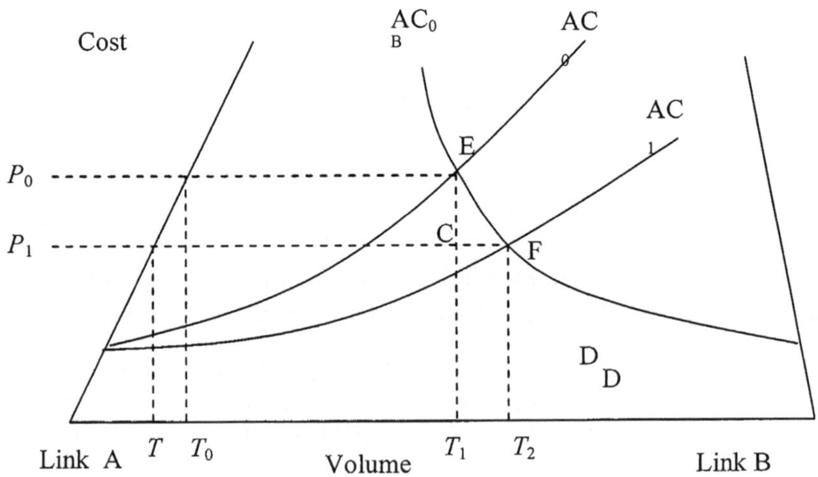

Figure 15.3 Two competing roads with elastic demand

VdZ is the change in the quantity of other goods and services consumed by the household (e.g., food, clothes, education, etc.).

The close connection between land and transportation, through the household's location budget, calls for a comprehensive analysis in order to evaluate the consequences of the transportation improvement. Several studies estimated the impact of a reduction in transportation cost on the consumption of transportation (Goodwin, 1996; Goodwin and Noland, 2003; Parthasaranthi et al., 2003). A significant difference between short-term elasticity (movement along the demand curve) and long-term elasticity (referring to a shift in the demand curve) was observed on some occasions (Oum, 1989; Oum et al., 1992; Oum and Walters, 2000; Lago and McEnroe, 1991; Goodwin, 1992; Goodwin et al., 2004; Graham and Glaister, 2002, 2004; Booz, Allen, Hamilton, 2003). Nevertheless, the elasticities do not exceed one in all cases, and hence a positive overall change in land value is to be expected.

Up to this point of the analysis, the distance from the CBD is the variable that generates transportation costs. In reality, cost is a combined function of travel distance and travel time (Mohring, 1965); thus, the cost per one unit of distance (e.g., 1 km) may be presented as

$$C = u + w/S \tag{15.13}$$

Where C is the cost per unit of distance, S is the vehicle speed, and u and w are cost factors (u is related to distance, and w is related to speed or time).

Equation (15.13) may also be rewritten as follows:

$$C = u + wK \tag{15.14}$$

Where K is the time required for traveling one unit of distance.

In urban areas, most transportation costs relate to travel time; thus, the component u may be ignored. Hence, in terms of equation (15.10), the term TP, which refers to the household's total transportation costs, can be replaced by Hb, where H is the total amount of time (hours) spent on trips and b is the per-hour unit cost (b being a strict linear multiplication of w from equation [15.14]). Note the implicit assumption of a fixed price per km or per trip, and that the household is permitted to be located anywhere in the urban area. Moreover, this form does not require all trips to go to and from the centre. It allows complete compatibility with the traffic-assignment procedure, in which trips are made from all origins to all destinations, whether the CBD or any other particular zone. Traffic-assignment results give the information needed for an estimation of total travel time (H), since in most cases the assignment-impedance function is trip time.

Using Hb as the total transport cost, equation (15.12) may be rewritten as

$$- b\mathrm{d}H = Q\mathrm{d}R + \mathrm{d}Q(R + \mathrm{d}R) + \mathrm{d}b(H + \mathrm{d}H) + V\mathrm{d}Z \qquad (15.15)$$

Equation (15.15) has exactly the same implication as equation (15.12). Note that the reduction in transportation costs is merely the value of time savings. Hence, the left-hand side of equation (15.15) represents the savings from the transportation project; i.e., the value of time multiplied by the time saved. The right-hand side of this equation presents the commodities on which the household will spend the saving that accrues to it. Note, also, that db stands for the change in the unit value of time and, therefore, may be considered as zero. Thus, equation (15.15) can be reduced to

$$- b\mathrm{d}H = Q\mathrm{d}R + \mathrm{d}Q(R + \mathrm{d}R) + V\mathrm{d}Z \qquad (15.16)$$

The expression dH, which describes the change in trip time, should be divided into two parts:

- $\mathrm{d}H_0$ = reduction in trip time owing to improved transportation;
- $\mathrm{d}H_1$ = increase in trip time from the upward shift in the transportation-demand curve.

Introducing these definitions into equation (15.16) and rearranging it, we obtain:

$$- b\mathrm{d}H_0 = Q\mathrm{d}R + \mathrm{d}Q(R + \mathrm{d}R) + b\mathrm{d}H_1 + V\mathrm{d}Z \qquad (15.17)$$

Equation (15.17) indicates that for every household, wherever located, the value of the saving in transportation time ($b\mathrm{d}H_0$) will be used as additional income for four possible purposes:

- Change in the price of land, which will tend to increase in value, at least in areas adjacent to the transportation project;
- Increase in the land area consumed by the household (in urban terms, this might be found as larger houses or apartments);
- Increase in the number of trips (and total time spent on travel) – induced traffic;
- Increase in the consumption of other goods.

Our study assumes that transportation improvement projects produce several impacts. They could increase land values, housing consumption, and expenditure on transportation. It is impossible to determine whether these are short-term effects that in the long term will reveal themselves fully in land values. Accordingly, we examined and compared the change in dwelling units and household car availability between highly impacted zones and less impacted zones.

EMPIRICAL RESULTS

The theoretical model presented above shows that if a transportation improvement reduces transportation costs in an urban area for certain users, then not only will unit land prices there increase, but also other goods and services may be impacted. In particular, according to the theoretical model, three main elements will be affected: number of trips (or total expenditure on transportation), amount of land consumed (or dwelling area) and consumption of other goods. Since it is quite difficult to determine the exact change in the consumption of "other goods," we will analyse the impact on the two other components: expenditure on transportation and dwelling area.

In order to find measurable and significant impacts on these variables, a large-scale project is needed, as well as a large number of observations. This was provided by the Ayalon Freeway, which traverses the Tel-Aviv Metropolitan area from north to south. Because of the long period of this Freeway's construction (it was built between 1975 and 1985) and the fact that the project entered into operation only gradually, section by section, it was expected that its impacts would last a long time. To analyse them, a long-term (from 1975 till 2000) observation was carried out (Aviram, 2001).

In order to determine the influence of the Ayalon project on the increase in the consumption of transportation, it was assumed that the rate of motorization would increase faster in the impacted areas than in the less impacted areas. The influence of the project was estimated according to the reduction in average travel time spent by travelers to all destinations. After the influence was defined, all traffic zones were divided (using cluster analysis) into two major groups: highly impacted areas and less impacted areas. For these two groups, a similar linear model was applied:

$$NV_{it} = a_i + b_i t \tag{15.18}$$

Where:

NV_{it} = average number of vehicles per household in group i in year t;
a_i and b_i = calibrated coefficients;
t = the calendar year;
i = type of zone: i = 1 for highly impacted zones and i = 2 for less impacted areas.

The results obtained, presented in Table 15.1, are quite significant. They indicate that the gradient of increase in the number of vehicles per household is significantly greater in the highly impacted areas than in the less impacted areas. Although this result means that transportation expenses increased more in the impacted areas, it implies more trips, too, since car ownership is an important, positive trigger for trip generation.

Table 15.1 Impact of the Ayalon Freeway on household car availability

	No. of Observations		Coefficient	t-value	Significance level
Impacted areas	967	intercept	0.368	10.4	0.0000
		slope	0.041	14.9	0.0000
Less impacted	5,418	intercept	0.277	20.1	0.0000
		slope	0.037	34.9	0.0000

Note: Other socio-economic variables may also affect the outcome. However, we are dealing here with a large number of traffic zones (observations) within a large area. Therefore, the influence of those other variables may not be so far reaching (the same comment applies to Table 15.2).

According to the model, another commodity that might increase, besides the number of vehicles per household, is the size of land consumption by the household. Hence, dwelling areas were analysed for the same period, between 1975 and 2000. According to the theoretical model, we would expect that the increase in disposable income that resulted from the project should be spent on the consumption of a larger lot (or dwelling size). Hence, two regression models, similar to those presented in Table 15.1, were preformed, one for the highly impacted areas and the other for the less impacted areas. In this case, the dependent variable was the average dwelling size while the independent variable was, as before, the calendar year. Results are presented in Table 15.2.

It is easy to notice that the average dwelling size seems to be increasing everywhere. However, the yearly rate of the increase (as well as the percentage of increase) is significantly higher in the highly impacted areas than in the less impacted areas.

The theoretical analysis indicated that transportation improvement had a mixed impact on land value, dwelling size and induced traffic. However, we did not define the relative magnitude of each impact, since it would depend heavily on the elasticities relative to the reduction in transport costs.

Aviram (2001) analysed the impacts of the Ayalon Freeway project on land values. His study evaluated changes in residential prices over an extended time period as a result of the construction of this urban highway, which radically changed the structure of the CBD and all measures of accessibility in the Tel Aviv Metropolitan Area. The study points out that during the period analysed the price of each square metre in highly impacted residential areas increased annually by US$40 more than in the less impacted areas (ibid., pp. 161–163). This shows that a higher proportion of household income was allocated to land consumption in the highly impacted areas. Less attention was given in the literature to changes in the dwelling size and transportation consumed by the household.

Table 15.2 Impact of the Ayalon Freeway on dwelling size

	No. of Observations		Coefficient	t-value	Significance Level
Impacted zones	1,072	intercept	79.03	49.3	0.0000
		slope	0.88	7.6	0.0000
Less impacted	6,268	intercept	75.89	128.0	0.0000
		slope	0.82	20.1	0.0000

CONCLUSIONS

The impact of investment in large-scale transportation projects within a metropolitan region is long lasting. The reason for this is that urban development and changes in land uses constitute a slow process, taking a long time to show its effects. It is, therefore, essential to investigate and measure these changes over a long period of time.

The impact of transportation projects on land values is controversial. It can be positive or negative, depending on the income effect and the substitution effect. Therefore, it is important to collect a large amount of data from as broad an area as possible in order to differentiate between relatively highly impacted areas and relatively less impacted areas. We would expect that land consumption per household will increase in the highly impacted areas.

The propensity to own cars increases faster in highly impacted areas, where accessibility improves relatively more than in less impacted areas.

The overall impact of the major Ayalon transportation project on total land values in the Tel-Aviv Metropolitan area was found to be positive and significant.

This study has shown that a reduction in transportation costs from an improvement in transportation may have several different impacts. Transportation improvement has a direct impact on the use of transportation itself through generating new traffic – induced demand. Even under the assumption that no equilibrium exists in the urban area, some of the classic results of urban economics still hold, in particular the impact of a transportation project on land values. Our results indicate that, at least in the short run, benefits derived from transportation improvements are not completely absorbed by land values. Other impacts, such as induced traffic and an increase in the consumption of dwelling size, also occur.

REFERENCES

Alcaly, R.E. (1976). Transportation and Urban Land Values: A Review of the Theoretical Literature, *Land Economics*, Vol. 52 (1), pp. 42–53.

Alonso, W. (1964). Equilibrium of the Household, Chapter 2, in W. Alonso: *Location and Land Use: Towards a General Theory*, Cambridge, MA: Harvard University Press, pp. 168–177.

Arnott, R. and J. Stiglitz (1979). Aggregate Land Rents, Expenditure on Public Goods, and Optimal City Size, *Quarterly Journal of Economics*, Vol. 93 (4), pp. 471–500.

Arnott, R. and J. Stiglitz (1981). Aggregate Land Rents and Aggregate Transport Cost, *The Economic Journal*, Vol. 91, pp. 331–347.

Aviram, H. (2001). *Evaluating the Impact of Urban Transportation Projects on Land Values*, Unpublished PhD Thesis, Technion-Israel Institute of Technology (in Hebrew).

Bajic, V. (1983). The Effects of a New Subway Line on Housing Pricing in Metropolitan Toronto, *Urban Studies*, Vol. 20, pp. 147–158.

Blum, U. (1997). Benefits and External Benefits of Transport: A Spatial View, in D.L. Greene, D.W. Jones, and M.A. Deluchi (Eds.), *The Full Costs and Benefits of Transportation, Contributions to Theory, Method and Measurement*, Berlin: Springer.

Booz, Allen, Hamilton (2003). ACT – Transport Demand Elasticities Study, Department of Urban Services, Canberra (Unpublished report, April).

Cervero, R. (1997). Transit Induced Accessibility and Agglomeration Benefits: A Land Market Evaluation, IURD Working Paper 691, University of California at Berkeley.

Chebreegziabiher, D., E. Pels and P. Rietveld (2007). The Impact of Railway Stations on Residential and Commercial Property Value: A Meta Analysis, *Journal of Real Estate Finance and Economics*, Vol. 35 (2), pp. 161–180.

Christensen, L.R. (1975). Substitution Effects in Consumer Expenditures, *Business Economics*, pp. 58–63.

Damm, D., S. Lerman, E. Lerner-Lam and J. Young (1980). Response of Urban Real Estate Values in Anticipation of the Washington Metro, *Journal of Transport Economics and Policy*, Vol. 14, pp. 315–335.

Dewees, D.N. (1976). The Effects of a Subway on Residential Property Values in Toronto, *Journal of Urban Economics*, Vol. 3 (4), pp. 357–369.

Ewing, R. (1997). Is Los Angeles Style Sprawl Desirable, *Journal of American Planning Association*, Vol. 3 (1), pp. 107–126.

Evans, W.A. (1983). The Determination of the Price of Land, *Urban Studies*, Vol. 20, pp. 119–129.

Gatzlaff, Dean H. and Marc T. Smith (1993). The Impact of the Miami Metrorail on the Value of Residences Near Station Locations, *Land Economics*, Vol. 69 (1), pp. 54–66.

Getz, M. (1975). A Model of the Impact of Transportation Investment on Land Rents, *Journal of Public Economics*, Vol. 4, pp. 57–74.

Goodwin, B.P. (1992). A Review of New Demand Elasticities with Special Reference to Short and Long Run Effects of Price Changes, *Journal of Transport Economics and Policy*, Vol. 26, pp. 155–169.

Goodwin, P.H. (1996). Empirical Evidence on Induced Traffic, A Review and Synthesis, *Transportation*, Vol. 23, pp. 35–54.

Goodwin P.H. and B.R. Noland (2003). Building New Roads Really Does Create Extra Traffic, *Applied Economics*, Vol. 35 (13), pp. 1451–1458.

Goodwin, P. H., J. Dargay and M. Hanly (2004). Elasticities of Road Traffic and Fuel Consumption with Respect to Price and Income: A Review, *Transport Reviews*, Vol. 24 (3), pp. 275–292.

Gordon, P. and H.W. Richardson. (1997). Are Compact Cities a Desirable Planning Goal? *Journal of American Planning Association*, Vol. 63 (1), pp. 95–106.

Graham, J.D. (2006). Wider Economic Benefits of Transport Improvements: Link Between Agglomeration and Productivity, Stage 2 Report, London DfT.

Graham, J.D. (2007). Agglomeration, Productivity, and Transport Investment, *Journal of Transport Economics and Policy*, Vol. 41 (3), pp. 317–343.

Graham, J.D. (2008). An Empirical Analytical Framework for Agglomeration Economies, *Annals of Regional Science*, Vol. 42, pp. 267–289.

Graham, J.D. and S. Glaister (2002). Review of Income and Price Elasticities of Demand for Road Traffic, Centre for Transport Studies, Imperial College, (working paper) (July).

Graham, J.D. and S. Glaister (2004). Road Traffic Demand Elasticity Estimates: A Review, *Transport Reviews*, Vol. 24 (3), pp. 261–274.

Herbert, J. and B. Stevens (1960). A Model for the Distribution of Residential Activity in Urban Areas, *Journal of Regional Science*, Vol. 2 (2), pp. 21–36.

Hills, P.J. (1996). What Is Induced Traffic? *Transportation*, Vol. 23, pp. 5–16.

Hongbo, Du and C. Mulley (2006). Relationship Between Transport Accessibility and Land Value. Local Model Approach with Geographically Weighted Regression, *Journal of the Transportation Research Board*, No. 1977 (2006), pp. 197–205.

Lago, A.M. and J.M. McEnroe (1991). Transit Service Elasticities, *Journal of Transport Economics and Policy*, May, pp. 44–51.

Mackett, R.L. (1991). LILT and MEPLAN: A Comparative Analysis of Land Use and Transport Policies for Leeds, *Transport Reviews*, Vol. 11 (2), pp. 131–144.

Mackett, R.L. (1993). Structure of Linkages between Transport and Land Use, *Transportation Research B*, Vol. 27B (3), pp. 189–206.

McDonald, F.J. and I.C. Osuji (1995). The Effect of Anticipated Transportation Improvement on Residential Land Values, *Regional Science and Urban Economics*, Vol. 25, pp. 261–278.

McDonald, John and P. McMillen (2004). Reaction of House Prices to a New Rapid Transit Line: Chicago's Midway Line, 1983–1999, *Real Estate Economics*, Vol. 32 (3), pp. 463–488.

Mills, S.E. (1969). The Value of Urban Land, in H.S. Perloff (ed.), *The Quality of the Urban Environment*, Baltimore: The Johns Hopkins Press, pp. 231–253.

Mohring, H. (1961). Land Values and the Measurement of Highway Benefit, *Journal of Political Economy*, pp. 236–249.

Mohring, H. (1965). Urban Highway Investment, in R. Dorfman (ed), *Measuring Benefits of Government Investment*, The Brookings Institution, pp. 231–291.

Mohring, H. (1993). Maximizing, Measuring and Not Double Counting Transportation-Improvement Benefits: A Primer on Closed and Open Economy Cost-Benefit Analysis, *Transportation Research B*, Vol. 27B, pp. 413–424.

Nash, C. and J. Preston (1991). Appraisal of Rail Investment Projects: Recent British Experience, *Transport Reviews*, Vol. 11 (4), pp. 295–309.

Oum H.T. (1989). Alternative Demand Models and Their Elasticity Estimates, *Journal of Transport Economics and Policy*, Vol. 23, May, pp. 163–187.

Oum, H.T. and G.W. Walters II (2000). Transport Demand Elasticities. Chapter 12 in D.A. Hensher and G.W.K.J. Button (eds), *Handbook of Transport Modeling*, Oxford: Elsevier Science.

Oum, T.H., G.W. Waters II and J.-S. Yong (1992). Concepts of Price Elasticities of Transport Demand and Recent Empirical Estimates, *Journal of Transport Economics and Policy*, Vol. 26, pp. 135–154.

Parthasaranthi P., M. Levinson and R. Kamachandra (2003). Induced Demand: A Microscopic Perspective, *Urban Studies*, Vol. 40 (7), pp. 1337–1351.

Pines, D. and Y. Weiss (1976). Land Improvement Projects and Land Values, *Journal of Urban Economics*, Vol. 3, pp. 1–13.

Pines D. and Y. Weiss. (1982). Land Improvement Projects and Land Values: An Addendum, *Journal of Urban Economics*, Vol. 2, pp. 198–204.

Ryan, S. (1999). Property Values and Transportation Facilities: Finding the Transportation-Land Use Connection, *Journal of Planning Literature*, Vol. 13 (4), pp. 412–427.

Ryan, S. (2005). The Value of Access to Highways and Light Rail Transit: Evidence for Industrial and Office Firms, *Urban Studies*, Vol. 42 (4), pp. 751–764.

Shefer, D. (1973). Localisation Economies in SMSAs: A Production Function Analysis, *Journal of Regional Science*, Vol. 15 (1), pp. 55–64.

Shefer, D. and H. Aviram (2005). Incorporating Agglomeration Economies in Cost Benefit Analysis of Transport Projects, No. 133, 45th Conference of the European Regional Science Association.

Venables, A. (2007). Evaluating Urban Transport Improvements: Cost Benefit Analysis in the Presence of Agglomeration and Income Taxation, *Journal of Transport Economics and Policy*, Vol. 41, pp. 173–188.

Van der Tak, H.G. and A. Ray (1971). The Economic Benefit of Road Transport Projects, World Bank Technical Paper, Baltimore: The Johns Hopkins Press..

Wegener, M. et al. (1991). One City, Three Models: Comparison of Land Use Transport Policy Simulation Models for Dortmund, *Transport Reviews*, Vol. 11 (2), pp. 107–129.

Wheaton, C.W. (1977). Residential Decentralisation, Land Rents and the Benefit of Urban Transportation Investment, *American Economic Review*, Vol. 67, pp. 138–143.

Yiu, C. Y. and S. K Wong. (2005). The Effects of Expected Transport Improvements on Housing Prices, *Urban Studies*, Vol. 42 (1), pp. 112–125.

Index